应用型本科国际经济与贸易专业精品系列规划教材

国际电子贸易

主　编　毛凤霞　秦成德

副主编　王轼颖　王　群

参　编　赵　青　刘　颖　李　鹏

　　　　霍建英　段　锞　冯青云

北京理工大学出版社

BEIJING INSTITUTE OF TECHNOLOGY PRESS

内 容 提 要

本书密切反映了在信息化的背景条件下,国际贸易对企业商务管理的新要求、新变化和新特点,突出理论创新和强化实务技能两个方面,学习外贸企业在电子化、信息化领域中最具先进性和实用性的理论与实践。通过对国际电子贸易理论和实践的学习、研究和分析,掌握国际贸易采用电子商务的现状和发展趋势,了解国际贸易电子商务的基本知识和基本原理,从而全面提高学生从事商务实践的理论素养和电子商务的应用能力。

本书共分15章,1~2章主要介绍国际电子商务、电子贸易理论;3~6章从国际贸易电子化设备角度,介绍国际电子贸易的网站建设、企业信息化、电子口岸的内容;7~11章介绍了国际电子贸易合同、国际电子支付与结算、国际电子物流、贸易程序简化及电子单证的内容;12~13章主要介绍国际网络营销及国际网络金融的内容;14~15章主要从制度法规方面,介绍国际电子贸易纠纷解决及国际电子商务法律制度。

本书每章内容在结构上分为5个部分:学习要点、导入案例、正文、本章小结、本章习题,以适应教师精讲、学生参与、师生互动、提高技能的新型教学理念和教学方法。

本书既可作为高等院校经管类国际贸易、电子商务、市场营销等专业和其他相关专业的教材,又可供企事业单位实际工作人员学习参考。

图书在版编目(CIP)数据

国际电子贸易/毛凤霞,秦成德主编 . —北京:北京理工大学出版社,2017.6(2017.7 重印)
ISBN 978 - 7 - 5682 - 4170 - 0

Ⅰ.①国… Ⅱ:①毛… ②秦… Ⅲ.①国际贸易 - 电子商务 Ⅳ.①F713.36

中国版本图书馆 CIP 数据核字(2017)第 134919 号

出版发行 / 北京理工大学出版社有限责任公司
社　　址 / 北京市海淀区中关村南大街 5 号
邮　　编 / 100081
电　　话 / (010)68914775(总编室)
　　　　　 (010)82562903(教材售后服务热线)
　　　　　 (010)68948351(其他图书服务热线)
网　　址 / http://www.bitpress.com.cn
经　　销 / 全国各地新华书店
印　　刷 / 三河市天利华印刷装订有限公司
开　　本 / 787 毫米×1092 毫米　1/16
印　　张 / 27　　　　　　　　　　　　　　　　　责任编辑 / 王俊洁
字　　数 / 625 千字　　　　　　　　　　　　　　文案编辑 / 王俊洁
版　　次 / 2017 年 6 月第 1 版　2017 年 7 月第 2 次印刷　责任校对 / 周瑞红
定　　价 / 54.80 元　　　　　　　　　　　　　　责任印制 / 施胜娟

前　言

21 世纪，网络经济的发展进入一个新时代，以互联网为依托的电子商务正在全方位改变企业运营模式。特别是在国际贸易领域，电子商务的发展及应用对节省交易费用、提高企业绩效、增加国际贸易交易量起到了重要作用。与电子商务应用相结合的国际贸易在市场竞争规则、经济增长方式等方面都将面临巨大变革。

国际电子贸易是一门实践性较强的应用学科，主要研究以计算机网络、通信技术为核心的 IT 技术在外贸企业信息化管理中的应用。是应用经济学与 IT 技术相结合的交叉学科，同时它也是国际经济与贸易专业的学生提高专业技能的重要课程。学生通过学习和实践，了解现代企业面对的社会经济环境系统，已经从一个工业化时代走向信息化时代，为适应这一变化的需要，加快以互联网为核心的现代信息技术在国际商务中的应用，对于提高企业的客户服务能力和水平以及提高企业绩效，特别是对于改善国际贸易工作流程具有重要的意义。

本教材密切反映了在信息化的背景条件下，国际贸易对企业商务流管理的新要求、新变化和新特点，突出理论创新和强化实务技能两个方面，学习外贸企业在电子化、信息化领域中最具先进性和实用性的理论与实践。通过对国际电子贸易的理论和实践的学习、研究和分析，掌握国际贸易采用电子商务的现状和发展趋势，了解国际贸易电子商务的基本知识和基本原理，从而全面提高学生从事商务实践的理论素养和电子商务的应用能力。

本教材每章内容在结构上分为 5 个部分：学习要点、导入案例、正文、本章小结、本章习题，以适应教师精讲、学生参与、师生互动、提高技能的新型教学理念和教学方法。

本书共分 15 章，主要由西安邮电大学经济与管理学院教师编写而成。毛凤霞教授担任第一主编，并负责拟定提纲、统稿和定稿；秦成德教授担任第二主编；王轶颖、王群担任副主编。具体编写分工：第 2、9 章由毛凤霞教授编写；第 1、5、6、15 章由秦成德教授编写；第 7、12 章由王群老师编写；第 8、10 章由王轶颖老师编写；第 3 章由段锞老师（西安培华学院）编写；第 4 章由赵青老师编写；第 6 章由冯青云（研究生）编写；第 11 章由刘颖老师编写；第 13 章由李鹏老师编写；第 14 章由霍建英老师编写。

本书既可作为高等院校经管类国际贸易、电子商务、市场营销等专业和其他相关专业的教材，又可供企事业单位实际工作人员学习参考。

本书在编写过程中得到了北京理工大学出版社、西安邮电大学及各位领导的大力支持和帮助，在此一并表示感谢。本书参考了大量相关领域的文献、网络资料，列示于书后的参考文献中，但仍可能有遗漏，在此谨向参考文献的作者表示诚挚的谢意，同时对未标注参考文献的作者表示由衷的歉意。

由于编者水平有限、编写时间仓促，加之国际贸易在电子商务实践领域发展变化十分迅速，本书难免存在疏漏和错误，恳请专家和读者批评指正。

编　者

2016 年 11 月

目 录

英文目录

中文目录

Preparations for International Trade

- Understand the significance of making an international trade plan.
- Know about the basic connotation of common used international trade laws and regulations.
- Master the availability of making international plan.
- Know the necessary conditions for carrying out trade.

International trade is also called foreign trade, it refers to the process of fair and deliberate exchange of goods or services between two or more countries. The statistics of success and failure for people who start international trade may not been seen, but they are probably similar to the statistics for business in general. Therefore, it is particularly important to carry out the preparatory work before doing import and/or export business. First of all, we need to be very familiar with the product and market, find a product with good market prospects, by conducting market research to analyze opportunities and threats, and to establish or seek to use an appropriate form of import and export organization, ultimately form a suitable international trade plan.

Second, importing and exporting are definitely business that require a broad and adequate international trade aspects of professional knowledge. In recent years, there are only few changes in the law relating to international trade contracts, however, due to the wide application of electronic commerce, the rapid development of means of transportation and communication, many procedures, rules and documents related to international trade have undergone great changes, especially the documents closely related to the Internet need to be updated in time. Therefore, we need to know the international trade conventions, customs and national laws and regulations etc.

Finally, after we have done the above preparations, intend to carry out international trade, we need to go through a number of procedures to obtain international trade qualifications, and for some goods, it is required to obtain the corresponding quota or license before doing import and/or export business.

1.1 Designing and Making an International Plan

Can I do international trade? The question is actually the same as "can I start my own business?", which are millions of people's dreams. Of the thousands of entrepreneurs who try, many fail, many succeed in some degree, and a few people become highly successful. So start any kind of business, we need to design and plan aspects of the preparatory work.

1.1.1 Preliminary Considerations before Business Plan

This chapter will discuss the preliminary considerations that anyone intending to export or import should consider. Before beginning to export or import, a number of considerations should be addressed to avoid costly mistakes and difficulties. Those companies that begin dong international trade or continue to do trade without any preparations will run into problems sooner or later. Therefore it is very important to have a solid business plan before you start your company. Such a plan will help your business better and determine whether it is likely to be profitable. Advice on preparing business plans is available from the Small Business Administration of Ministry of Industry and Information Technology of the People's Republic of China and from Small Business Development Center, books, software, and the Internet.

Usually the contents of trade plan include the types of trade and products; potential markets, market segments, and competition; promotion and sales plan; organization and personnel; import/export logistics; schedule of start-up activities and so forth. To develop such a complete plan, the first thing we should do is the preparatory work before carrying out international trade, also need to conduct market research, and then consider how to finance, how to find a supplier and set forth the agreement, understand the regulations of foreign trade, how to avoid import and export fraud and so on.

The Basic Conditions for Carrying Out Trade

Importing and exporting are definitely business that require knowledge of purchasing, marketing/selling, finance, and other functions that go along with being your own boss. It is necessary to take a course for entrepreneurs or at least to read some of the many books on how to start and manage a small and medium business.

Doing business no doubt requires some capital. You normally need some money to register your business and for basic office equipment, communications, and so forth. Then, if you plan to buy merchandise for import or export, you will probably have to pay for it before you can sell it and collect from your buyer, in addition, you may have to lay out cash for transportation, storage, and other services. Even some people make a living acting as agents or brokers and do not take title to or possession of merchandise. Doing business this way limits their need for capital, but even so, they usually set up small offices and have communication and other expenses. Also, if you plan to quit your job to start this business, you will need money to live on for several months. You may be able to borrow part of the money from bank but only if you have business experience, cash, and collateral.

Then doing international trade requires knowledge of the world. Especially the world geography, it really helps to know where countries are located, which ones are friendly with our country, which have strong currencies, and so forth. We can obtain these information by reading newspapers, such as International Business Daily, or watching TV online or listening to the BBC news, etc.

Doing import and export also requires the familiarity with foreign cultures and languages. In the international trade, the influence of language and cultural differences should be paied attention to. Because of the differences of thinking mode, value concept, customs and language phenomenon, there are often obstacles in cross cultural communication. Therefore, to understand other countries' culture, values and customs, try to use simple, clear and definite terms in communication, do not use easy to cause misunderstanding of the ambiguous words, pun, slang, idioms, also do not use easy to cause the other offensive words. In international trade activities, comply with international practice and a certain courtesy. We suggest trying to learn a bit local language of the country that you expect to do business with. If you need to employ local people as translators (for written work) or interpreters (for oral conversations), please be sure that the persons are qualified, and do not use the translation directly by computer or some website translation machine, the translation should be checked by the qualified translator, so as to avoid making mistakes.

To be engaged in international trade, it is necessary to have some characteristics. Persistence is usually a good virtue in business activities. Usually at the start stage, there is no order probably in a few months or half a year, this is the time to keep calm, we must overcome the psychological anxie-ty, calm down step by step to lay the foundation, long-term accumulation will bring harvest. On the other hand, if you are losing money or barely breaking even, you may judge that the time has come to change your business model or to close for a while. A word of advice here is to take stock at least every month of customers you are working on and the status of each and decide whether, and if so how, to continue your pursuit of them. In addition, attention to detail is also engaged in the interna-tional trade needs of good quality. International trade is much more complex than commercial activity in China. For example, the product from one country to another, it involves the issue of transport, most of the transportation is by ship, then we need a boat, the management of the ship is the ship company, shipping company will arrange the containers to the freight forwarding, we will contact the freight forwarding to arrange the shipment. Before goods shipped out of the customs territory of one country, we need to report to the government, so there is the customs, we need to declare to the customs, some products also need to apply for inspection, the completion of these links need to be careful, other matters such as understanding the needs of customers and the product itself also need to pay attention to the details. Details can be especially critical when they relate to payment. And also in international trade, as with any business, you have to give the buyer exactly what he or she asks for. It is certainly possible to start an import or export business by finding an attractive product and then looking for buyers, and this is the way it is usually done. Selling, however, is usually much harder than buying. If you have or can find someone who wants to buy a product from you, and will give you the specifications, you can probably find and supply the item.

Market Research

Market research is vital to carry out international trade. Even if you have some experience in international trade, it's unwise to rush into the market that has not been investigated. Most small business are started with inadequate market studies or with none at all. Yet a market study is almost the only way to support your income projections and your marketing plan. Market research can help us analyze the realization of income, can help us complete the market research program. Market study before trade will help to analyze how to market them and to whom to sell them.

The following figure 1. 1 is a typical outline for a report on the market for any product being exported from one country to another. If you can produce or buy this kind of information before you begin, you might decide to try another product or another market. If you decide to go ahead, your chances of success will be increased considerably. Moreover, a solid market study looks very good to potential investors and lenders.

Basic Informations
• Product name and HS Code
• Country of origin, exporting country, importing country
Regulations
• Exporting country controls and taxes
• Importing country controls and taxes
• Import restrictions, quotas, and so forth
• Import duties
• Marketing and labeling laws
• Other regulations
Supply and Demand
• Availability of supply in the exporting country
• Domestic production less imports in the importing country (five years' statistics)
• Imports for consumption (five years' statistics with trend calculation)
• Percent of product for consumption that is imported
• Industry experts' perception of current and future supply and demand in the importing country
• The perception of selected buyers with the same business scope
Competition
• Survey of producers in the importing country
• Description of selected producers
• Sources of imports (countries), with import market share of each
• Average FOB, CFR and CIF prices from each country
Target Market

• The market and market segments
• Characteristics of important market
• Kinds of industrial users
• Main industrial users and a brief description
Product Description
• Main types/varieties of the product
• Required or desired product characteristics
• Required or desired packaging and labeling
Distribution
• Normal distribution system in the importing country
• Principal importers and wholesalers
• Principal industrial distributors
Pricing
• Representative prices and markups at each level in the channel
• Price trends
• Discounts used in the trade
Promotion
• Methods of promotion used in the trade
• Promotional assistance usually provided to the exporters
• Approximate costs of this assistance
Logistics
• Steamship lines and airlines serving the route
• The usual transport mode for this kind of product
• Availability of vessels (planes) and cost of shipping
• Shipping term normally used
• Payment term normally used
Other Considerations
• Local laws on the product, label, distribution, pricing, promotion
• The distributor's and consumer's attitudes of acceptance to new suppliers
• Image of exporting country in the importing country

Figure 1. 1　Example of market study on an exporting product

1. 1. 2　Main Contents of Trade Plan

Issues to be Considered in Making Trade Plans

The first issue to be considered is the objectives of international trade. Why are you thinking of starting a business? Are you interested in starting your own business and are thinking that it could be an import and/or export business? What are your objectives? If your answer is "to make money", you should probably spend extra time on good planning with realistic financial projections. If you basically want to experiment with international trade and perhaps be better off getting more work experience, you are at a lower level of commitment and can start your business in a less formal way, even at low salaries.

The second issue is the necessary conditions for international trade. Do you have the following things? If you have business training and experience, capital, product knowledge, contacts with buyers, travel experience, or foreign languages, you'll be in a much stronger position than if you lack these things. And then make sure you plan to import, export, or both. If you just start your business, we suggest picking one or the other, at least at first. It's hard enough to start one business; you don't want to be starting two at the same time.

The third issue is the types of trade. If a company plans to import or export by itself, that is to say, they own the self-support imports and exports rights, they can buy goods in China and sell goods to the other countries directly. This model usually requires quite a bit of money and involves risk. If a company plans to begin as commission agents, and they can be a selling or a buying agent for a China or a foreign company. There is usually a formal written agreement by which the principal designates a firm or individual to act his agent in return for a commission on completed transactions. If you just try to broker, or arrange, deals between unrelated parties. You can try to collect a commission from either party to the transaction, or from both. This model does not always require written contracts. No matter what model to take, a new company should consider registering what kind of name, whether existing in the form of company, taking which forms of company, etc. Answers to these questions will be related to liability and taxation.

The fourth issue to be considered is to find out your target market. This is the most important issue. Select a target market should have practical significance, that is a market you can identify and contact. Instead of trying to export suitcases to anyone in the world, try selling them to department stores in some selected countries, as they will help to sell your suitcases to the middle-income consumers who like to travel around the world. Therefore you can't put everyone in a country as your target customer, but these people who are in the appropriate area, with appropriate income levels, taking on the appropriate lifestyle will become your target customers. The next is how to meet the target customers? That is, chose what kind of method for promotion, and what channel to sell the products to the customers. Firstly, we can use some B2B platforms for promotion, such as Alibaba, Made in China, Global Sources, etc. Secondly, we can find the potential customer by attending trading fairs and exhibitions, such as China Export Commodities Fair (Canton Fair for short) and other foreign professional exhibitions. At last, we could take direct selling, avoiding the intermediate processes, such as doing cross-border electronic commerce by using of a variety of platforms such as Amazon, eBay, and some other local websites in foreign countries etc. Our purpose is to sell the goods combining with the logistics and all kinds of media.

The fifth issue to be considered is the type (s) of product (s) you plan to deal in. The best answer to this question is that you will deal in a product for which you already have a customer. If you can not do that try looking for a product that you like and know a lot about, that is available in sufficient quantity, and that can be transported to and entered (through customs) into your market country. Whichever product you choose, you will probably find that domestic company import it, export it, or both. Almost every kind of product can be traded internationally, and you may find the right way to carry on.

The sixth issue includes the sources of supply companies and countries. Both importers and exporters usually try to deal with companies that produce merchandise, rather than buying from intermediaries. Middlemen have a role to play, but each one has a profit and costs that increase the price to the consumer. Therefore, how to find manufacturers in China and abroad is also the knowledge we need to master. Some factors to consider in selecting a country from which to buy are its reputation for quality, its cost structure, transportation to China, and customs duties.

The seventh issue is the methods of shipping. There are several different methods of shipping. Small items can be handled by couriers. You might use airfreight for larger shipments and sea or land for still larger ones, but the decision also varies with such factors as the value and fragility of the cargo. Intermodal shipments are very common, for example Chinese Fuling Pickled mustard from China to Indonesia might be moved to its destination by truck, ship, train, and possibly oxcart to some small stores. In international shipping there are a number of standardized terms such as FOB and CIF. It is important that you understand these. The main ones are explained later in this book.

The eighth issue to be considered is the method of international payment you plan to use. Inexpensive and secure payment is vital to any business. In international transactions there is an extra element because money must cross international boundaries. Different countries use the different monetary system, currency exchange rates will also change frequently, so it is essential to adopt appropriate and safe payment methods. In addition to the traditional payment methods such as letter of credit and document against payment, etc. , we also should pay attention to and master the new payment methods such as PayPal, international Alipay, etc. Unfortunately, in every country there may be profiteers who try to order merchandise from overseas and not pay for it.

The ninth issue is the regulations of Chinese and foreign government. The international trade operators are subject to the laws of multiple jurisdictions, including the national and local laws and regulations of the exporting and the importing countries. There are also bilateral and multilateral agreements that come into play with regard to many transactions. The main regulations are explained later in this book and the information about how to find out about many of the others are given.

The tenth issue to be considered is having a certain business premises. Many people start business in their homes, and that has become easier and more acceptable in the Internet Era than it used to be. Equipping a home office is not difficult, a computer and several office equipment including telephone, fax machine, printer, and so on, but also it can save a lot of money on rent expense. But this is only applicable to those small company that just have started, it is too close to the family, which may bring some annoyance.

The eleventh issue to be considered is how to select the service companies. You may want to use the services of an accountant and/or an attorney to help set up your business. Then you will need a bank, an Internet provider, a telephone company, and an insurance broker. When you start doing business, you will need a freight forwarder and perhaps a customs broker, a courier service, and others. When your business is done to a certain extent, you might also want to look for help and guidance from a number of professional organizations or industry associations.

The twelfth issue to be considered is the source of your capital investment as well as the source of income and profitability. No matter how you start the business, you will have to make some investments, including start-up costs, such as equipment investment, the cost of your time and operating costs before you start to make a profit. Once you have a bottom-line number, you will know whether you have enough money. If you don't have enough money, you can determine how much money you need and can look at possible source. After you figure out the start-up costs, you should project sales and then project your income and your expenses.

Making Trade Plan

As mentioned before, it is very important to make a useful and reliable business plan before starting a business. Such a plan will help you plan your business better and determine whether it is likely to be profitable. The preparatory work before carrying out the international trade is a very complicated process. It involves a lot of research and analysis about their company and target markets. The following figure1. 2 is a suggested outline for an import/export business plan, with some annotations. We can find other outlines in books on business planning and on Internet.

Cover Page
Table of Contents
Executive Summary (This should give the highlight of each of the sections that follows. Most people who read business plans look through the summary and then read other sections only as needed for clarification or additional details.)
Narrative Section (This section of the plan describes in words, with graphs if you wish, how your business will be set up and operated. It can be from 12 to 15 pages in length.)
• Type of trade and products
Whether you plan to import or export and your sources of supply, types of products, and product characteristics including packages, labels, and brands.
• Potential markets, market segments, and competition
Figure 1. 2 (continued)
A description of the likely consumers of your product, the channels of distribution through which you will reach them, and your main competitors.
• Promotion and sales plan
Your anticipated strategy with regard to pricing, promotion and selling, that is, how you actually cause your customers to buy your product.
• Organization and personnel

A description of the company's managers, employees, and sources of assistance, such as your advisers, accountant, banker, and attorney.
• Import/export logistics How you expect to handle the functions of packing, shipping, insurance, documentation, and so forth.
• Schedule of start-up activities
A list of steps in starting the business with a target date for completion of each, and a final target date for opening the doors and beginning to operate. Financial Section
This section of the plan shows, in numbers, what you expect in the way of sales, income, expenses, and other financial aspects of the business.
• Schedule of start-up costs
• Proforma income and expense statements
• Proforma cash flow statements
• Proforma balance sheets
• Main assumptions used in preparing the financial projections
• Financial analysis
Supporting Documents
This section can contain any document that will establish that the information in your plan is correct or that will be useful to the reader, for example:
• Pictures of your products
• A summary of your market study
• Sample promotional literature
• Owner's resumes
• Detailed costs for a sample import or export transaction
• Lease agreement for office space
• Detailed sales forecasts
• Letters of intent from suppliers

Figure 1. 2   Examples of Trade Plan

1. 2   Be Familiar with International Trade Laws and Regulations

The legal system of international trade mainly consists of three parts: international trade conventions, international trade customs and practices, national commercial laws and regulations. In the process of international trade, the parties wish to adopt the law applicable to the country that they are familiar with. But because people come from different countries or regions, it may be controversial on which national law should be used. Therefore, in the long-term trade practice, we are willing to adopt a generally accepted international trade contract or practice, and sometimes willing to apply the law of a certain country in the contract. So international traders should be familiar with international trade conventions, international trade customs and practices, national business laws and regulations.

1. 2. 1　National Business Laws and Regulations

The existing international commercial conventions and practices, still are not complete uniform rules about all the fields of international trade, and even the existing conventions and rules of international trade have not yet been recognized and adopted by all countries and regions. When a person or enterprise is engaged in economic and trade activities across the border, it may also choose to a national law applicable to the settlement of contract, dispute according to the rules of conflict of laws. Therefore, the national law has a great influence on the international sale of goods.

Understanding Anglo-American Law System and Continental Law System

At present, the common law system and civil law system have great influence on legal formulation of international trade and economic cooperation. In the countries of continental law system, the commercial law is mainly used in the form of codification. The layout is divided into two: one is the separation of civil and commercial form, that is, the civil law and commercial law were compiled into two laws, in such a country generally compiled the sale of goods acts into the civil code, included a special section or in the chapter on debt which to be explicitly stipulated in the civil law. In addition, it is also specified the special matters of transaction in the code of commercial law. Countries that adopt this approach are France, Germany, Japan and other countries. The other is combination of the civil law and commercial law which separated in some states. Namely, there is no separate commercial code, and the commercial law, including the content of the civil code is incorporated into the civil code, such as Italy and other countries. In the countries of Anglo-American law system, commercial law mainly take the form of specific regulations. Precedent is an important source of commercial law in the countries of Anglo-American legal system.

Chinese National Law on International Business

In China, at present, a number of laws and regulations regulating commercial activities are scattered in the General Principles of the Civil Law of the People's Republic of China (1986) and other separate regulations. There are some other important laws and regulations about international economic and trade activities, which have covered a very wide range of fields, such as the Contract law of the People's Republic of China (1999), Maritime law of P. R. C. (1992), Company law of the People's Republic of China revised in 2013, the law of the People's Republic of China Note revised in 2004, Insurance law of P. R. C. revised in 2015, etc.

The Legal System of Chinese Foreign Trade Control

China's foreign trade control system is mainly composed of customs supervising system, tariff system, foreign trade operators management system, import and export licensing system, the Entry-Exit inspection and quarantine system, foreign change management system, trade remedy system, etc. In order to ensure the implementation of the system of trade control, China has basically established and gradually perfected the legal system of foreign trade, which take the Foreign Trade law of the People's Republic of China as the core, promulgated and implemented in 1994, and implements the foreign trade control independently in accordance with the relevant provisions of these laws, administrative regulations, departmental rules and China's implementation of international conventions.

Currently there are some laws on management of foreign trade in China, such as the Foreign Trade law of the People's Republic of China, the Customs Law, the law of the People's Republic of China on Import and Export Commodity Inspection Law, the law of the People's Republic of China on the Entry and Exit Animal and Plant Quarantine, the law of the People's Republic of China on the Prevention and Control of Solid Waste, the Frontier Health Quarantine Law of the P. R. C. , the law of Wildlife Protection of People's Republic of China, the Drug Administration law of the People's Republic of China, the law of the People's Republic of China on the Protection of Cultural Relics, the Food Hygiene Law of the People's Republic of China.

Moreover, there are some administrative rules and regulations related to foreign trade control in China, such as Regulations on Technology Import and Export Administration of the People's Republic of China, Regulations on Technology Import and Export Administration of the People's Republic of China, Regulations of the People's Republic of China on Import and Export Duties, Regulations on Customs Protection of Intellectual Property Rights of P. R. C. , Regulations of the People's Republic of China on the Protection of Wild Plants, Regulations on Foreign Exchange Control of the People's Republic of China, Regulation on Countervailing of the People's Republic of China, Regulations of the People's Republic of China on Anti-dumping, Regulation on Safeguard Measures of the People's Republic of China, etc.

Furthermore, there are a lot of ministerial rules about the management of international trade in China, such as Management Rules of Goods Import License, Management Rules of Goods Export License, Management Measures on Cargo Automatic Import License, Management Measures on Verification of Export Proceeds, Measures for the Administration of Imported Drugs, Measures for the Administration of Radioactive Drugs, Measures for the Administration of Import and Export Licenses for Dual-Use Items and Technologies, etc.

All kinds of international treaties or agreements to which the People's Republic of China is a contracting party or participating party, although they do not belong to the category of domestic law of our country, in terms of its effectiveness they can be regarded as one of the sources of law in our country. Mainly include: the relevant bilateral or multilateral trade agreements signed by China during the accession to the World Trade Organization, International Convention on the Simplification and Harmonization of Customs Systems (also known as the Kyoto Convention) , Convention on International Trade in Endangered Species of Wild Fauna and Flora (Montreal protocol for short) , International Convention on Substances that Deplete the Ozone Layer, International Convention on Psychotropic Substances, London Guidelines for the Exchange of Information on Chemicals in International Trade, International Convention on the Prior Informed Consent Procedure for Certain Hazardous Chemicals and Pesticides in International Trade (also known as the Rotterdam Convention) , the Basel Convention on the Control of Transboundary Movements of Hazardous Wastes and Their Disposal, Convention on Establishing the World Intellectual Property Organization, etc.

1. 2. 2　International Trade Conventions

Convention Relating to Contracts of International Sale of Goods

- United Nations Convention on Contracts for the International Sale of Goods (CISG)

The United Nations Convention on Contracts for the International Sale of Goods (CISG; the Vienna Convention for short) is a treaty that is a uniform international sales law. As of the end of 2015, it has been ratified by 84 states that account for a significant proportion of world trade, making it one of the most successful international uniform laws. Vietnam was the most recent state to ratify the Convention, having acceded to it on December 18, 2015.

The CISG was developed by the United Nations Commission on International Trade Law (UNCITRAL), and was signed in Vienna in 1980. The CISG is sometimes referred to as the Vienna Convention (but is not to be confused with other treaties signed in Vienna). It came into force as a multilateral treaty on January 1, 1988, after being ratified by 11 countries.

Generally, the CISG is deemed to be incorporated into (and supplant) any otherwise applicable domestic law (s) with respect to a transaction in goods between parties from different Contracting States.

The CISG has been regarded as a success for the UNCITRAL, as the Convention has been accepted by states from "every geographical region, every stage of economic development and every major legal, social and economic system".

The objectives of CISG are: To adopt uniform rules governing contracts for the international sale of goods; To adopt uniform rules that account for different social, economic, and legal system; To contribute to the removal of legal barriers in international trade; To promote the development of international trade.

The CISG is divided into four parts: Part I: Sphere of Application and General Provisions; Part II: Formation of the Contract (Articles 14 – 24); Part III: sale of goods, obligations of the seller, obligations of the buyer, passing of risk, obligations common to both buyer and seller; Part IV: Final Provisions (Articles 89 – 101) include how and when the Convention comes into force, permitted reservations and declarations, and the application of the Convention to international sales where both States concerned have the same or similar law on the subject.

The CISG applies to contracts of the sale of goods between parties whose places of business are in different States, when the States are Contracting States (Article 1 (1) (a)). Given the significant number of Contracting States, this is the usual path to the CISG's applicability.

The CISG also applies if the parties are situated in different countries (which need not be Contracting States) and the conflict of law rules lead to the application of the law of a Contracting State. For example, a contract between a Chinese trader and a Thailand trader may contain a clause that arbitration will be in Sydney under Australian law with the consequence that the CISG would apply. A number of States have declared they will not be bound by this condition.

The CISG is intended to apply to commercial goods and products only. With some limited exceptions, the CISG does not apply to personal, family, or household goods, nor does it apply to auctions, ships, aircraft, or intangibles and services. The position of computer software is "controversial" and will depend upon various conditions and situations.

China acceded to the CISG in 1986. But, China excluded two stipulations of CISG when submitting the instruments of ratification and accession, i. e. article 11 and (b) of (1) of article 1. The Chinese government formally notified the UN secretary general to withdraw a declaration of the United Nations Convention on Contracts for the international sale of goods in January 2013, which is not subject to the provisions of article eleventh and the provisions relating to the provisions of article eleventh. The withdrawal has officially entered into force.

- The Two Hague Convention on the Unification of the International Sale of Goods

The two Hague convention referred to Convention Relating to a Uniform Law on the International Sale of Goods in 1964 (ULIS for short) and Convention Relating to a Uniform Law on the Formation of the Contract for International Sale of Goods in 1964 (ULFIS for short), made by the International Institute for the Unification of Private Law (UNIDROIT for short).

After 30 years of efforts, 28 countries participated in the Diplomatic Conference held in Hague in April 25, 1964 and adopted ULIS and its annex Unified law of International Sale of Goods, which signed in July 1 st of the same year, come into effective in August 18, 1972. At the same time, the Hague conference in 1964 adopted ULFIS. But till now, only 9 nations have signed, ratified and acceded to the ULIS and ULFIS, including Belgium, Gambia, Germany, Israel, Italy, the Netherlands, San Marino, the United Kingdom, and Luxembourg.

- Convention on the Limitation Period of 1974 and 1980

The Convention on the Limitation Period in the International Sale of Goods (the "Limitation Convention" for short) is a uniform law treaty prepared by the United Nations Commission on International Trade Law (UNCITRAL). It deals with the prescription of actions relating to contracts for the international sale of goods due to the passage of time.

As of February 2016, 30 states have ratified, acceded to, succeeded to, or participated under the Limitation Convention. The Limitation Convention has been signed, but not yet ratified, acceded to, succeeded to, or participated under, by Brazil, Bulgaria, Costa Rica, Mongolia, Nicaragua, and the Russian Federation.

The Limitation Convention was originally prepared as a chapter of a broader treaty on contracts for international sale of goods. Adopted in 1974, it was amended in 1980 to be fully aligned, especially with respect to scope of application, to the United Nations Convention on Contracts for the International Sale of Goods (CISG), adopted in the same year.

United Nation Convention on the Limitation Period in the International Sale of Goods (1974 Limitation for short) was made and approved by UNCITRAL at New York on June 14 1974. It is Substantive law Convention on the period of extinguishing right to be relevant to the international sale of goods. The Convention establishes a unified rule on the time limit for legal proceedings arising from international sales contracts. In order to to make 1970 Limitation more adapt to the United Nations Convention on Contracts for the international sale of goods in 1980 (CISG for short), In April 1980 at Vienna, the United Nations Diplomatic Conference adopted CISG, at the same time, approved the the Protocol Amending the Convention on the Limitation Period in the International Sale of goods (1980 Protocol for short). Convention on Limitation Period in the International sale of goods as Amended by 1980 Protocol (1980 Limitation Convention for short) has come into force from August 1st 1988. By the end of 2005, the former has 25 participating countries, and the latter has 18 participating countries.

These two limitation Convention, are only applicable to international transactions, and the contract within its scope of application shall be exempted from appealing to the rules of international private law. International contracts outside the scope of application of the Convention and other contracts subject to application of other effective chosen law, are not affected by the Limitation Conventions. The purely domestic sales contract is not affected by the Time Limitation Conventions, but is subject to the domestic laws and regulations. China hasn't acceded to either the 1974 Limitation Convention or the 1980 Limitation Convention.

Conventions Relating to International Payment

● Geneva Convention on the Unification of laws Relating to Bills of Exchange and Promissory Notes

Convention on the Unification of the Law Relating to Bills of Exchange and Promissory Notes is also known as Convention Providing a Uniform of Law for Bills of Exchange and Promissory Notes 1930 (ULB), was made and approved by the International Union on June 7, 1930 at the first uniform meeting of the negotiable instruments law in Geneva, which came into effect from January 1, 1934.

Geneva unified law system only solves the two bill law system conflict between France and Germany, and many European countries, Japan and some Latin American countries have adopted Geneva Conventions on the laws of negotiable instruments. Because of the provisions of the Geneva Convention are in conflict with the traditions and practices of the notes, the Anglo American law system countries refused to accede to these conventions. Therefore, two major legal systems in the international law of negotiable instruments have been formed, common law system and Geneva unified bill law system.

● Convention on International Bill of Exchange and International Promissory Note of the United Nations

The United Nations Convention on International Bills of Exchange and International Promissory Notes was adopted by the United Nations General Assembly, forty-third session in 1988, and open for signature. In accordance with the relevant provisions of the Convention, the Convention shall come into force upon approval or accession by at least 10 states, but till now, there are only 5 countries which have ratified this convention, the Convention has not yet entered into force.

● Convention on International Factoring (Ottawa, 1988)

Convention on International Factoring was approved by UNIDROIT in 1988 at Ottawa and came into force in 1995. And till now, more than 130 countries has acceded to it.

International Conventions Relating to the Carriage of Goods

(Ⅰ) Conventions relating to international marine transport

There are three international conventions which plays an important role in the unification of the relevant laws and regulations of the bill of lading, such as International Convention for the Unification of Certain Rules of Law Relating to Bill of Lading (Hague Rules for short), Protocol to Amend the International Convention for the Unification of Certain Rules of Law Relating to Bill of Lading (Hague-Visby Rules for short), United Nations Convention of the Carriage of Goods by Sea, 1978 (Hamburg Rules for short), and United Nations Convention on Contracts for the International Carriage of Goods Wholly or Partly by Sea (Rotterdam Rules for short).

● Hague Rules

The Hague Rules of 1924 is an international convention to impose minimum standards upon commercial carriers of goods by sea. Previously, only the common law provided protection to cargo-owners; but the Hague Rules should not be seen as a "consumers' charter" for shippers because the 1924 Convention actually favored carriers and reduced some of their obligations to shippers.

The Hague Rules represented the first attempt by the international community to find a workable and uniform means of dealing with the problem of ship owners regularly excluding themselves from all liability for loss or damage to cargo.

The Hague Rules form the basis of national legislation in almost all of the world's major trading nations, and probably cover more than 90 percent of world trade. The Hague Rules have been updated by two protocols, but neither addressed the basic liability provisions, which remain unchanged.

● Hague-Visby Rules

The Hague-Visby Rules is a set of international rules for the international carriage of goods by sea. They are a slightly updated version of the original Hague Rules which were drafted in Brussels in 1924.

The official title of the Hague Rules is the "International Convention for the Unification of Certain Rules of Law relating to Bills of Lading". After being amended by the Brussels Amendments (officially the "Protocol to Amend the International Convention for the Unification of Certain Rules of Law Relating to Bills of Lading") in 1968, the Rules became known colloquially as the Hague-Visby Rules.

The premise of the Hague-Visby Rules (and of the earlier English common law from which the Rules are drawn) was that a carrier typically has far greater bargaining power than the shipper, and that to protect the interests of the shipper/cargo-owner, the law should impose some minimum affreightment obligations upon the carrier. However, the Hague and Hague-Visby Rules were hardly a charter of new protections for cargo-owners; the English common law prior to 1924 provided more protection for cargo-owners, and imposed more liabilities upon "common carriers".

A final amendment was made in the SDR Protocol in 1979. Many countries declined to adopt the Hague-Visby Rules and stayed with the 1924 Hague Rules. Some other countries which upgraded to Hague-Visby subsequently failed to adopt the 1979 SDR protocol.

- Hamburg Rules

The Hamburg Rules are a set of rules governing the international shipment of goods, resulting from the United Nations International Convention on the Carriage of Goods by Sea adopted in Hamburg on March 31 1978. The Convention was an attempt to form a uniform legal base for the transportation of goods on oceangoing ships. A driving force behind the convention was the attempt of developing countries' to level the playing field. It came into force on November 1 1992. As of October 2014, the convention had been ratified by 34 countries.

Some countries did not ratify any of the above mentioned conventions, but drafted their own Maritime Law by reference to the contents of the three conventions, such as China. In addition, some countries did not join any convention, and there is no relevant domestic law clearly in their countries. The inconsistent transport rules have brought many inconvenience to international trade, affected the free transfer of goods, increased transaction costs. This phenomenon has aroused great attention of the international community, the voice of building a unified rule is rising day by day in the field of international maritime transport of goods.

- Rotterdam Rules

The "Rotterdam Rules" is a treaty proposing new international rules to revise the legal framework for maritime affreightment and carriage of goods by sea. The Rules primarily address the legal relationship between carriers and cargo-owners. As of October 2015, the Rules are not yet in force as they have been ratified by only three states.

The aim of the convention is to extend and modernize existing international rules and achieve uniformity of International trade law in the field of maritime carriage, updating and/or replacing many provisions in the Hague Rules, Hague-Visby Rules and Hamburg Rules. The convention establishes a modern, comprehensive, uniform legal regime governing the rights and obligations of shippers, carriers and consignees under a contract for door-to-door shipments that involve international sea transport.

The Rotterdam Rules will enter into effect a year after 20 countries have ratified that treaty. As of August 9 2011, there are 24 signatories to the treaty. The most recent country to sign the treaty was Sweden, which signed on July 20 2011. Spain was the first country to ratify the convention in January 2011. China, Japan, the shipping and trading countries such as Germany, the United Kingdom, Italy, Canada, Australia have not signed.

(Ⅱ) *Conventions relating to international land transport*

Conventions relating to international land transport mainly are Berne Convention concerning International Carriage by Rail (COTIF for short), Agreement concerning International Carriage of Goods by Rail (CMIC for short) . COTIF and CMIC are the two main rules of international railway freight transportation. COTIF is the consolidation of International Convention concerning International Carriage by Rail (CIM for short) and International Convention concerning Carriage of Passengers and Luggage by Rail (CIV for short) .

CIM was signed in 1961 at Bern and come into effective from January 1, 1975. Its member states include 28 countries, such as France, Germany, Belgium, Italy, Sweden, Spain Iran, Iraq, Syria, the northwest of Algeria, the northwest of Morocco, Tunisia etc. CMIC was entered into in 1951 at Warsaw, China has acceded to CMIC in 1953. Revised on July 1, 1974, there are 12 member states such as the former Soviet Union, Eastern Europe countries, China, North Korea, Vietnam. The Eastern European countries of CMIC are members of CIM, In this way, the import and export goods of CMIC countries can be transported to the member countries of CIM through the railway, which provides a more favorable condition for the communication of carriage of goods by international railway. China is a member of CMIC, and all import and export goods by rail transport shall be handled according to the provisions of CIMC.

(Ⅲ) *Conventions relating to international air transport*

The Convention for the Unification of certain rules relating to international carriage by air, commonly known as the Warsaw Convention, is an international convention which regulates liability for international carriage of persons, luggage, or goods performed by aircraft for reward. Originally signed in 1929 in Warsaw Poland and amended several times, China joined the convention in 1957 and the convention entered into force in China in October 1958.

Warsaw Convention was revised by its Hague Protocol in 1955 and by Guadalajara Convention in 1960. The Hague Protocol, officially the Protocol to Amend the Convention for the Unification of Certain Rules Relating to International Carriage by Air, is a treaty signed in the Hague. It has come into force on August 3 rd, 1963 and China has acceded to the Hague Protocol in 1975. But China hasn't acceded to Guadalajara Convention. They serve to amend the Warsaw Convention. While officially the Hague Protocol is intended to become a single entity with the Warsaw Convention, it has only be ratified by 137 of the original 152 parties to the Warsaw Convention. The binding version of the treaty is written in French, but certified versions also exist in English and Spanish. The official depository of the treaty is the Government of Poland.

The Montreal Convention, formally the Convention for the Unification of Certain Rules for International Carriage by Air (Montreal Convention for short), is a multilateral treaty adopted by a diplomatic meeting of ICAO member states in 1999. It amended important provisions of the Warsaw Convention's regime concerning compensation for the victims of air disasters. The Convention attempts to re-establish uniformity and predictability of rules relating to the international carriage of passengers, baggage and cargo. Whilst maintaining the core provisions which have served the international air community for several decades (i. e. , the Warsaw regime), the new treaty achieves modernization in a number of key areas. As of the end of 2015, there are 119 parties to the Convention. 118 of the 191 ICAO Member States have ratified the Montreal Convention. including Argentina, Australia, Brazil, Canada, China, all member states of the European Union, India, Israel, Japan, South Korea, Malaysia, Mexico, New Zealand, Norway, Pakistan, Saudi Arabia, Singapore, South Africa, Switzerland, Turkey, Ukraine, the United Arab Emirates, and the United States.

(Ⅳ) *Conventions relating to international multimodal transport*

The United Nations Convention on international multimodal transport of goods was adopted by the 84 members of UNCTAD unanimously at the second meeting of the United Nations Conference on international multimodal transport in Geneva in May 24, 1980. The Convention consists of 40 articles and an annex. The structure is divided into eight parts, such as general, documents, multimodal transport operator liability, shipper's liability, claims and litigation, supplement regulations, customs matters and final provisions and so on. China has signed it. But it did not attract the necessary number of ratifications and thus has not entered into force.

1. 2. 3　International Trade Customs and Practices

The international trade practice, also known as the "international business practice", "international trade regulations", are widely recognized and accepted habitual practices, rules and interpretations by both buyers and sellers and others engaged in related international trade activities, and to serve as a dispute resolution regulations if it doesn't breach the laws, public welfare, bone fides or other contract items. Because international trade covers a very wide field, the range of international trade customs and practice is relatively wide. The following are the main types of international trade practices.

International Customs and Practices Relating to Trade Terms

In order to standardize the understanding of the seller and buyer relating to their obligations in international sales agreement, various nomenclatures have been developed that use abbreviations, such as ex-works, FOB, CIF, landed, and so on. While these shorthand abbreviations can be useful, they can also be sources of confusion. The International Chamber [INCO] has developed the "Incoterms," which were revised in 2010. There are also the Revised American Foreign Trade Definitions and the Warsaw Terms. Although these abbreviated terms of sales are similar, they also differ from nomenclature to nomenclature, and it is important to specify in the sales agreement which nomenclature is being used when an abbreviation is utilized.

- Warsaw-Oxford Rules 1932

In order to make uniform provisions and interpretation on the responsibilities of the two parties under CIF terms, the International Law Association held a meeting in Warsaw in 1928, and worked out the Uniform Rules for CIF Sales Contracts, which was called Warsaw Rules 1928, and renamed Warsaw-Oxford Rules 1932 at the Oxford Convention and includes 21 clauses. The rules provide a set of uniform regulations applicable for those transactions between buyers and sellers under CIF trade terms, it would be used for the parties that wished to adopt it. When both parties lack the standard form of contract or general trading conditions, they can agree to adopt this plan. Warsaw-Oxford rules has been in use ever since published in 1932, and has become a very influential international trade practice. It is mainly used to indicate the nature and characteristic of the CIF contract and also to stipulate the responsibilities of the two parties under CIF terms.

- Revised American Foreign Trade Definitions 1941

In 1919, nine American Foreign commercial groups drew up the US Export Quotations and Abbreviations in 1919, then revised in 1941 and renamed Revised American Foreign Trade Definitions

1941. It was adopted by the American Chamber of Commerce, the National Importers Association and the American Foreign Trade Association in the same year. It defines six trade terms, i. e. , Ex-point of origin, FOB, FAS, C&F, CIF and Ex-Dock. Notably, The Definition is divided FOB into six types, except the fifth kinds, namely FOB (FOB Vessel), the other trade terms are explained quite differently from those in INCOTERMS, and the Definitions requires the buyer to bear the costs and responsibilities related to the export customs formalities under all the FOB terms. It varies from the definition of Incoterms greatly. In order to specify respective obligations of the buyers and sellers under every trade term, the various terms of trade listed in the Definitions are in general followed by a series of explanatory note. These notes are an inseparable part of the definitions of trade terms. The Definitions also have a certain influence in international trade practices, these trade terms are often adopted in the United States of America, Canada and some other countries in America.

- International Rules for the Interpretation of Trade Terms

The Incoterms Rules or international Commercial Terms are a series of pre-defined commercial terms published by the International Chamber of Commerce (ICC) . First published in 1936, the Incoterms rules have been periodically updated, with the eighth version—Incoterms 2010—having been published on January 1, 2011. "Incoterms" is a registered trademark of the ICC.

The Incoterms are widely used in International commercial transactions or procurement processes, and accepted by governments, legal authorities, and practitioners worldwide for the interpretation of most commonly used terms in international trade. The purpose of Incoterms is to reduce or remove altogether uncertainties arising from different interpretation of the rules in different countries. As such they are regularly incorporated into sales contracts worldwide. The trade terms of the International Chamber of Commerce have adapted themselves to the changes of the development of international trade, and based on the principle of party autonomy with the nature of arbitrary law. Therefore, while negotiating the contract, the seller and the buyer can directly apply to certain practices, can also change, modify the rules or add any other terms, whether it is necessary to apply to the above mentioned rules, depending on the parties voluntarily. The Incoterms is widely used in the foreign trade practices of China, if the parties conclude the business on CIF basis they may also apply to the provisions of the "Warsaw – Oxford rules", if import from the United States and Canada on FOB basis, they should also consider the special interpretation and application of FOB terms of "1941 revised American foreign trade definitions" upon the signing of the contract and performing the contract.

Customs and Practices Relating to International Settlement

- Uniform Customs and Practice for Documentary Credits

The Uniform Customs and Practice for Documentary Credits (UCP) is a set of rules on the issuance and use of letters of credit. To clear the rights, responsibilities of all parties concerned, payment terms and definitions in Credit operations, to reconcile the contradictions between the parties concerned, to alleviate the confusion caused by individual countries' promoting their own national rules on letter of credit practice, ICC (International Chamber of Commerce) firstly published the

UCP in 1933 and subsequently updated it throughout the years. The ICC has developed and moulded the UCP by regular revisions, the current version being the UCP600. It was approved by the Banking Commission of the ICC at its meeting in Paris on October 25 2006. This latest version formally commenced on July 1 2007.

The UCP is utilized by bankers and commercial parties in more than 175 countries in trade finance. Some 11% – 15% of international trade utilizes letters of credit, totaling over a trillion dollars (US) each year. The UCP has become the international customs of the credit.

- The Uniform Rules for Collection

In order to unify the collection business practices, to alleviate the contradictions and disputes caused by all parties involved, ICC (International Chamber of Commerce) made the Uniform Rules for Collection (URC for short) in 1958. And then the URC was revised in 1978 and 1995. Its latest version is URC522 (1995 Revision) (ICC Publication No. 522) . The URC is widely used by bankers from many countries, since its publication, it has become the international customs in collection business.

Customs and Practices Relating to International Transportation Insurance

- Institute Cargo Clauses

In the international transport insurance business, the British insurance regulations system, especially the insurance policy and insurance clauses, has a great effect on the insurance clauses of the world. At present, most countries in the world in marine insurance business directly adopt the London Institute Cargo Clauses (I. C. C for short) made by the Institute of London Underwriters.

In the 19 th century, Lloyd's and the Institute of London Underwriters (a grouping of London company insurers) developed between them standardized clauses for the use of marine insurance, and these have been maintained since. These are known as the Institute Clauses because the Institute covered the cost of their publication. I. C. C. first established in 1912, after several revisions, the latest version was completed in 2008, which has come into force since January 1st, 2009.

- York-Antwerp Rules

"York-Antwerp rules" were first promulgated in 1890 and have been amended several times, last amended from May 2004 to June. International Maritime Committee in Vancouver, Canada revised York Antwerp Rules 1994, and the revised rules called York Antwerp Rules 2004, having come into force 1 st January 2005. Every amendment to the rules will not lead to the abolition of the old rules, in 1974 and 2004 of the rules of international common "York-Antwerp rules", and choose to use. York-Antwerp Rules 1974 and York-Antwerp Rules 2004 are commonly used in the world at present, providing two alternative provision, the choice among which is left to each adopting parties concerned.

Although the rules are not international conventions, there are usually clauses regarding the adjustment of general average in the contract of charter companies or on the bill of lading in accordance with York-Antwerp rules, this rules in practice have managed to avoid the negative impact caused the differences of general average system, have become international customs and practices and widely accepted by the international shipping and insurance industry.

Customs and Practices Relating to International Arbitration

The UNCITRAL Arbitration Rules were made by the UNCITRAL in 1976 and revised in 2010, which have been effective since August 15 2010. The rules are selected by parties either as part of their contract, or after a dispute arises, to govern the conduct of an arbitration intended to resolve a dispute or disputes between themselves.

The framework of commercial arbitration on the level of legislation is composed of Convention on the Recognition and Enforcement of Foreign Arbitral Awards (New York, 1958) (the "New York Convention" for short), UNCITRAL Arbitration Rules in 1976, and UNCITRAL Model Law on International Commercial Arbitration constitute in 1985. And among them UNCITRAL Arbitration Rules has been recognized as one of the most successful contractual text of international documents since its adoption, and has been widely used in Ad hoc arbitration, investment disputes arbitration between investor and the host country, inter-state arbitration and commercial arbitration under the management of Standing Arbitration Institutions.

1.3　Obtaining the International Trade Qualification

1.3.1　Organizing for Export and Import Operations

As we discussed in the previous section, exporting and importing requires that certain personnel must have specialized knowledge. The personnel involved and their organization vary from company to company, and sometimes the same personnel have roles in both exporting and importing. In small companies, one person may perform all of exports or imports, while in large companies or companies with a large amount of exports or imports, the number of personnel may be large. As business increases, specialties may develop within the department, and the duties performed by any one person may become narrower.

Export Department

For many companies, the export department begins in the sales or marketing department. That department may develop leads or identify customs located in other countries. Because the export order may require special procedures in manufacturing, credit checking, insuring, packing, shipping, and collection, It will be necessary for specific personnel to interface with freight forwarders, couriers, banks, packing companies, steamship lines, airlines, translators, government agencies, domestic transportation companies, and attorneys. Because most manufacturers have personnel who must interface with domestic transportation companies, often additional personnel will be assigned to that department to manage export shipments and interface with other outside services. The number of personnel needed and the assignment of responsibilities depend upon the size of the company and the volume of exports involved.

Import Department

A manufacturer's import department often grows out of the purchasing department, whose personnel have been assigned the responsibility of procuring raw materials of components for the manufacturing process. For importers or trading companies that deal in finished goods, the import department may begin as the result of being appointed as the China distributor for a foreign manufacturer or from purchasing a product produced by a foreign manufacturer that has China sales potential. Increasingly, a number of China manufacturers are moving their manufacturing operations overseas to cheaper labor regions and importing products they formerly manufactured in China. it will be necessary for the import department to be in contact with foreign freight forwarders, China customs brokers, banks, China Customs, marine insurance companies, and other service companies.

Combined Export and Import

In many companies, some or all of the functions of the export and import departments are combined in some way. In smaller companies, where the volume of exports or imports does justify more personnel, one or two persons may have responsibility for both export and import procedures and documentations. As companies grow larger or the volume of export/import business increases, these functions tend to be separated more into export departments and import departments. Some activities such as being in contact with some of the same outside parties (typically banks, those freight forwarders that are also customs brokers, or domestic transportation companies) may be consolidated specific persons for both export and import, while other personnel will work exclusively on exports or on imports.

1.3.2　Obtaining the Foreign Trade Dealer Qualification

To carry out international trade in China needs to have the license to run import and export business trade. According to the provisions of the State, the export consists of self-support export and principal-agent exports. If a company plans to import or export by itself, it needs to go through all the procedures and formalities in this section. But if a company plans to appoint an agency to import or export, it will be the agency's responsibilities to do all the procedures and formalities in this section.

Foreign trade dealers refer to the legal entities, other organizations and individuals engaging in foreign trade dealings in compliance with the provisions of the Foreign Trade Law of the P. R. C. (2004) and other administrative regulations, departmental rules, the rules and other related regulations. The management system of foreign trade dealers is one of China's foreign trade management systems.

At present, Chinese governments implement the system of the Record and Registration of Foreign Trade Operators. According to Article 9 of the Foreign Trade Law of the P. R. C. (2004), a foreign trade dealer who intends to engage in the import and export of goods or technologies shall register with the department for foreign trade under the State Council or the body it entrusts with the registration, unless otherwise prescribed by laws, administrative regulations or by the said department. Where a foreign trade dealer fails to register as required by regulations, the Customs shall not process the procedures of declaration, inspection and release for the import or export of goods. Foreign trade dealers may act as an agent for others carrying out foreign trade business within the scope of business. Procedures for opening a trade company and obtaining the international trade qualification usually include the following several steps.

Handling the Business Registration Formalities

(I) *For new entrants*

● Apply to the local Industrial and Commercial Administration Bureau for business registration. After registration, the applicant shall get a legal business license with "Import and Export of Goods" in its business scope.

● Apply to local Quality and Technical Supervision Security for organization code certificate.

● Open an account in a bank with business license and organization code certificate.

● Apply to tax authority for tax registration.

(II) *For registered enterprises*

Apply to the local Industrial and Commercial Administration Bureau for expanding business scope. After registration, the applicant shall get a legal business license with "Import and Export of Goods" in its business scope.

● Registering with the Authority Responsible for Foreign Trade

● Applying to Tax Authority for Tax Registration of Alteration within 30 Days after Getting the Above Registration Form.

● Applying to Local Customs Office for Registration in Order to Obtain the Declaration Rights.

● Applying to General Administration of Quality Super vision, Inspection and Quarantine of P. R. C. for Registration in Order to Get the Right of Applying for Quarantine and Inspection.

● Applying to the Local Office of State Administration of Foreign Exchange for Obtaining the Right of Settling Foreign Exchange Payment and Receipt and Ensuring Compliance with SAFE Requirements.

● Applying to China E-port for Registration.

1.3.3   Obtaining the International Trade Quota or License

As one of the trade measures of non-tariff barrier, import and export licensing system is a common means of import and export trade management all over the world. It is an administrative measure which charged by the relevant departments of foreign trade through licensing the import and export control of goods. It's an integral part of an important component in National Foreign Trade Policy.

The import and export licensing systems of goods and technology are the main parts of the management system of China's import and export license, the range of management systems including three categories such as prohibiting, limiting, free import and export of goods and technology. If you want to tell whether one good that you are going to export or import belongs to one of the prohibiting, limiting, free categories, you must determine according to its tariff categories or Harmonized System number (HS code). Furthermore, the kinds of these goods are adjusted constantly by our government according to the changing conditions of international trade. So the foreign trade dealers shall always pay attention to the new policies and regulations on the websites of Ministry of Commerce of P. R. C. , General Administration of Customs of P. R. C. and other relevant authorities.

Applying for the Import or Export Quota or License

(I) If the applicant shall apply for the quota by Internet

• The applicant shall enter the online application system (http: //egov. mofcom. gov. cn) through the website of the Ministry of Commerce of P. R. C. . Then the applicant choose to enter the specific system of specific kind of quota or license.

• The applicant fills in and submits the application form online according to the instructions of the system and relevant laws and regulations.

• The applicant delivers the required printed documents to the nominated authority after receiving notice from the online system.

(II) If the applicant shall apply for the quota or license by written, he/she shall directly deliver the printed application form and all other required documents to the nominated authority. For instance, if the applicant applies for some pesticides import permit, he/she may need to apply to China's Agriculture Ministry by written; if the applicant wants to import some specific artwork, he/she may need to apply to Ministry of Culture of P. R. C for approval by written.

• The nominated authority examines the application and relevant documents delivered by the applicant according to the relevant laws and regulations, then notifies the applicant of the result by the online system or by announcement.

• The applicant gets the quota certificate or license from the nominated authority after obtaining the approved notice or announcement from the authority.

International Trade Terms

- Understand the nature for international trade terms, and the purposes of using trade terms in international business.
- Know three basic rules of international trade terms.
- Know the interpretations of trade terms in Incoterms® 2010.
- Master the six main trade terms in Incoterms® 2010.

The seller and buyer in international trade, in general, rarely use cash on delivery. Normally, they will use international trade terms, such as "FOB" "CFR" or "CIF", to identify where to deliver the goods, what price is, and how to clarify the risk, responsibilities and the expenses between the seller and the buyer. Different trade terms define different delivery place, different rights and obligations of the buyer and seller. When the buyer and seller agree to use certain trade term in the contract, other terms of the contract should keep in line with it, take "FOB" term as an example, under FOB term, the buyer should take the responsibility of booking and chartering a ship, as well as paying for the ocean freight, in addition, the buyer needs to effect the insurance and pays the premium. Therefore, even for the same batch of goods, the price will be different when adapting different trade terms. Therefore, it is the trade terms that determine the nature of the contract, as well as the rights and obligations of the buyer and seller.

2. 1　International Trade Terms and Its Basic Rules

2. 1. 1　The Nature of International Trade Terms

Trade Terms Are the Terms of Delivery

Trade terms define the division of the responsibilities, expenses and risks between buyers and sellers during the delivery of goods, which we called terms of delivery, that is the point of which delivery occurs, i. e. , the point at which the risk of loss or damage transfers from the seller to the buyer. For example, under the FOB (Free on Board) term, and DDP (Delivered Duty Paid) term, the rights and the obligations of both parties also differ from the place of passage of title and risk of loss.

Trade Terms Specify the Component of Commodity Price

Different trade terms contain different ancillary expenses, so even the same transaction will quote differently under different trade terms. For example, FOB price does not include freight and premium from the port of shipment to the port of destination, while the CIF price does, that is why FOB price is always lower than CIF price. Thus, the final price of a contract determined by which trade term is chosen by both parties.

2. 1. 2　The Purpose of Trade Terms

Simplify Transaction Procedures and Promote the Business

In international trade, trade terms can significantly simplify the business negotiation, and promote the trade transaction. International trade always involves many procedures, which should be completed either by the seller or buyer. For example, who should effect marine cargo insurance? Who should book shipping space? Who should apply for the import and/or export license? And so on. It is very time-consuming for the exporter and importer to negotiate the obligations and the cost is involved in every procedure listed above for transaction. But so long as the buyers and sellers agree on trade terms by which the transaction goes, it can easily specify the respective responsibility, expenses and risk that should be borne by the buyer and seller, in this way the trade terms can simplify trade procedures, help buyers and sellers save time consultation, then make a deal quickly.

Facilitate the Buyers and Sellers of Cost Accounting and Pricing

Since trade terms stand for the components of the international price of a commodity, different trade terms including different costs, the price quoted will vary due to different trade terms. While quoting a price, it is necessary for the buyer and seller to consider the costs that contain in the trade terms, such as freight, insurance, handling charges, tariffs, which is conducive to both buyers and sellers on price comparisons and cost accounting.

2. 1. 3　**Three Sets of Rules**

International trade rules, refer to the practices and interpretations that is generally recognized and gradually formed with international trade practice. It includes: some interpretations or rules that use to explain some aspects of international trade, such as trade terms, payment terms etc. by some international organization; traditional practices on some of the major international ports and terminals; practices from different industries. In addition, the typical cases or judgments awarded by judicial authority or arbitration organization, also regarded as part of international trade practices. There are three main international trade practices relating to trade terms: "*Warsaw-Oxford Rules 1932*", "*Revised American Foreign Trade Definitions 1941*", "*International Rules for the Interpretation of Trade Terms 2010*".

Warsaw-Oxford Rules 1932

This rule is designed by International Law Association to explain the CIF (Cost, Insurance and Freight) contract. During the 1928 convention of International Law Association held in Warsaw, Poland, a set of rules relating to CIF contract were established— "Warsaw rules, 1928", which was amended on Oxford Conference in 1932, still in use now. It combined into 21 provisions. The rule mainly describes the characters of a CIF contract, and specifies the division of responsibilities and methods of transfer of cargo ownership between buyer and seller, while using CIF trade terms. When the parties conclude a contract base on "Warsaw-Oxford Rules", it means that they agree to make a CIF contract, and "Warsaw Oxford Rules" will have a binding. However, if these rules conflict with the stipulations of the contract, the contract shall prevail. As the "Warsaw Oxford Rules" only provide interpretations and regulations for CIF terms, it rarely used by traders in practice.

Revised American Foreign Trade Definitions 1941

In 1919, 9 business communities in the United States devised "the U. S. Export Quotation and Abbreviations", which was revised in 1941, and then the "*Revised American Foreign Trade Definitions 1941*" was adopted in the same year by American Chamber of Commerce, the American Export-Import Association and the American National Foreign Trade Association. The definition provide explanation to six trade terms:

Ex point of origin

FOB (Free on Board)

FAS (Free Along Side)

C&F (Cost and Freight)

CIF (Cost, Insurance and Freight)

Ex dock (named port of importation)

The above definitions are most commonly used by the United States, Canada and some other American countries, but they are rarely used internationally due to their content have great differences from other general interpretation. In recent years, American business communities or trade organizations consider abandoning this "definition", and use "International Incoterms", which was devised by International Chamber of Commerce. It should be noted that the rule is still used in the Americas, so when these countries in international trade, the parties agree to adopt this rule, it is necessary to specify that the contract is subject to "Revised American Foreign Trade Definitions 1941", otherwise it is non-binding.

International Rules for the Interpretation of Trade Terms 2010

Incoterms rules was made by the International Chamber of Commerce in 1936 firstly. Till now, it has been revised for seven times and so has eight versions totally. Nowadays, Incoterms as an international trade practices, is more and more universally recognized and applicable by international trade practice, it has become an important rule for both parties to sign and fulfill a contract, as well as to solve business disputes.

The main reason for successive revisions of Incoterms is to adapt them to contemporary commercial practice. "Incoterms" contained 8 trade terms as amended in year 1953, two another trade terms were added in 1967 version, that is the "Delivered at Frontier" (DAF) and "Delivered Duty Paid" (DDP); later the "Departure airports delivery" (FOA) was added in 1976 version; in 1980 the "Free Carrier (FRC)" and " freight, insurance paid to (destination) (CIP)" were added, until then, Incoterms 1980 has already contained 14 trade terms. Further, revision in 1990 was to adapt itself to the increased use of Electronic Data Interchange (EDI) in business transaction, Incoterms 1990 has 13 trade terms, and it came into force on 1st July, 1990. Revision in 2000 took account of the spread of customs-free zones in the 1990s.

The latest version, Incoterms® 2010, has come into force since 1 st January, 2011. This revision takes account of the continued spread of customs-free zones, the increased use of electronic communications in business transactions, heightened concern about security in the movement of goods and consolidates in transport practices. Incoterms® 2010 deleted four trade terms in Group D of Incoterms 2000, that is "Delivered Duty Unpaid (DDU)" "Delivered at Frontier (DAF)" "Delivered ex Ship (DES)" "Delivered ex Quay (DEQ)", but added two new trade terms for Group D "Delivered at Terminal (DAT)" and "Delivered at Place (DAP)" for substitution.

Needless to say, every revision of Incoterms is to improve all the trade terms to facilitate their practical implementation. It is notable that even though the new Incoterms entered into force, the Incoterms 2000 can still be used in doing international trade. It is important however to clearly specify the chosen version Incoterms 2010, Incoterms 2000 or any earlier version, through such words as, "the chosen Incoterms 2010 including the named place, followed by Incoterms 2010".

Although international trade practices play a role in the settlement of trade disputes, we should pay attention to the following questions:

● International trade practice is not legal, so it is not binding on both parties, it may be or may not be used.

● If the buyer and seller in the contract agree to adopt certain practice, then the practice will be binding for both parties.

• If the contract specify that it adopts to certain practice, when which is contrary to the stipulation in the contract, as long as the stipulation do not conflict with the national law, it will be recognized and protected by the relevant laws. In other words, in such case, both parties should be binding by the stipulation of the contract.

• If the contract neither specifies on an issue clearly, nor designates certain practice adopted, when disputes happen and submitted for litigation or arbitration, the court or arbitration commission can make a judgment base on the relevant practices.

While doing import and export business, we need to understand and master some international trade rules, which are essential while negotiating business, signing a contract, fulfilling the contract, as well as resolving the disputes. When disputes arise, we can struggle for our rights base on these rules.

2.2　International Rules for the Interpretation of Trade Terms 2010

Incoterms® 2010 updates and consolidates the delivered rules, reduce the total number of rules from 13 to 11, and offer a simpler and clearer presentation of all the rules. Incoterms® 2010 is also the fist version of the Incoterms rules to make all references to buyers and sellers gender-neutral.

2.2.1　Classification of Incoterms 2010

Incoterms® 2010 is a bit different from Incoterm 2000. Incoterms 2000 divides the thirteen trade terms into four categories of group E, F, C, D. The first group is "E" -terms (EXW), the seller only makes the goods available to the buyer at the seller's own premises. The second group is "F" -terms (FCA, FAS and FOB), the seller is called upon to deliver the goods to a carrier appointed by the buyer; contracts under F-terms are shipment contract. The third group is "C" -terms (CFR, CIF, CPT and CIP), the seller has to contract for major carriage, but without assuming the risk of loss or damage to the goods or additional costs due to events occurring after shipment or dispatch. Contracts under C-terms are shipment contract as well. The most important feature for group C is that the division of relevant risk and costs are separated. The fourth group is "D" -terms (DAF, DES, DEQ, DDU and DDP), the seller has to bear all risk and costs needed to bring the goods to the place of destination; Contracts under "D" -terms are arrival contract.

Incoterms® 2010 grouped 11 three-letter trade terms into two categories according to the mode of transport (maritime vs. any other mode [s] s).

Group 1: Incoterms that apply to any mode of transport

They are EXW (Ex Works), FCA (Free Carrier), CPT (Carriage Paid to), CIP (Carriage and Insurance Paid to), DAT (Delivered at Terminal), DAP (Delivered at Place) and DDP (Delivered Duty Paid) .

These seven kinds of trade terms not only apply to any mode of transport, but also apply to where more than one mode of transport. In addition, they can be used when the vessel carries just part of the consignment.

Group 2: Incoterms that apply to sea and inland waterway transport only

They are FAS (free alongside ship), FOB (free on board), CFR (cost and freight) and CIF (cost, insurance and freight).

Under these trade terms, the sellers should deliver the goods to buyers at the port, therefore, they are classified as "suitable terms of maritime and inland waterway". As for terms FOB, CFR, and CIF in Incoterms® 2010, "delivery" means when the goods have been put "on board of vessel", not when the goods "across the ship's rail" as stipulated in Incoterms 2000. Such modifications are more practical in international business.

Figure 2.1 illustrates the critical points for the division of relevant risks and costs under the eleven terms. When using this chart, it needs to be born in mind that all the issues are discussed from the seller's perspective.

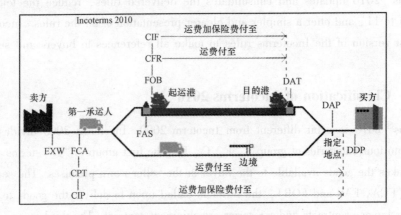

Figure2.1   Incoterms Chart

2.2.2   The Main Features of Incoterms® 2010

(I) Incoterms® 2010 reduces the total number of trade terms from 13 to 11. This change is accomplished by using two new trade terms DAT (Delivery at Terminal) and DAP (Deliver at Place) which are applied to any mode of transport, instead of terms DAF (Delivered at Frontier), DES (Delivered Ex Ship), DEQ (Delivered Ex Quay) and DDU (Delivered Duty Unpaid) in Incoterms 2000.

For these two new terms, delivery occurs at the nominated place. While using DAT terms, the goods arrive at the nominated destination by the disposal of buyer, and unloaded from the transporting vehicle (the same as the previous terms DEQ). While using DAP terms, the goods also arrive at the nominated place by the disposal of the buyer, but only well prepared for discharge (the same as previous DAF, DES and DDU).

（Ⅱ）It is required that in Incoterms® 2010 the parties, including the buyer, the seller and the transport contractors should provide information whether the goods involved can meet the security requirements. This will help the shipping company to know whether or not the cargo carried by the ship violating the Dangerous Goods Rules, in this way, they could prevent the dangerous cargo loaded onto the ship's hold without the relevant security documents. With the development of electronic transport in the current international trade market, Incoterms® 2010 specifies that electronic documents can replace paper documents within the agreement of buyer and seller, they have the same legal force.

（Ⅲ）It is noteworthy that there is no longer any "ship's rail" concept in Incoterms® 2010. In the interpretation of FOB, CFR and CIF terms, the concept of "ship's rail" is deleted and replaced by "loading on board". That is in Incoterms® 2010, the sellers will bear all the risk before "the goods have been loaded on board of the vessel", not as "the goods passing the "ship's rail" in Incoterms 2000.

（Ⅳ）Taking into account the intra-trade among large regional trading blocs, it is specified that the Incoterms® 2010 are available for application to both international domestic sales contracts. And it will further convergence with the "United Nations Convention on Contracts for the International Sale of Goods（CISG）" and "Rotterdam Rules".

The new international Incoterms will further promote the development of international trade, and help in resolving international disputes.

2.2.3　Main Trade Terms in Incoterms 2010

FOB Free on Board （...named port of shipment）

Definition

FOB is an Abbreviation for "free on board", when using this term, the port of shipment shall be nominated behind FOB, for example, if Guangzhou is the port of shipment, said the FOB Guangzhou. FOB means that the seller delivers the goods on board the vessel nominated by the buyer at the named port of shipment or procures the goods already so delivered. The risks of loss of or damage to the goods passes when the goods are on board the vessel, and the buyer bear all costs from that moment onwards. This rule is to be used only for sea or inland waterway transport.

The obligations of buyer and seller

（Ⅰ）The seller's obligations

● The seller must deliver the goods on board the vessel nominated by buyer at the agreed port of loading within the agreed period, and notify the buyer timely.

● The seller must obtain, on its own risks and expense, any export license or other official documents and carry out all the customs formalities necessary for export.

● The seller must bear all the costs and risks of loss of or damage to the goods until they have been loaded on board the vessel.

● The seller must provide the evidence at its own expense that he has complete the delivery obligation. The required documents include commercial invoice, clean on board shipping documents and other documents as contracted or some electronic information with equivalent force.

(II) The buyer's obligations

• The buyer must contract, at its own expense for the carriage of the goods from the named port of shipment, take responsibility to charter a ship and booking shipping space, as well as to pay the freight. The buyer must give the seller sufficient notice of deliver time and vessel name, so as to make the seller ready for shipment.

• The buyer must bear all the costs and the risks of loss of or damage to the goods from the time that the goods have been loaded on board the vessel.

• The buyer must obtain, on its own risk and expense, any import license or other official documents and carry out all customs formalities for the import of goods and for their transport through any country.

• The seller must accept the shipping documents provided by the seller, and make the payment accordingly.

Some notable issues under the FOB term

(I) The "delivery point" and "risk-transfer point"

The seller must deliver, within the contracted period, the goods either by placing them on board the vessel nominated by buyer at the loading point, if any, indicated by the buyer at the named port of shipment or by procuring the goods so delivered. It is the obligation of "delivery" for the seller. The buyer must bear all the risks and costs at the moment when the goods have been delivered. The risk here means the risk of loss of or damage to the goods, and the cost is the cost excluding the normal freight.

(II) Link-up of vessel and goods

Under FOB term, it is the buyer's duty to arrange shipment of goods, therefore both parties must try to ensure the transfer of goods meet the shipment schedule. On one hand, the buyer should send the ship just in time, on the other hand, the seller should deliver the goods to the nominated port. In order to meet both ends well, the buyer should give the seller prompt notice of the vessel's name, loading berth, and ETD after he has contract for the carriage. (Known as "Shipping Instruction").

If the buyer designates the vessel, but fails to notify timely the seller of the vessel's name, loading berth and shipment time; or the vessel nominated by the buyer fails to arrive on time, is unable to take the goods, or closes for cargo earlier than the time notified. Then, the buyer bears all risks of loss or damage to the goods, provided that the goods have been clearly identified as the contract goods. As sending a ship by buyer is a precondition for seller to fulfill the contract, if the buyer fails to designate a vessel within the stipulated time, the seller may claim damages, or may revoke the contract if serious consequences occur. In contrast, the seller should pay dead freight or demurrage, in case of he did not get ready for the goods within the stipulated time.

The seller shall give the buyer sufficient notice promptly after goods have been loaded on board the vessel. As under FOB term, it is the buyer to effect insurance and the risks of loss of or damage to the goods are transferred from seller to buyer when the goods have been loaded on board the vessel. If the seller does not notify the buyer of the shipment promptly, the buyer may fail to make insurance timely, and the goods may not be insured properly. Therefore, the shipping notice sent by buyer can facilitate the seller to take delivery of goods and other matters.

Under FOB term, the seller can provide assistance to the buyer, at the buyer's request, such as chartering a ship or booking shipping space, as well as obtaining bill of lading or other transport documents, but the buyer should bear all the risks and expenses on his account.

(Ⅲ) Fees for loading the goods on board

Under FOB, the seller bears all the costs until the goods have been loaded on board the vessel, and the buyer bears afterward. However, loading is a continuous process, it is hard to divide all the loading costs between buyer and the seller. As to liner transport, no disputes will arise, because the liner freight paid by the buyer has already included all the loading charges. But under voyage charter, according to the usual practice, the charter party will normally stipulate that the ship owner will not bear the loading charges. In this case, both buyer and seller should clarify in their contract that who should cover the loading, trimming, or stowing charges. To prevent disputes arising between the buyer and seller, variants of FOB have been created to specify the division of the loading charges. The five variations of FOB are as follows:

• *FOB Liner Terms*　This term follows the practice of liner. That is sellers under this terms are not paying the loading, it is the party who contracts for carriage to pay for the loading charges, this party is the buyer.

• *FOB Under Tackle*　This variant allows the seller to cover the cost until the goods are placed somewhere within the reach of the tackle of the vessel nominated by the buyer. Other costs will be bore by the buyer.

• *FOB Stowed*　Under this term the seller has to cover the loading charges as well as the stowage charge when the goods are placed inside the hold of the ship.

• *FOB Trimmed*　Besides loading the goods on board the vessel, the seller should also trim the goods to make the vessel evenly balanced. The seller pays all the loading and trimming expenses.

• *FOB Stowed and Trimmed*　The seller should pay all the loading, stowing and trimming charges.

The above five variants of FOB usually used by charter transport, it is no need to use the variants by liner transport, since the freight under liner transport includes the loading charges which are borne by the charterers. Generally speaking, the variants only make the division of cost, the risk-transfer point is still the same that the risk transferred from the seller to buyer once the goods are loaded on board the vessel.

Some different interpretations for FOB

The interpretations and the charges for relevant export formalities for FOB terms in " Revised

American Foreign Trade Definitions 1941" vary from that in "International Incoterms" greatly. There are six interpretations for FOB in "Revised American Foreign Trade Definitions 1941", only the fifth one is similar with Incoterms regarding the obligations of the seller and the buyer. When signing FOB contract with businessmen in American or other Americas, in order to make the obligations similar to the Incoterms, we should use "FOB vessel" instead of " FOB", and indicate the port of shipment, such as "FOB vessel New York". Since FOB New York means the seller is responsible only for shipping the goods to anywhere in New York City, not responsible for shipping the goods to the port of New York and loading on board the vessel. Especially in import trade, even if under "FOB vessel", it is the buyer not the seller should make the export customs formalities, the seller only provides, at the request of the buyer, assistance for the buyer to get the require exporting documents issued by exporting country, what's more, all the relevant costs, such as export tariff and other taxes and fees should be borne by the buyer as well. Therefore, while doing import trade under FOB terms with the businessmen from the United States, Canada and other countries in North American, we should use "FOB vessel" and specify that the buyer is responsible for export customs formalities. As to transfer of risk of the Definitions, the seller should bear the risk of loss of or damage to the goods until the goods are placed inside the hole of the ship.

CIF Cost, Insurance and Freight (...named port of destination)

Definition

CIF is an Abbreviation for "cost, insurance and freight". When using this term, the port of destination shall be nominated behind CIF, for example, if Qingdao is the port of destination, said CIF Qingdao. CIF means that the seller is responsible for chartering a ship or booking shipping space, delivering the contracted goods on board the vessel at the named port of shipment on schedule, as well as effecting the insurance and paying the premium and freight necessary to bring the goods to the named port of destination. The seller only bears the risks of loss of or damage to the goods before the goods have been loaded on board the vessel, which is the same as FOB. This rule is to be used only for sea or inland waterway transport.

The obligations of buyer and seller

(I) The seller's obligations

• The seller is responsible for chartering a ship and booking shipping space, delivering the goods on board the vessel at the named port of loading within the agreed period, paying the freight necessary to bring the goods to the named port of destination, and notify the buyer of the shipment.

• The seller must bear all the costs and risks of loss of or damage to the goods until they have been loaded on board the vessel.

• The seller should effect insurance and pay the premium.

• The seller should make all the customs formalities necessary for export.

• The seller must provide the relevant transport documents, such as commercial invoice, insurance policy, the normal transport documents for transporting the goods to the port of destination, or some electronic information with equivalent force.

(II) The buyer's obligations

• The buyer bears all the costs and risks of loss of or damage to the goods from the time they have been loaded on board the vessel.

• The buyer must accept the transport documents provided by seller, and make the payment as contracted.

• The buyer must take delivery of the goods at the port of destination and carry out all customs formalities for the import of goods.

• The buyer must pay all of the additional costs while in transit until the goods arrival at the port of destination, excluding the normal freight and premium.

Some notable issues under CIF term

(I) Shipment contract

CIF contracts are shipment contracts. Under CIF term, the seller fulfills his delivery obligation when he has put the goods on board the vessel at the named port of shipment within the agreed period. That is why we define the CIF contract as a "shipment contract", not an "arrival contract". Risk of loss transfers to buyer when the goods are placed on board. Under CIF term, although the seller is responsible for carriage and insurance, but he does need to guarantee the safety of the goods to the port of destination. In whatever case, there should not be any provision or clause guaranteeing the arrival time of the shipment.

(II) Issues on Insurance Coverage

Under CIF terms, the seller is responsible for effecting the marine cargo insurance for the goods from port of loading to port of destination, and paying the premium. Since the risks after the goods being loaded on board the vessel are borne by the buyer, but the insurance during the transit is effected by the seller, to some extent, the seller procures the insurance on behalf of the buyer. The seller is required to obtain cargo insurance only on minimum cover if there is no other special stipulation in the sales contract. The minimum insurance amount shall cover the price provided in the contract plus 10%, and shall be provided in the currency of the contract. In practice, in order to avoid the disputes arisen between both parties, they usually specify in their contract what risks to cover and how much the insured amount is.

(III) Symbolic Delivery

CIF is a special type of trade terms that documents submission are more important than goods delivery, thus, it known as "symbolic delivery" or "Payment Against Documents". Symbolic delivery means that once the seller puts the contracted goods on board the vessel at the port of shipment, and submits the relating shipping documents to the buyer, then, the seller has completed his delivery obligation, he doesn't need to guarantee of arrival of goods.

Payment against documents means that the buyer once accepts the shipping documents, he must pay the price as provided in the contract, even if at that time the goods had been lost or damaged. He can lodge a claim after making payments, against the relevant party, such as the insurance company, therefore, CIF is also called documents transaction. Under CIF terms, the seller makes delivery against documents, and the buyer makes payment against documents. As long as the seller has submitted the full set of qualified documents stipulated in the contract within the scheduled time, regardless of whether the goods are damaged or lost, the buyer must pay the price. On the contrary, in case the seller's documents are not in conformity with what required in the contract, even if the goods arrived at the port of destination in good condition, the buyer still has the right to dishonor the payment.

(Ⅳ) Issue about contract for carriage

Under CIF contract, one basic obligation for the seller is to arrange the shipment from the port of shipment to the port of destination. According to Incoterms® 2010, the contract of carriage must be made on usual terms at the seller's expense and provide for carriage by the usual route in a vessel of type normally used for the transport of the type of goods sold. Therefore, if the seller fails to arrange the shipment accordingly, and could not deliver the goods on time, which will lead to a breach of contract and may bear the legal responsibility. Usually the buyer has no right to designate the vessel or shipping company, however, the seller could accept buyer's requirements in the case of they are easy to handle and without additional expenses, even those these requirements are not specified in the contract, such as the nationality of the ship, ship age, ship class, etc. But to the terms specified clearly in the contract, it must be strictly enforced.

(Ⅴ) Unloading Charges Issue

Under CIF contract, it is the seller's obligation to make shipment and pay the ocean freight. While adopting liner transportation, the unloading charges are included in liner freight, thus, it is the seller who take responsibility for unloading charges in liner transportation. However, for charter transportation, it usually specifies in the charter party that the ship-owner doesn't need to bear the loading and unloading charges. Under CIF terms, the loading charges at the port of shipment will be borne by the seller, but disputes arisen in who should bear the unloading charges at the port of destination. CIF variants aim at settling the miscellaneous charges occurred when the vessel arrives at the port of destination.

● CIF Liner Terms　This term follows the practice of liners which requires the party contracting for carriage, that is the seller to pay for all the unloading charges.

● CIF Landed　This variant requires the seller to cover the necessary handling costs until the goods are placed upon the dock, including the barge fees and wharfage charges.

● CIF Ex Tackle　Under this term the seller will pay for moving the goods from the vessel to a location within the reach of the crane's tackle. The unloading point can be somewhere at the port or in an incoming lighter. That is when the vessel can not reach the quay, the seller only needs to unload the goods to a lighter, and the charge for renting a lighter should be borne by the buyer.

● CIF Ex Ship's Hold　Under this term the buyer should pay the unloading charges moving the goods from ship's hold to the unloading dock.

CFR Cost and Freight (...named port of destination)

Definition

CFR is an Abbreviation for "cost and freight", the cost here means FOB price. when using this term, the port of destination shall be nominated behind CFR, for example, if Rotterdam is the port of destination, said CFR Rotterdam. CFR means that the seller is responsible for chartering a ship or booking shipping space, delivering the contracted goods on board the vessel at the named port of shipment on schedule. The seller only bears the risks of loss of or damage to the goods before the goods have been loaded on board the vessel, and pays the ocean freight. This rule is to be used only for sea or inland waterway transport.

The obligations of buyer and seller

(I) The seller's obligations

• The seller must contract for the carriage and pay for the freight, as well as deliver the goods on board the vessel at the port of loading within the agreed period and notify the buyer of the shipment.

• The seller must bear all the costs and risks of loss of or damage to the goods until the goods have been loaded on board the vessel.

• The seller should effect the export customs clearance.

• The seller must provide the relevant documents, such as transport documents or some electronic information with equivalent force.

(II) The buyer's obligations

• The buyer bears all the costs and risks of loss of or damage to the goods from the time they have been loaded on board the vessel.

• The buyer must take delivery of the goods at the port of destination and carry out all customs formalities for the import of goods.

• The buyer must accept the transport documents provided by the seller, and make the payment as contracted.

Some notable issues under the CIF term

(I) Unloading Charges Issue.

CFR variants are the same with CIF variants, they also aim at settling the miscellaneous unloading charges at the port of destination. The CFR variants are: CIF Liner Terms, CIF Landed, CIF Ex Tackle, CIF Ex Ship's Hold.

(II) Shipment Notice

Under CFR contract the buyer is responsible for arranging cargo insurance, when the goods have been loaded on board the vessel, the seller should send shipment notice to the buyer immediately by e-mail, fax, etc. so that the buyer can make the insurance timely. If the seller fails to provide shipment details to the buyer after the shipment is made, the buyer will not be able to obtain insurance against his risk of loss of or damage to the goods during the ocean transport, then the seller should bear the risk himself. Therefore, it is significant for seller to provide shipment notice under CFR term.

(Ⅲ) Fraud issues of CFR term

While doing import business, we should be cautious about using CFR term, because more fraud would occur under this term. Under CFR contract, it is the seller in foreign country to effect shipment, and we make insurance, it is possible that the seller colludes with the shipping company to issue false bill of lading, or the seller may charter unsuitable vessel, issue a fake documents, such as quality certificates and certificates of origin, etc. In this case, we will suffer losses. Therefore, when adopting CFR terms in import transaction, we should do business with customers with good reputation, and specify the vessel details.

The above terms FOB, CIF and CFR are all commonly used trade terms that deliver the goods at the port of shipment, and the risk-transfer points are the same, the risk transfers from seller to buyer when the goods have been loaded on board the vessel. However, the expenses and the obligations for seller and buyer are different.

FCA Free Carrier (...named place of delivery)

Definition

FCA is an abbreviation for "free carrier". When using this term, CFR shall be followed by a named place, for example, if the named place is Changsha, said FCA Changsha. FCA means that the seller complete delivery when the goods have been handed over to the carrier nominated by the buyer at the agreed time and place, and requires the seller to clear the goods for export. FCA applies to all kinds of mode of transport, including multimodal transport, but the seller is only responsible for handing over the goods to the first carrier. This trade term is widely used in international trade and plays a great important role in international practice.

The obligations of buyer and seller

(Ⅰ) The seller's obligations

• The seller must hand over the goods in conformity with the contract to the carrier nominated by the buyer at the named place within the agreed period.

• The seller is responsible for obtaining export license or other official authorization and carry out all customs formalities necessary for export of the goods.

• The seller must bear all the costs and risks of loss of or damage to the goods until the goods have been handed over to the carrier.

• The seller must provide commercial invoice and the usual proof that the goods have been delivered to the nominated carrier, or some electronic information with equivalent force.

(Ⅱ) The buyer's obligations

• The buyer must contract at its own expense for the carriage of the goods, and notify the seller of the name of carrier, delivery time and place promptly.

• The buyer must bears all the costs and risks of loss of or damage to the goods from the time that the goods have been delivered to the carrier.

• The buyer must obtain the import license and other official authorization on his own expenses and risk, carry out all the necessary customs formalities for the import of goods, including the transit formalities when necessary.

- The buyer must pay the price of the goods as provided in the contract, take delivery of the goods as contracted, and accept the qualified documents provided by the seller.

Some notable issues under the FCA term

(I) Delivery Place

Since the FCA applies to all modes of transport, its delivery place varies due to different modes of transport and different place designated for delivery.

Delivery is completed:

- If the named place is the seller's premises, when the goods have been loaded on the means of transport provided by the buyer.

- In any other case, when the goods are placed at the disposal of the carrier or another person nominated by the buyer on the seller's means of transport ready for unloading.

In general, the loading and unloading obligations under FCA are: If delivery occurs at the seller's premises, the seller is responsible for loading. If delivery occurs at any other place, the seller is not responsible for loading and unloading.

(II) Contract of Carriage

Under FCA term, it is the buyer's obligation to make a contract of carriage at its own expense, and notify the seller of the name of carrier, delivery time and place promptly. However, if it is more convenient for the seller to make carriage stuff, he may help the buyer to do so, but the risks and expenses should be borne by the buyer himself.

(III) Risk-Transfer Point

The seller must bear all the costs and risks of loss of or damage to the goods until the goods are placed at the disposal of the carrier. When using the FCA term, the goods are always grouped together to form a large collection, which is called "group shipping packing", such as containers and pallets are common collective packing. In this case, the seller should take such collection costs into account while pricing under the FCA term.

CPT Carriage Paid to (*...named place of destination*)

Definition

"Carriage Paid to..." means that the seller delivers the goods to the carrier or another person nominated by the seller itself, and that the seller must contract for and pay the costs of carriage necessary to bring the goods to the named place of destination, however, the buyer should bear all risks any other costs occurring after the goods have been so delivered. CPT applies to any mode of transport, including multimodal transport.

The obligations of buyer and seller

(I) The seller's obligations

- The seller is responsible for carrying out all of the export customs clearances, and contracting for and pay the costs of carriage necessary to bring the goods to the named place of destination, as well as notifying the buyer once the goods have been so delivered.

● The seller must bear all the costs and risks of loss of or damage to the goods until the goods have been delivered to the carrier.

● The seller must provide the relevant agreed documents, or some electronic information with equivalent force.

(Ⅱ) The buyer's obligations

● The buyer must bear all the costs and risks of loss of or damage to the goods from the time that the goods have been delivered to the carrier.

● The buyer must bear all of the expenses and unloading charges occurring in the transit of goods, except the normal freight paid by the seller.

● The buyer must accept delivery of the goods at the place of dispatch and receive them from the carrier at the place of destination, accept the qualified documents provided by the seller and pay the price of the goods.

Some notable issues under the CPT term

(Ⅰ) Place of delivery and the transfer of risk

Under CPT term, the seller is responsible for making contract of carriage and paying the normal freight bringing the goods from the place of shipment to the agreed place of destination, but other expenses are borne by the buyer. CPT is similar to CFR that the seller bears the cost of carriage but the risk transferred from the seller to the buyer at the time when the goods have been delivered to the carrier, that is all the risks during the transit of goods are borne by the buyer.

(Ⅱ) Shipping notice

Under CPT term, the seller should notify the buyer promptly when the goods have been so delivered to the carrier, so as to ensure the buyer can make the insurance in time and take delivery of the goods from the carrier at the named place of destination. If a specific point is not agreed or is not determined by practice, the seller may select the point of delivery and the point at the named place of destination that best suit its purpose. Shipment notice is as important as that for FOB and CFR terms. When CPT, CIP, CFR or CIF are used, the seller fulfills its obligation to deliver when it hands the goods over to the carrier and not when the goods reach the place of destination.

CIP Carriage and Insurance Paid to (...*named place of destination*)

Definition

CIP is an abbreviation for "carriage and insurance paid to". CIP means that the seller delivers the goods to the carrier or another person nominated by the seller at an agreed place (if any such place is agreed between the parties), and that the seller must contract for and pay the costs of carriage necessary to bring the goods to the named place of destination. The seller also contracts for insurance cover against the buyer's risk of loss of or damage to the goods during the carriage. That is while using CIP term, the seller should carry out both the carriage contract and the insurance contract, pay for the carriage cost and the premium. however, the buyer should note that under CIP the seller is required to obtain insurance only on minimum cover, the buyer should undertake all the risks and additional expenses occurring after the goods have been so delivered. CIP applies to any mode of transport, including multimodal transport, which is the same as CPT.

The obligations of buyer and seller

(I) The seller's obligations

● The seller is responsible for carrying out all of the export customs clearances, making the carriage contract and insurance contract on its own expense, delivering the goods to the carrier, as well as notifying the buyer once the goods have been so delivered.

● The seller must bear all the costs and risks of loss of or damage to the goods until the goods have been delivered to the carrier.

● The seller must provide the relevant agreed documents, or some electronic information with equivalent force.

(II) The buyer's obligations

● The buyer must bear all the costs and risks of loss of or damage to the goods from the time that the goods have been delivered to the carrier.

● The buyer must bear all of the expenses and unloading charges occurring in the transit of goods, except the normal freight been paid by the seller.

● The buyer must take delivery of the goods at the place of destination, accept the qualified documents provided by the seller and pay the price of the goods.

Some notable issues under the CIP term

Under CIP term, the seller is responsible for contracting for insurance cover against the buyer's risk of loss of or damage to the goods during the carriage, which is the same as CIF term, the seller makes the insurance on behalf of the buyer due to the premium contained in the CIP price. Generally speaking, the seller should only cover the agreed coverage against the goods. Otherwise, the insurance shall cover, at a minimum, the price provided in the contract plus 10% (i. e. , 110%) and shall be in the currency of the contract. For CIP contract, the seller has no obligation to cover additional risks, such as war risk, strike risk and SRCC risk, however when required by the buyer, the seller shall provide, but the additional insurance costs of such insurance should be borne by the buyer.

2.2.4　FOB, CFR, CIF VS. FCA, CPT, CIP

Points in common:

● They are all symbolic delivery and shipment contracts.

● The export clearances are borne by the exporters, and the import clearances are borne by the importers.

● The relationship among the trade terms in each group are the same: the seller is responsible for arranging carriage under terms FOB and FCA; while the buyer is responsible for arranging carriage under terms CFR and CPT; for terms CIF and CIP, the seller undertakes the obligations of arranging carriage and insurance. Other respective notable issues are the same.

The differences:

● Applicable modes of transport are different

FOB, CFR, CIF apply to water transport (including sea and inland waterway transport), the carrier generally is a shipping company. FCA, CPT, CIP apply to any mode of transport, such as water transport, land transport and air transport, etc. In addition, they can be used irrespective of whether one or more than one mode transport is employed, the carrier can be a shipping company, railway company, airlines and multimodal transport operators.

- Delivery point and risk-transfer point are different

For terms FOB, CFR and CIF, the delivery point and risk-transfer point are that when the goods have been loaded on board the vessel. While for terms FCA, CPT and CIP are that when the goods are placed at the disposal of the carrier.

- The responsibility of loading and unloading charges are differential.

Under FOB, CFR and CIF terms, for liner transport, it is the party who responsible for transport is responsible for loading and unloading charges, while for charter transport, the loading and unloading charges will be specified by terms variants. Under terms FCA, CPT and CIP, it is generally the carrier who bears the loading and unloading duties, so there is no variants in this group.

- The related transport documents are different

The transport documents under terms FOB, CFR and CIF are generally clean on board bills of lading, which is also a document of title to the goods. While for terms FCA, CPT and CIP, what kind of transport documents submitted depends on the transport mode selected, but one thing to note is that the air waybill and rail waybill are not documents of title to the goods.

- The types of cargo insurance are different.

When using FOB, CFR and CIF terms, the goods are covered by Marine Cargo Insurance. When using FCA, CPT and CIP terms, what type of transport insurance should be covered is depend on what kind of transport mode adopted, it can be Marine Cargo Insurance, Overland Transportation Insurance and Air Transportation Insurance.

2. 2. 5　Other Rrade Terms in Incoterms® 2010

EXW Ex works (...named place of delivery)

"Ex Works" means that the seller delivers when it places the goods at the disposal of the buyer at the seller's premises or at another named place (i. e. , works, factory, warehouse, etc.) . The seller does not need to load the goods on any collecting vehicle, nor does it need to clear the goods for export, where such clearance is applicable. The buyer bears all costs and risks involved in taking the goods from the agreed point, if any, at the named place of delivery. This term thus imposes the minimum obligations on the seller, and imposes the maximum obligation on the buyer.

Under EXW term, export clearance procedure is handled by the buyer instead of the seller, in this case, the buyer needs to be aware that whether the exporting country could accept the person or the representative from another county to handle the export clearance formalities. If the buyer cannot directly or indirectly obtain export clearance, he is advised not to use this term. FCA may be preferable, especially for the countries whose land are neighboring.

FAS Free alongside ship (...named port of shipment)

"Free Alongside Ship" means that the seller delivers when the goods are placed alongside the vessel (e. g. , on a quay or a barge) nominated by the buyer at the named port of shipment. The risk of loss of or damage to the goods passes when the goods are alongside the ship, and the buyer bears all costs and risks (including the risks and costs that happen during the barging) from that moment onwards. FAS term requires the seller to clear the goods for export. It can be used only for sea or inland waterway transport.

DAT Delivered at Terminal (...named terminal at port or place of destination)

The term DAT is one of the two new terms introduced by Incoterms® 2010. "Delivered at Terminal" means that the seller delivers when the goods, once unloaded from the arriving means of transport, are placed at the disposal of the buyer at a named terminal at the named port or place of destination. "Terminal" includes any place, whether covered or not, such as a quay, warehouse, container yard or road, rail or air cargo terminal. The seller bears all risks involved in bringing the goods to and unloading them at the terminal at the named port or place of destination. This term may be used to any mode of transport or multimodal transport.

DAP Delivered at Place (...named place of destination)

The term DAP takes place of DAF (Delivered at Frontier), DES (Delivered Ex Ship) and DDU (Delivered Duty Unpaid) in Incoterms 2000. "Delivered at Place" means that the seller delivers when the goods are placed at the disposal of the buyer on the arriving means of transport ready for unloading at the named place of destination. This term is very similar to CPT, that the seller will cover the charges and handle the operations necessary to send the goods to the named place of destination. However, one significant difference from CPT term is that, DAP requires the seller to bear all risks involved in bring the goods to the named destination. Under DAP at the time of delivery the risk and cost related to unloading the goods are for the account of the buyer. This term may be used to any mode of transport or multimodal transport.

DDP delivered duty paid (... named place destination)

"Delivered Duty Paid" means that the seller delivers the goods when the goods are placed at the disposal of the buyer, cleared for import on the arriving means of transport ready for unloading at the named place of destination. The seller bears all the costs and risks involved in bringing the goods to the place of destination and has an obligation to clear the goods not only for export but also for import, to pay any duty for both export and import and to carry out all customs formalities. DDP represents the maximum obligation for the seller.

If the seller cannot directly or indirectly obtain import clearance, he is advised not to use this term. If the parties wish the buyer to bear all risks and costs of import clearance, the DAP should be used. Any VAT or other taxes payable upon import are for the seller's account unless expressly agreed otherwise in the sales contract. For example, if the parties wish the buyer to bear certain kind of expense, such as value-added tax, then it should stipulate "delivered duty paid, VAT. unpaid （…named place destination）" in the contract.

This term may be used to any mode of transport or multimodal transport.

The following Table 2.1 shows the responsibilities, risks and expenses between the buyer and seller.

Table 2.1 Incoterms® 2010 Comparison Table

Trade Termas	Place of Delivery	Divison of Risks	Ex. clearance	Contract of carriage	Contract of insurance	Im. clearance	Mode of transport
EXW	seller's	Goods at the disposal of buyer	B	B	B	B	Any mode
FCA	Seller's premises or inland place or port of shipment	Goods at the disposal of carrier	S	B	B	B	Any mode
FAS	Port of shipment	Goods place alongside the vessel	S	B	B	B	Sca or inland waterway
FOB	port of shipment	Goods on board the vessel	S	B	B	B	Sca or inland waterway
CFR	port of shipment	Goods on board the vessel	S	S	B	B	Sca or inland waterway
CIF	port of shipment	Goods on board the vessel	S	S	S	B	Sca or inland waterway
CPT	Inland place or port of shipment	Goods at the disposal of carrier	S	S	B	B	Any mode
CIP	Inland place or port of shipment	Goods at the disposal of carrier	S	S	S	B	Any mode
DAT	Terminal at port or place of destination	Goods unloaded, at the disposal of buyer	S	S	S	B	Any mode
DAP	Inland place or port of destination	Goods not unloaded, at the disposal of buyer	S	S	S	B	Any mode
DDP	Inland place in the importing county	Goods not unloaded, at the disposal of buyer	S	S	S*	S Any mode	

NOTE: * indicates that in chosen occasions when insurance is not obliged by Incoterms® 2010, however applicable, this is the party who will obtain insurance.

(B = Buyer, S = Seller)

2.3 Clauses Commonly Used about the Trade Terms in Contract

2.3.1 Some examples for international trade terms of sales contract

1. USD100 per dozen CFR New York.

2. USD150 per M/T CIF London including our 2% commission.

3. RMB25 per case CFR Singapore less 1% discount.

4. USD4.5 per dozen FOB net Shanghai.

5. Unless otherwise specified, prices are FOB/Ex Works with freight allowed to US port of Westinghouse choice.

6. In the case of EXW contract, insurance is to be effected by the end users after loading. In the case of CIF contract, insurance is to be effected by the seller for 110% of invoice value on All Risks basis.

7. The Letter of Credit shall permit payment of 100% of the CIF price on presentation of ship-

ping documents stipulated within the delivery date specified in the Letter of Credit.

8. USD1230. 00 per ton CIF Shanghai including 3% commission. The commission shall be payable only after seller has received the full amount of all payment due to seller.

9. Fluctuation in the freight and contingent imposition of export levies and changes therein, after the date of sale, to be for buyer's account regardless of CIF or FOB term.

2. 3. 2   Determinants of Choice of Trade terms

The main objective of the Incoterms is to give the two trading parties a guideline of the obligations they need to fulfill. In an actual transaction, which term to choose is eventually determined by the trader's acceptance of cost, obligation and risk. It is very important to consider all the possible aspects while choosing the Incoterms. Traders should not accept any term which requires a performance beyond his ability. Other criteria for selecting terms may include:

- Exporting and importing possibilities
- Pricing purpose
- Availability of transport
- Competitive position on the transport market
- Ability of risk control

It is a noticeable fact that in spite of its detailed definition about relevant issues, the Incoterms® 2010 has a very restricted scope of governance. It cannot replace any provision in a sales contract which regulates an aspect other than those governed by it. The familiarity with the Incoterms® 2010 enables the traders to communicate more easily, but other skills obtained through long time of practice and experience will be indispensable.

Chapter 3

The Calculation of Export Price

★LEARNING OBJECTIVES

- Know the major factors in pricing decision.
- Understand the function and calculation of commission and discount.
- Master calculating the export price.
- Know the conversion of the different trade terms.
- Master the skills of negotiating the price terms of sales contract.

In international business, price clause is the core terms and conditions of a contract. In last section we have introduced the 11 kinds of trade terms in Incoterms 2010, the price term of sales contract is connected with trade terms, the international price varies due to different trade term adopted. In this section we will explain how to calculate the export price of commodity and how to stipulate the price terms in international sales contracts.

3. 1   The Components of the Export Price

3. 1. 1   Expression of export price

In international trade, the price term of sales contract includes the unit price and total price. Total price is the total amount of a deal. The price of a commodity usually refers to the unit price. The standard format of a price in international trade has four components: a code of currency, price per unit, measurement unit and trade terms. They typically look like the following expression.

USD100 per piece CIF New York or FOB Guangzhou EUR25 per dozen

It is up to the exporter if they put the trade term at the beginning or the end of the price. But all terms should be followed by the name of an appropriate place as defined by Incoterms 2010.

3.1.2　Pricing strategy

In international sales of goods, there are following pricing methods can be used:

Fixed pricing

The seller delivers and the buyer accepts the commodities at a fixed price agreed by both parties, neither party shall have the right to change the agreed price.

Flexible pricing

The pricing time and the pricing method are specified in the price terms, for instance: "the price will be negotiated and decided by both parties 60 days before the shipment according to the international price level". Or only the pricing time is fixed, for instance: "To be priced on July 1, 2010 by both parties".

Partially fixed price and partially unfixed price

The parties concerned only fix the price for the commodities to be delivered recently, and leave the price of the commodities to be delivered in the long term open.

Floating pricing

At the time of pricing, the price adjustment is also stipulated, for instance: "If the concluded price for other buyers is 5% higher or lower than the contract price, both parties will negotiate to adjust the contract price for the quantity of the contract."

3.1.3　Pricing methods

In order to master the pricing well, four pricing methods are illustrated here.

Cost-plus pricing

It involves adding to a basic cost figure, an amount to cover profit and other unassigned costs, then to arrive at a selling price. Cost-plus pricing is the easiest way of export pricing.

Marginal cost pricing

It is one that making the incremented cost of unit product for export is lower than the earlier average production cost for the domestic market. When using this method, it is important to find out the break-even point—the minimum quantity required by which the exporter can sell at a particular price without a loss. The further the sales are above the break-even point, the higher the profits. Sales below the break-even point result in a loss to the seller.

Buyer-based approach

Prices are set according to purchasing power of the buyer and the perceived value in the target market. It needs a good understanding of the market place if this approach is to be adopted.

Competition-based approach

If competition is fierce, the exporter has to provide prices benchmarked to competitors or market average so as to stay in business. In this case profit margins could be lowered.

Generally speaking, no matter what techniques is used, the export price must include cost and expected profit. Otherwise, the export does not make sense.

3.1.4 Pricing Considerations

Pricing for foreign markets is usually different from that for the domestic market. Without adequate research into the build-up of an export price and unforeseen cost components and contingencies, transactions that initially appear attractive may prove unprofitable or unexpectedly resource-intensive. The following part will provide you with details of the major components determining the export price. An export price consists of the production cost, the expenses and the profits.

Production Cost

The calculation of actual cost of producing a product is the core element in pricing. Production cost in a narrow sense normally includes material cost, labor cost, allocation of fixed and packing cost. Other administrative costs also considered as part of the overall production costs. If the exporter is not a manufacturer, it is then not necessary for him to concern the details. He can simply conclude all these costs into a "factory price" or "purchasing cost" . Of course, He should consider the overhead costs as well.

It is noteworthy that, in order to promote the development of export trade, China implements the export rebates system, it refers to the refund of export products, domestic production and circulation in the actual payment of the product tax, value added tax, business tax and special consumption tax. Export tax rebate system balance domestic tax mainly through the refund of export tax already paid the tax burden on domestic products, so that the cost of their products is not included into the international market and competing with foreign product under the same conditions, thereby enhancing competitiveness and expanding exports foreign exchange. Therefore, if there is export tax refund, it should be deducted when calculating the actual cost of an export commodity. The steps to calculate the export tax refunds: firstly, to determine the actual price of export commodities; and then multiplied by the tax refund rate with the actual price of the commodity. which is:

The actual price of export commodity = purchase price (including VAT) / (1 + VAT rate)

Export tax refunds = actual price of export commodity × export tax refund rate

Actual cost of export commodity = purchase price (including VAT) – export tax refunds

Expenses

From the factory workshop until the place of destination in the importing country internationally traded products may incur five categories of expenses.

- Sales Expenses

This refers to all the costs related to international marketing and sales activities. In order to promote the product in the international market, exports often have to take part in international exhibitions, print product catalogs, or set up company websites. To enter certain markets exporters may be willing to pay the local intermediaries for their cooperation and assistance. In addition, the sales commission and discount are regarded as a kind of sales expenses. These expenses can be substantial and should not be ignored.

- Delivery Expenses

In order to physically move the goods from one location to another, especially across country borders and over oceans, traders need to cover local and overseas warehousing and transportation costs, insurance premiums, tariffs and duties, customs clearance charges and necessary documentation costs such as applying for export or import licenses. When the goods arrive at the port of destination, there will be unloading charges, terminal costs, which includes handing, wharfage and harbor dues that must be paid by the exporter to the port authorities. The complex composition of delivery cost is one of the major reasons why export prices differ greatly from domestic selling prices.

- Financing Expenses

Another difference between export price and domestic selling price come from the different financing arrangements. Export business usually involves long time production and overseas transport. It is very common for a transaction to take three to six months to complete, sometimes even longer. To keep operating, exporters have to use various financing channels to obtain funds. This may lead to financing expenses; such as bank interest. In addition, banking charges will occur because of the different payment terms adopted, among which banking charges for L/C payment is relatively higher. To minimize such expenses exporters have to well consider the choice of financing channels and payment methods.

- Other Expenses

Here, space is left for the inclusion of unexpected additional expenses such as the cost of overseas telegrams or phone calls, fax charges, extra storage charges, and even "gifts" to the foreign customers.

The above separate costs are now available for consideration into a properly printed price list. These are the ones being used by most of the Chinese trading companies. The export price is a multidimensional variable. Whether or not you can export your merchandise with price calculated from the above costing sheet, the foreign consumer is always the final controller of your price. Accurate calculation does not imply right exporting. That is dependent upon your pricing strategies and policies. There are still some factors that should be considered, for example, the quality and classification of the goods, the quantity for transaction, the distances of the transportation, the delivery destination, the seasonal demand alteration.

Anticipated Profit

Another essential part of the export price is the anticipated profit. How much an exporter wants to make out of a particular transaction directly impacts on the price level of the product. This has a direct relationship with the company's marketing objective in the foreign market. For example, some companies may attempt to launch into a new market; some may look for long-term market growth; some just want to set up an outlet for surplus production or outmoded products; and many firms view the foreign market as a secondary market and consequently have lower expectation regarding market share and sales volume. All these different objectives naturally affect pricing decision.

There are two ways to calculate the anticipated profit: one is to calculate by an absolute number, for example, the profit is USD10 per M/T. The other is to calculate by a percentage of a certain amount, for example, the profit is 10% of the "factory price".

3.2 Calculation of Export Price

3.2.1 Calculation for the main trade terms

we take the cost-plus approach as the basic tool for pricing. The calculation requires one to analyze in detail cost incurred. Using a worksheet can make the process easier and clear to understand. Table 3.1 provides a sample calculating chart for the nine Incoterms terms if a waterway transport is employed. Here only some typical items are listed. The actual application of such a worksheet may be subject to specific cost variations among different transactions.

Table 3.1   Costing Worksheet

Item	Sub – total	Total
Manufacturing cost		
+ Export packing (optional, depending on mode of transport)		
+ Profit margin		
+ Administration overhead		
− Possible discounts/rebates/ sales commission		
= Selling price ex works (EXW)		
+ Local transport cost from plant to place of loading (train/truck)		
+ Costs for export clearance		
= Selling price free carrier (FCA)		
+ Local transport costs to shipping port		
+ Local transport insurance to shipping port if applicable		
= Selling price free alongside ship (FAS)		
+ Storage costs, terminal handling charge (THC), loading onto ship		
= Selling price on board (FOB)		
+ Main ocean freight to port of destination		
Selling price cost and freight (CFR)		
+ Minimum marine cargo insurance premium		
Selling price cost, insurance, freight (CIF)		
+ Additional costs for full transport insurance		
+ Unloading at terminal, THC if applicable		

Continued

Item	Sub – total	Total
Selling price delivered at terminal (DAT)		
+ Local transport costs to nominated destination		
Selling price delivered at place (DAP)		
+ Cost of import customs clearance		
+ Import duties, any VAT or other taxes payable upon import		
Price delivered duty paid (DDP)		

After having a general idea about the cost elements responding to Incoterms, a trader is advised to focus his attention to the particulars of the calculation of the most commonly adopted trade terms: FOB, CFR and CIF. The following part will explore those details related to the price conversion of the three terms.

FOB

$$FOB = \text{Actual cost} + \text{expenses at home} + \text{profit}$$

(I) Cost

For trading companies, the cost indicates the purchasing price of the export products; for manufacturer, the cost is the manufacturer price of the commodities. If there is tax rebate, it should be deducted when calculating the actual cost of the export commodity. If there is no a tax rebate, then the actual cost is the purchasing price or the manufacturer price of the commodity (the specific formulas please refer to the previous section).

Generally, our VAT rate is 17% or 13%, the rebate rate for each export commodity can be checked from the State Administration of Taxation, or can be obtained by searching the internet.

(II) Expenses

Export commodity expenses mainly refer to the costs of commodity circulation. Typically include packaging, storage charges, domestic freight, certification fees, port charges, inspection and quarantine expenses, taxes, interest, operating expenses, bank charges and so on.

Generally, there are two ways to calculate the expenses: Firstly, the exporter adds each domestic expense and overseas expense one by one, and then share to each unit of export commodity, this method is called Cost-plus Pricing. Secondly, set a basic amount, and then multiplied by a certain percentage, which is generally 3% to 8% of the purchase price, this method is called the proportional method.

The following describes the calculation formula for certain expense :

- Customs Tariff

$$\text{Export Tariff} = \text{duty-paid price of export product} \times \text{export tariff rate}$$
$$\text{Export Tariff} = FOB/ (1 + \text{export tariff rate}) \times \text{export tariff rate}$$

- Bank Interest

Bank Interest = purchasing price (purchasing cost) × loan interest rate (annual) × (the number of days loan/ the number of days in one year). For example, the purchasing cost of one commodity is RMB100, the bank's load interest rate is 9%, the loan period is two months, then the bank's load interest for this unit commodity is $100 \times 9\% \times (2/12) = 1.5$ (RMB)

- Bank Charges

Which refer to the international settlement fees, such as transfer fees, collection fees, and credit fees. Usually charge in two ways: pay-per-transaction (to be shared equally on each unit of goods) or by a certain percentage of the amount charged (usually by the offer price).

(Ⅲ) Profit

As sating in the previous section, the profit is calculate by two ways: one is to calculate by an absolute number, the other is to calculate by a percentage of a certain amount.

Normally the price will be calculated in the exporter's currency. If he decides to quote the price in a foreign currency, at the end of the calculation he then has to apply a valid exchange rate to the result and covert it into foreign currency, which is:

FOB in foreign currency = FOB in local currency/Exchange rat

CFR

If an FOB price is available, it is much easier to calculate the CFR price. According to the definition in the Incoterms 2010, the difference between FOB and CFR, in the sense of cost, is the ocean freight for sending the goods from the port of shipment to the port of destination.

$$CFR = FOB + Ocean\ Freight$$

Ocean freight quotations can be provided by shipping lines or shipping forwarder. The amount of ocean freight depends on mode of transport and the calculation method adopted, the details of calculation of ocean freight refer to chapter 5.

CIF

In a similar pattern the CIF price can be calculated given the FOB or CFR price.

$$CIF = FOB + Ocean\ Freight + Insurance\ Premium\ or$$
$$CIF = CFR + Insurance\ Premium$$

The key here is to find out the insurance premium charges. It is the normal practice of the insurance company to calculate the insurance premium based on the face value of a contract. In addition, it is also a standard industry norm that a markup, which is normally 10%, will be added on top of the contract value covering the incidental costs like processing the claim, survey costs, and the possibly inflated costs of replacements. Consequently, the formula for calculating the insurance premium will be (details refer to Chapter7):

$$Insurance\ Premium\ (I)\ = CIF \times (1 + 10\%) \times Premium\ Rate\ (R)$$

If the two formulas are combined, the result will be:

$$CIF = CFR + CIF \times (1 + 10\%) \times Premium\ Rate\ (R)\ or$$
$$CIF = CFR/ (1 - 110\% \times R)$$

The premium rate for a product can be obtained from any insurance company. When CFR price is available, the formula can help calculate the CIF price directly.

3.2.2   Price Conversion

When the buyer and the seller negotiate about the price, often they will change their trade terms they offered according to the requirement of the other party. This concerns the price conversion. Formulas for price conversion are as follows:

FOB converts into CFR or CIF

$$CFR = FOB + F$$

$$CIF = (FOB + F) / [1 - (1 + \text{insurance markup}) \times \text{Premium Rate}]$$

CFR converts into FOB or CIF

$$FOB = CFR - F$$

$$CIF = CFR / [1 - (1 + \text{insurance markup}) \times \text{Premium Rate}]$$

CIF converts into FOB or CFR

$$FOB = CIF \times [1 - (1 + \text{insurance markup}) \times \text{Premium Rate}] - F$$

$$CFR = CIF \times [1 - (1 + \text{insurance markup}) \times \text{Premium Rate}]$$

Note: $F = freight$

3.2.3   Price Including Commission

If a price shown in the contract directly comes from the calculation of basic costs and profit, it is called a "net price". But occasionally traders have to make some adjustments to the net prices to achieve the goal of promoting sales. These adjustments include commission and discount.

Commission is an incentive payment made to the middle persons or brokers for their intermediary service. The commission can be paid by exporter to the sales agent, or paid by importer to the purchase agent.

Expression

Commission is normally expressed by mentioning a percentage as commission rate at the end of the price with commission. For example:

USD200 per yard CFR Hongkong including 2.5% commission

A more frequently adopted way of quoting price including commission is to use a capitalized "C" to indicate commission behind the trade terms. For example:

USD2000 per dozen CIFC 5 Singapore or

USD2000 per dozen CIFC 5% Singapore

Sometimes the commission can be expressed in specific amount, such as:

USD25 commission per M/T

A price which contains a proportion as commission payment is called a "price including commission", it can be made openly or implicitly in the contract.

Calculation

It is internationally customary to calculate commission by some percentage of the amount of transaction. There are two ways introduced for commission calculation:

- Commission calculated based on FOB price or FCA price

This means that if the deal is finalized by CIF, CIP or any other terms, costs like ocean freight and insurance should be deducted before the commission is calculated. The reason is that the freight and insurance are not sales revenue to the seller, the seller does not need to pay commission for the freight charge, insurance or any costs of the same nature, but only for the value of the goods itself.

For example: the CIF price is USD1000, the ocean freight is USD100 and the premium is USD10, the commission rate is 2.5%. Then, the commission = $(1000 - 100 - 10) \times 2.5\%$ = 22.25 (USD)

This method is rarely used in practice, because it is neither good to seller nor to the middlemen. To the seller, the freight charges and premium have deducted directly from the total amount in advance, the seller will pay less commission, but receive less foreign exchange income as well. To the middlemen, this way will reduce his commission income and discourage their enthusiasm.

- Commission calculated based on the invoice value or contract value of a transaction

This way is most commonly used for commission calculation. Commission can be calculated based on pricing including commission, and the net price can be obtained by deducted to commission from price including commission.

$$\text{Commission} = \text{contract value} \times \text{commission rate}$$

For example, if the contract value of a transaction based on CIFC2.5% London is USD1000, the commission rate is 2.5%, then commission value will then be $1000 \times 2.5\% = 25$ (USD). Sometimes one needs to calculate the price including commission from a net price. The formula will be:

$$\text{Price including commission} = \text{net price}/ (1 - \text{commission rate})$$
$$\text{Commission} = \text{Price including commission} \times \text{commission rate}$$
$$\text{Net price} = \text{Price including commission} - \text{Commission}$$

Net price refers to the contract price that without commission and discount, such as "USD1000 per M/T net FOB Guangzhou".

For example: If the net CIF price of one export good is 1000 US dollars, the commission rate is 5%, then its

CIF price including commission = net price/ $(1 - \text{commission rate})$ = $1000/ (1 - 5\%)$ = 1052.64 (USD)

Commission = Price including commission × commission rate = $1052.64 \times 5\%$ = 52.64 (USD), or

Net price = Price including commission-Commission = $1052.64 - 1000 = 52.64$ (USD)

Payment

Commission usually paid by the exporter when he receives the full payment. However it is better for the involved parties to reach an agreement in advance if the commission is paid in this way, so as to avoid disputes. Otherwise, the middlemen may require the payment of commission once the contract concluded, in this case, the fulfillment of the contract may be more risky because of the lack of guarantee of the middlemen.

3.2.4　Discount

Discount is the price deduction allowed by the seller to buyer. What discount allowed depends on the mutual relationship between the seller and buyer.

Expression

The discount is normally expressed in the contract as a percentage of the total value or a fixed amount. For example:

USD200 per dozen CIF New York less 1.5% discount

Or expressed in specific amount, such as:

USD3 discount per dozen

Calculation

The calculation of discount is simple, it is not necessary to consider the FOB value or CIF value, normally it based on the contract price. The discount rate multiplied by the contract value is the discount amount. That is,

$$Discount = contract\ price \times discount\ rate$$

Then the actual price of the product will be:

$$Actual\ price = contract\ price\text{-}discount$$
$$= contract\ price \times (1 - discount\ rate)$$

In addition, discount may be calculated based on the quantity of the commodity. For example, if the discount for a commodity is USD 5 per piece, and the total amount is 500 pieces, then the discount $= 5 \times 500 = 2500$ (USD)

Payment

It is normally deducted by the seller directly during the payment of goods.

E. g. Guangdong Xingguang Exp-Imp company purchases one lot of lead ingot（铅锭）from Sunshine Steel factory. The purchasing price including VAT is RMB15000/MT, the VAT rate is 17%, and the export tax rebate rate is 13%, the total export expenses are 6% of purchasing value, the profit is 11% of purchasing value. The ocean freight from Guangzhou to London is USD115/MT, and it covers All Risk for 110% of invoice value, the premium rate is 0.7%. Please calculate the CIF London price. If customer wants the CIFC5% price, then what price shall Guangdong Xingguang Exp-Imp company quote? Suppose the exchange rate is 1 USD = 6.42 RMB

$$FOB = [15000 - 15000 \div (1 + 17\%) \times 13\% + 15000 \times 6\% + 15000 \times 11\%] \div 6.42$$
$$= (15000 - 1666.67 + 900 + 1650) \div 6.42$$
$$= USD\ 2474.00$$

$$CFR = FOB + F = 2474.04 + 115 = USD\ 2589.04$$
$$CIF = CFR \div (1 - 1.1 \times 0.7\%) = 2589.04 \div 0.9923 = USD\ 2609.13$$
$$CIFC5\% = CIF \div (1 - 5\%) = 2609.13 \div (1 - 5\%) = USD\ 2746.45$$

3.2.5 Understand the price

Profit-loss rate for export goods

Profit-loss rate for export goods is probably the best known and most widely used ratio for measuring the profitability of a company or a particular transaction. It refers to the ratio between export-loss amount and export total cost. The formulas are as follows:

Profit-loss rate for export goods = (export net RMB income (FOB) – export total cost) /export total cost × 100%

Export profit-loss amount = export net RMB income (FOB) – export total cost

NOTE: All the figures normally would be calculated in the exporter's local currency.

The total export cost is the sum of purchasing cost (including Value Added Tax) and domestic charges, excluding the taxes-drawback of export taxes. The export net RMB income refers to FOB price, which is exchanged into RMB by the current exchange rate. If the profit-loss rate is greater than zero, there will be benefits, otherwise loss. One thing should be noted that the calculation of both export income and cost must exclude any overseas transportation charge and insurance cost since exporters are not supposed to make profit from the transportation charge and insurance payment should the real revenue and cost be considered.

Exchange cost of export products

Exchange cost of export products is also an important index to show the profits and losses of the export business. It refers to the ratio between total cost of export goods and the foreign exchange net income. That is to say how much RMB can be exchanged for 1 US dollars, or the total RMB cost needed for exporting goods worthy of net 1 US dollar. The formula is as follows:

The exchange cost of export products = total export cost (RMB) /export net income (USD) × 100%

If the exchange cost of export goods is greater than the exchange rate quotation when making settlement, there will be losses, conversely, there will be profits. For export company, thus, the exchange cost of export goods is the lower the better.

There is an inner link between the exchange cost and the profit-loss rate, the greater the export loss ratio is, the higher the exchange cost will be. On the contrary, the smaller the loss rate or a little profit is, the lower the exchange cost will be.

Profit-loss ratio = conversion rate-exchange cost/exchange cost × 100%

3. 3   Price Terms in Sales Contract

Price term is one of the main terms and conditions of the international sales contract, it used to determine the price of the contract, thus, the price term should be complete, clear, specific and correct. Generally a price term includes two basic elements, which are unit price and total amount of goods, and the involved currency should be the same.

3. 3. 1   Examples for prices terms in the contract

Unit price: USD12. 00 per M/T FOB stowed and trimmed Singapore.

Total Amount: USD1200 000. 00 （ SAY US DOLLARS ONE MILLION AND TWO HUNDRED THOUSAND ONLY）

Unit price: JPY3 000 per set CFR Tokyo including 2% discount

Total Amount: JPY5880 000 （SAY JAPANESE YEN FIVE MILLION EIGHT HUNDERED AND EIGHTY THOUSAND ONLY）

3. 3. 2   Some notable issues while stipulating the price terms

• Select the appropriate trade terms according to transport mode and marketing intent to be adopted.

• Fighting for favorable currency in international settlement, and adding hedging terms if necessary.

• Adopt flexible pricing methods, so as to avoid the risks from price fluctuation.

• With reference to the international trade practices, using commissions and discounts reasonably.

• If there are quality latitude and "more or less" clauses stipulated in terms of quality and terms of quantity, as to the difference in quality or quantity, the pricing of this part should be set forth in the contract in full detail.

• It should be clearly stipulated in the contract referring the measurement unit, involved currency, port of loading, port of destination, etc. so as to execute the contract well.

Terms of Commodity

- Know the different ways of quality stipulation.
- Master the quantity measurement units and systems.
- Understand the approaches to weight calculation.
- Understand the functions and features of different types of export packaging.

Commodity is the subject matter of international sales contracts. An international sales contract is dealing with the transfer of ownership of the subject matter from one party (the seller) in one country to the other (the buyer) in another country. According to CISG Article 35 (1), the seller must deliver goods which are of the quantity, quality and description required by the contract and which are contained or packaged in the manner required by the contract. The failure to deliver goods in strict conformity with the contract will result in disputes. Therefore, terms and conditions concerning commodity must be explicitly stipulated in the sales contract to avoid subsequent disputes between the seller and the buyer.

4.1　Name of Commodity

When conducting business negotiations and making contracts, both the buyer and seller should first reach an agreement concerning what commodity or goods are under transaction and describe the goods in the sales contract exactly. The description of commodity comprises two components: name and quality of the commodity.

4.1.1　The meaning of Name of Commodity

Name of commodity is the commodity name, it is a title or concept which can make some kind of goods different from other commodities. Commodity name should be able to highly reflect the natural attribute of the commodity, purpose and main performance. We usually use raw materials, main component, production process, the appearance and the names of people and places to name the commodity.

4. 1. 2　Name of Commodity in the contract

The name of a commodity is an indispensable part of a sales contract. In most cases, it may be simply put under the contract item "Name of Commodity" (e. g. Name of Commodity: China Sesame Seed). However, many goods under the same name are of different kinds. It is necessary to provide further description such as specifications, grade, etc. Hence, the name and the quality description are often merged into one (e. g. Description of Commodity: Jasmine Flower green Tea, Grade One). In such a case, the name of commodity may constitute part of "the Description of Commodity" or " Description of Goods " in the contract.

4. 1. 3　Issues to be considered in the Name of Commodity

Name of commodity, as a basis for the delivery of goods, has a bearing on the interests and rights of both importers and exporters. Therefore, the name should be clearly and properly specified in a sales contract and the name of the goods delivered should exactly conform to the contract. To avoid subsequent disputes, the following three issues need to be considered:

Being Clear, Specific & Precise

The expression of the commodity name must be specific and precise, avoiding vagueness and ambiguity. For example, the name "rice" is too general as there are different kinds of rice in the market. In addition, the same type of rice produced in different places may be of different quality. Therefore, proper description of commodities like rice should also include details such as type, name of origin and some necessary specifications.

Being Practical

The wording of name of commodity should avoid unnecessary modifiers, especially those restrictive ones adding to the difficulty in the execution of the contract. Take "pure cotton T-shirts" for example. The word "pure" is an unnecessary modifier which may make it extremely difficult for the seller to fulfill the contract unless he has the intention and capability to deliver T-shirts which are of 100% cotton, or the buyer and seller have reached an agreement concerning the definition of "pure cotton" .

Adopting Widely Accepted Names

Sometimes a product is named differently in different countries and regions. As a result, different named may refer to the same commodity or the same name may mean different products. For example, "Coke" may refer to the soft drink Coca-Cola, or coke, a kind of solid fuel. To protect the interests of both parties, commodity names used on contracts should bear common interpretation by the seller and the buyer. Possible disputes can be further avoided by adopting internationally standardized and widely accepted names such as those listed in the Harmonized System.

In some cases, the appropriate choice of commodity names may facilitate the flow of import and export and reduce costs of transaction in terms of reducing customs tariff, avoiding non-tariff trade barriers and lowering transportation costs. It happens that when the same product is imported or exported under different names or categories of commodities, differential tariff rates or freight rates and trade policies may apply. Traders, therefore, should give some consideration to the tactical employment of the commodity names so as to benefit the most.

4. 2   Specifying Quality

The importance of quality is self-evident. It is the core element of the whole business. If the quality of products can not be guaranteed, the ground for transaction has been undermined. In addition, it determines the price value of the goods. Good quality always allows products to enjoy premium prices. Moreover, it concerns the image of goods. Products of high quality are more competitive in the world market. Consequently the task of defining the quality details on a business contract becomes critical as it involves the fundamental interests of both parties . The seller depends on it for setting production or purchase standards: and the buyer uses it as the criteria for accepting the goods. If quality requirements are not presented properly, disputes are likely to occur.

4. 2. 1   Meaning of Specification

Quality refers to the intrinsic elements of commodities including the internal properties or ingredients as well as the external appearance. Hence description of the quality may provide information related to the shape, structure, form, color, flavor, chemical composition, physical and mechanical property, biological features and other aspects of a commodity or product.

4. 2. 2   Quality Stipulations

Owing to the fact that large portions of international commodity transactions are conducted for goods normally unavailable at the time of contracting, two major ways have been developed to describe the quality of goods in a contract: sale by description and sale by sample. In some rare cases, a third way may also be adopted for the deal of ready goods —sale by actual commodity or sale by actual quality. In the following section (Table 4. 1) we will introduce the details of the methods and discuss some issues concerning quality clauses in the contract.

Table 4. 1 Quality Stipulations

Categories	Types
	Quality Stipulations
Sale by Description	Sale by Specification
	Sale by Grade
	Sale by Standard
	Sale by Brand Name or Trade Mark
	Sale by origin
	Sale by Descriptions or Illustrations
Sale by Sample	Sale by Sell's Sample
	Sale by Buyer's Sample
	Sale by Counter Sample
Sale by Actual Commodity/ Sale by Actual Quality	

Sale by Sample

A sample refers to a single item of a consignment of goods or a part of a whole product possibly selected from a whole consignment or specially designed or processed. It is a usual practice that a sample can be regarded as the representative of the whole shipment to be delivered. In trading, samples are frequently offered to potential buyers as the evidence of quality.

A sale is made by sample when the seller and buyer agree that samples are used as reference of quality and condition of the goods to be delivered. This method is used when it is difficult to describe quality of the commodity by words. Some products contain some properties or features that are far beyond the scope of any scientific or technical description. When words are futile, the product can talk. A lot of light industrial products, agricultural native produce, arts and crafts, and garments rely on samples for quality confirmation. In the case of a sale by s ample, the reference number and the date of sampling should be indicated. Typical expression on the contract may read as "Pikachu Plush Toys Size 22, Quality as Per Sample A031316 X ".

Samples can be provided by either the seller or buyer. According to the supplier of the sample, there are three cases under s ale by sample: sale by seller's sample, sale by buyer's sample and sale by counter sample.

- Sale by Seller's Sample

When a sale is made based on the sample provided by the seller, it is a sale by seller's sample. In international trade it often happens that the seller presents collection of samples and the buyer makes his selection. Since the buyer's purchasing decision is based on the samples presented by the seller, it is rather natural to stick to the same samples for the purpose of quality specification. This is the most common one among the three.

If this method is used, it should be clearly stipulated in the contract that quality is "as Per Seller's Sample " and the quality of the actual goods should conform to that of the sample.

- Sale by Buyer's Sample

Sometimes buyers may provide samples to sellers, requesting supply of the same good. In this case, samples from buyers are referred to as the benchmark of quality requirement. This is a sale by buyer's sample.

Under such circumstances, wording like "quality as Per Buyer's Sample" should be clearly stipulated in the contract and the seller should provide goods of the same quality as that of the buyer's sample.

In case of sale by buyer's sample, the seller has to study the samples thoroughly, ensuring that all details are covered. A frequent mistake sellers may make is that they are able to identify the external features, but ignore the intrinsic characteristics of the sample.

In addition, he might need to pay special attention to factors such as raw material supply, processing techniques and equipment available, managing to achieve the equivalent quality of the samples. Considering the probability of miscomprehension of product properties and the limitation of local conditions sometimes, sellers are not encouraged to use this approach.

- Sale by Counter Sample

A counter sample is a replica made by the seller of the sample provided, normally by the buyer. As the buyer is usually required to return the sample back to the seller with his confirmation if he approves of the sample, a counter sample is also called a returned sample or a confirmed sample.

This method is a good substitute to sale by buyer's sample because it removes the risk sellers have to bear under other approaches. By sending a replica to the buyer, the seller is free from the uncertainty of getting the right picture of the product. If the counter sample is not in conformity with the original one, it will be disapproved by the buyer. The seller hence will not have the chance to provide the wrong mass products due to his misunderstanding. On the other hand, once the counter sample is confirmed by the buyer, it will replace the original sample and become the final standard of quality of the transaction. The seller would be more comfortable to prepare the mass products according to a sample provided by him s elf. Even it the worst scenario that the buyer later finds the counter sample does not match with the original, the seller will not carry any responsibility as the counter sample has been confirmed by the buyer.

Sale Actual Commodity or Sale by Actual Quality

In some occasions, traders may also deal with goods already available at the time of contracting. The quality of the actual goods will then form the condition of transaction. This is called sale by actual commodity or sale by actual quality.

Trading under this term, buyers normally inspect the goods before their purchase is made and once the sale is finalized sellers do not provide any guarantee on the quality or condition of the goods sold, nor hold any liability of compensation in case of defects . Sale by actual commodity or sale by actual quality applies to items such as jewelry, paintings, arts and crafts and most products in stock. They mainly take place in some special forms like sale by consignment, sale by auction, exhibition sale, etc.

It should be noted that all approaches discussed above are not independent of each other. If necessary, two or more ways can be used simultaneously in order to specify the quality with clarity. In the meantime, traders, especially sellers, should also be aware of the problem of "double standard". When one specifying method is sufficient to serve the purpose, it is advisable not to employ any others because sellers are obliged to provide goods in strict conformity with the requirements contracted. As CISG Article 35 states, if a transaction is made both by description and by sample, the seller must deliver goods not only conforming to quality descriptions in the contract, but also possessing the equivalent quality of the sample. The more criteria are set, the greater the difficulties would be for the sellers to fulfill the their obligation.

Sale by Description

Sale by description is a way to specify the quality of most commodities in international trade. Sale by description may take the form of sale by specification; sale by grade; sale by standard; sale by brand name or trade mark; sale by origin and sale by descriptions or illustrations. Different forms may be chosen depending on the attributes, nature, and characteristics of a commodity. Generally speaking, sale by description is applicable to commodities of which quality can be expressed by some scientific indices.

● Sale by Specification

The specifications of a commodity comprise some important indicators such composition, content, purity, length and size. Defining quality by specification is simple and accurate, therefore is widely used in international trade (see Figure 4.1).

Examples
1. Color Lamps, Candle Type, 110v, 28w
2. China Soybean, Oil Content   20% min.
Moisture   12% max.
Admixture   2% max.
Imperfect Grains   8% max.
3. C708 Chinese Grey Duck Feather Down Content 18%, 1% more or less

Figure 4.1   Examples of Sale by Specification

● Sale by Grade

Based on some industry customary practice or traditions, some products are classified into different grades, such as Grade A, B, C; or Grade 1, 2, 3. Under a sale by grade, the quality of a product may be indicated simply by stating its grade, presumably the seller and the buyer have reached a consensus on the implication of grades. To avoid misunderstanding and subsequent disputes, however, it is recommended to lay down some major specifications apart from the use of grade (see Figure 4.2).

Example	
Fresh Hen Eggs, shell light brown and clean, even in size	
Grade AA:	60 – 65gm per egg
Grade A:	55 – 60gm per egg
Grade B:	50 – 55gm per egg
Grade C:	45 – 50gm per egg

Figure 4. 2   Examples of Sale by Grade

- Sale by Standard

When specifications or grades are laid down and proclaimed in a unified way, they become standards. Standards are formulated either by governments or by commercial organizations. Some apply to individual country; others are used internationally. Many countries have their own standards, for example, BS in Britain, ANSI in the USA, JIS in Japan and GB in China. The typical international standard is ISO (International Standards Organization) Standard.

As standards of commodities are subject to changes or amendments over time, it is important to mark the year in which the standard is created to avoid ambiguities.

Different categories of products have different standards. Some special standards are designed for particular reasons. Fair Average Quality (F. A. Q.) and Good Merchantable Quality (G. M. Q.) are good examples of this type. Generally speaking, it is difficult to establish fixed quality standards for agricultural products since they can be easily affected by all sorts of external factors. Thus F. A. Q. is used to indicate that the quality of the product is about equal to the average quality level of the same crop within a certain period of time (e. g. a year) . G. M. Q. , on the other hand, means that the goods are of a quality sufficiently good to satisfy the purpose of use or consumption which are mutually understood by both the buyer and the seller. It is used sometimes as the bottom-line of quality requirement.

Similar to s ale by grade, standards are usually supplemented by some detailed specifications for clearer presentation of information (see Figure 4. 3)

Tetracycline HCL Tablets (Sugar Coated) 250mg. B. P. 1973 (Note : BP refers to British Pharmacopoeia)	
China Northeast Soybean 2006 New Crop, F. A. Q :	
Moisture	15% max.
Admixture	1% max.
Imperfect granules	7% max.
Oil content	17% min.

Figure 4. 3   Examples of Sale by Standard

- Sale by Brand Name or Trade Mark

It is possible to define the quality of some products by simply referring to their brand names or trademarks. Many consumption products adopt this approach . Typical examples include Sony Televisions, Haier Refrigerators, Tigerhead Batteries and Triangle Tires. A brand name is a company-specific name for a particular product or a group of products, usually used to differentiate that product from competitor offerings . It usually forms a part of an easily recognizable design on packing or advertising material . A trademark, on the other hand, refers to a name or a symbol that is unique to one product, representing a commercial enterprise. Trade marks are usually registered officially thereupon protected. Both brand names and trade marks are signs for distinguishing the products from other competing goods of the same line.

This approach, however, can only apply to those widely recognized brands or trademarks . Only when a brand name or a trademark is established in the market successfully, can it be considered the indication of certain level of quality. It is also believed that in most cases goods of the same brand or trade mark are of unified and stable quality. As a result, other products of the same company may benefit from the spillover effect of the brand as well. The importance of branding has been increasingly highlighted and more and more enterprises are taking proactive measures to build up their brands in the international market. In practice, due to the varieties and complexities of some brand products, some detailed quality particulars must be specially and legibly stated in the contract as well, like Panasonic Television, TH – 42 PV65 C.

- Sale by Origin

Sale by origin refers to the sale of goods by using the name of the place of origin as the indication of quality. This is more suitable for agricultural products or by-products. Owing to the unique and favorable natural conditions or traditional production techniques in some areas the native products are renowned for their speciality and excellence in quality. Sale by origin is usually used in company with other necessary quality indices such as specifications, grade or brands. The way to specify the tea as "West-lake Longjing Tea, Grade 1, Zhejiang Origin" is a typical example.

- Sale by Descriptions or Illustrations

Sale by descriptions or illustrations is especially applicable to full sets of equipment or instruments. These commodities are usually complicated in property and structure . It is difficult to use simple indicator such as data or parameters to describe the quality . In addition, the installation, usage and maintenance of such equipment or instruments shall follow certain procedures. Therefore specific descriptions, sometimes with illustrations, are necessary for specifying the quality.

In the case of sale by descriptions or illustrations, clauses such as "quality and technical data to be in conformity with the description submitted by the seller" are to be stipulated the contract and relevant technical manuals, booklets of directions, drawings or diagrams will be attached as well to serve the purpose (see Figure 4. 4) .

Example

- Samsung FHD Monitor, CF591, quality and technical data to be strictly in conformity with the instructions attached.
- Haier Portable Washer, Model HLP24E, quality and technical data to be strictly in conformity with the descriptions submitted by the seller.

Figure 4. 4　Examples of Sale by Descriptions or Illustrations

4. 2. 3   Other Quality Clauses

A part from defining the basic quality criteria, quality clauses in a sales contract usually include some other elements as well, which constitute the condition to apply the quality standards formerly identified. One of them is the statement of quality latitude or quality tolerance. If it is a sale by buyer's sample or a sale by counter sample, a safeguard clause is also inserted.

Quality Latitude or Quality Tolerance

Absolute equivalence of quality is not practically feasible in most cases. Therefore, the quality latitude or quality tolerance clause is introduced to address the issue.

Quality latitude means the permissible range within which the quality of the goods delivered by the seller may be flexibly controlled. Quality tolerance refers to the quality deviation recognized, which allows the quality of the goods delivered to have certain difference within a range. According to international practice, quality deviation recognized which allows the quality of the goods delivered to have certain difference within a range. According to international practice, quality with some deviation within a certain range is still considered to meet the quality stipulation in a contract.

Evidently a quality latitude or tolerance clause is necessary in the contract. It is normally set by stating the flexible quality range or scope, the maximum or minimum requirements, or a deviation allowance for certain quality indexes (see Figure 4. 5).

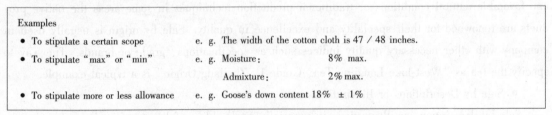

Examples
- To stipulate a certain scope   e. g. The width of the cotton cloth is 47 / 48 inches.
- To stipulate "max" or "min"   e. g. Moisture: 8% max.

                                      Admixture: 2% max.
- To stipulate more or less allowance   e. g. Goose's down content 18% ± 1%

Figure 4. 5   Example of Quality Latitude or Tolerance

When a sale is made by sample, it is also impossible for the seller to deliver goods of quality identical to that of the sample. Therefore, wording such as " about " should be added to the quality clause to allow quality flexibility (see Figure 4. 6).

Examples
- "Quality shall be about equal to the sample. "
- "Quality is nearly the same as the sample. "
- "The quality of the goods shipped to be about equal to the sample. "
- "Shipment shall be similar to the sample. "

Figure 4. 6   Examples of Quality Latitude Concerning Sale by Sample

In most cases, prices of the commodities remain unchanged when the quality varies within the tolerance or latitude unless otherwise specified in the contract, However, when the difference in quality is big enough to constitute fundamental change to the quality of goods, prices are to be adjusted accordingly and it should be explicitly stipulated in the contract (see Figure 4. 7).

Examples

- "Soybean, if the oil content of the goods shipped is 1% higher, the price will be accordingly increased by 2%"
- Chinese Sesame seed F. A. Q. 2008 Crop, Moisture (max.) 8%; Admixture (max.) 2%; Oil Content 52% basis. Should the oil content of the goods actually shipped be 1% higher or lower, the price will be accordingly increased or decreased by 1%, and any fraction will be proportionally calculated.

Figure 4 . 7   Examples of Price Adjustment

Safeguard Clause

According to CISG Article 42, the seller must deliver goods that are free from any right or claim of a third party based on industrial property or other intellectual property. In the case of sale by buyer's sample or sale by counter sample however, there exists the possibility that the seller might provide goods alleged to have infringed the copy right of a third party without his privities. Therefore, a safeguard clause is usually stipulated in the contract. It can protect the seller from any undeserved liabilities (see Figure 4. 8).

Examples

- "It is the buyer rather than the seller who shall be responsible for any disputes arising from the infringement of the third party's intellectual properties."
- For any cotton piece goods produced with the design, trade marks, brands, and/or stampings provided by the buyer, should there be any disputes arising from the infringement upon the third party's industrial property rights, it is the buyer to be held responsible for it.
- In case the buyer's sample results in any disputes of infringement of industrial property, the seller will have nothing to do with it.

Figure 4 . 8   Examples of Safeguard Clause

4. 3   Measuring Quantity

Quantity is one of the essential conditions in a business contract in international trade. It is obligatory for sellers to deliver the quantity of goods that is identical to what is called for in the contract.

According to CISG Art. 52 (2), if the quantity delivered is more than that agreed upon, the buyer is entitled to reject delivery of the excess quantity, or he may take delivery of all or part of the excess quantity, he must pay for it at the contract rate. If the quantity delivered is less than that agreed upon, sellers should make up for the shortage within the time limit without rendering buyers any unreasonable inconvenience and cost. However, buyers are still entitled to lodge a claim for compensation.

4. 3. 1   Use of Unit

In international trade, the quantity of commodities is always shown as a specific amount in different measurement units such as weight number, length, area and volume. Since different commodities have different natures and characteristics, the adoption of measurement units varies. The commonly used measurement units are as follows:

Weight

Kilogram (kg.); ton (t); metric ton (m/t); quintal (q.); gram (gm.); pound (lb); ounce (oz.); long ton (l/t); and short ton (s/t.) are units of weight. These are widely used for the measurement of raw materials and agricultural products such as minerals, raw chemicals, coal, oil, wool, cotton, corns and medicine.

Capacity

Units of capacity are liter (l.); gallon (gal.) bushel (bu.), etc. They are applicable to products such as wheat, corn, gas, gasoline and beer.

Number

There are various measurement units by means of number such as piece (pc.), package (pkg.), pair, set, dozen (doz.), gross (gr.), great gross (g. gr.), ream (rm.), roll or coil, unit, head, case, bale, barrel or drum, bag etc. They are for daily industrial and general products, such as stationery, paper, toys, ready-made clothes, vehicles and live animals.

Length

Yard (yd.); meter (m.); foot (ft.); centimeter (cm.), etc. are measurement units of length. They are often used for textiles, ropes and wires and so on.

Area

Square yard (yd^2); square meter (m^2); square foot (ft^2) and square inch, etc. are measurement units of area. These units are for leatherwear and plastic products such as plastic sheet and floor, leather, and wire meshes.

Volume

Measurement units of volume are cubic yard (yd^3); cubic meter (m^3); cubic foot (ft^3); cubic inch, etc. They are for chemical gas, wood, etc.

When adopting a particular measurement unit to define the quantity, traders need to be aware of the consistency of system. Currently, the measurement systems in use include the Metric System, the U. S. System, the British System and the International System of Units (S. I.), among which the metric system and S. I. Units are universally accepted and adopted all countries. A notable fact is that some units in different systems carry the same name though; they are indicating standards of measurement with significant difference. In addition, it is true that due to the local background and customary practice, different countries adopt traders, while having their business different systems of measurement. As a result, traders, while having their business communications, need to clarify the use of unit [1] and measurement system to avoid unnecessary disputes.

4. 3. 2   Interpretation of Weight

In international trade, weight is a very important concept. Besides serving as the most frequently used unit for defining quantity of commodity in transaction, it constitutes an indispensable part of information about each consignment, frequently demanded by different parties with different intentions. Moreover, when used in various occasions, the data would be treated differently, resulting in a series of different names. Being able to interpret the concepts properly will add value to the operational proficiency of the practitioner . The next part will focus on explaining the application of weight data in different scenarios.

Gross Weight

Gross weight refers to the weight of the commodity plus the weight of the packing. Gross weight is applicable to commodities of comparatively low value.

Net Weight

Net weight means the actual weight of a commodity itself excluding the weight of the packing. According to CISG Article 56, the weight of a commodity is calculated by its net weight unless otherwise stated in the contract .

However, some products can be weighed only when they are packed. If net weight is used, the weight of packing, i. e. the tare, must be deducted. Thus the relationship between net weight and gross weight is as follows:

$$\text{Net weight} = \text{Gross weight} - \text{Tare}$$

Sometimes the packing may become an indivisible part of the product, such as tobacco flakes; or the packing material is almost of the same value as that of the goods, like grain and fodder. If it is inconvenient to measure the net weight of the commodity, the " gross for net " practice will be adopted. In the case of " gross for net", the goods are priced by their gross weight instead of the net.

Conditioned Weight

Some commodities like wool, cotton and raw silk, not usually packed in a vacuum container, tend to absorb moisture. The weight of these commodities is likely to be unstable due to the fluctuation of their actual moisture content, which varies greatly from time to time, and from place to place. When these products are of high value, it becomes important for the difference between " dry goods" (i. e. goods with little moisture) and " wet goods" (i. e. goods with much moisture) and

---

1   Same measurement unit indicating different quantity

Ton maybe interpreted as 1 metric ton, 1 long ton or 1 short ton

But : 1 metric ton = 1 000 KG

1 short ton = 907. 18 KG

1 long ton = 1 016. 05 KG

the slightest difference may cause considerable difference in the price. In that case, conditioned weight is used. Conditioned weight equals to the dry net weight of a commodity plus the standard moisture content. It can be calculated by first deducting the actual moisture from the actual net weight of the commodity and then adding the standard moisture. The formula of calculating the conditioned weight is explained as below :

    Conditioned Weight

    = Dried Net Weight × (1 + Standard Regain Rate)

    = Actual Net Weight / (1 + Actual Regain Rate) × (1 + Standard Regain Rate)

    = Actual Net Weight × (1 + Standard Regain Rate) / (1 + Actual Regain Rate)

Note:

Dried net weight: the weight of the goods when all the water content is abstracted by a scientific method from the goods.

Regain rate: the ratio between the water content and the dry weight of the goods.

Standard regain rate: the ratio between the water content and the dry weight of the goods which is accepted in the world market or agreed upon by the seller and the buyer.

Actual regain rate: the ratio between the actual regaining water content in the goods and the actual dry weight.

Eg. An exporter has signed a contract to export 10 metric tons (5% more or less) of raw silk to Spain. It is stipulated in the contract that conditioned weight is used and the standard regain rate is 10%. If the actual regain is 21% when shipment is made, how many metric tons of raw silk should the exporter deliver in order to fulfill the contract?

Calculation:

Actual Net weight = Conditioned weight × (1 + actual regain rate / 1 + standard regain rate)

Actual Net weight = 10 M/T × (1 + 21% / 1 + 10%) = 11 MT

Answer: In order to fulfill the contract, the exporter should deliver 11 metric tons of raw silk, 5% more or less.

Theoretical Weight

The weight is a theoretical weight when the total weight of the product is calculated by multiplying the total quantity and the unit weight, rather than measured actually. Theoretical weight is applicable to commodities such as galvanized iron, tinplate and steel plate. These commodities are of identical or standardized sizes and specifications ; the weight of each unit is almost the same, thus the total can be obtained by means of theoretical calculation.

Legal Weight

Legal weight is the weight of the goods including the immediate, inner, or direct packing of the goods. According to the customs laws and regulations in some countries, legal weight is usually used as the basis for tariff calculation.

4.3.3 More or Less Clause

In practice, sometimes it is hard to strictly control the quantity of goods supplied, especially bulk cargoes such as agricultural products and mineral products, owing to contingencies in goods preparation. Besides, the differences in transport facilities may also lead to the inconsistency between the actual shippable quantity and the contracted quantity that might be the result of estimation. For the sake of efficient shipment and less complexity in contract execution it is common to allow the seller to deliver the goods with a certain percentage of more or less in quantity accordingly. This kind of stipulation on the contract is usually referred to as the "more or less clause".

When drafting the more or less clause the following issues are to be concerned: how much more or less should be allowed, which party is entitled to make the decision, and how should the more or less portion on of the goods be priced?

How much "more or less" should be allowed?

There is no fixed rule to decide how much the range of quantity difference should be. Traders normally make their decisions based on customs or result of negotiation. The ratio also varies among different lines of products. Most traders prefer to express the flexible portion by percentage; and the common range for the ratio is between 3% and 10%. An example may read as "1,000 metric tons, with 5% more or less". Other traders, on the other hand, would like to set an absolute number as the tolerable quantity disparity. Whatever way chosen, the stipulation should be clear and specific. Ambiguous terms like " about " "circa" "approximately" should be avoided, as they are easy to be interpreted differently.

To reduce disputes, UCP600 Article 30 stipulates that the words " about " or "approximately" used in connection with the amount or the quantity or the unit price are interpreted as allowing a tolerance not exceeding 10% more or 10% less. In absence of any more or less specifying the quantity, a tolerance not exceeding 5% more or 5% less is allowed, provided that the quantity is not stipulated by number (of packing units or individual items) and provided that the total amount of the goods does not exceed the amount of the credit.

Which Party Is to Make the Decision?

Who should determine how much more or less would be allowed? This should also be legibly stipulated in the contract to avoid possible disputes. In practice, as the quantity of goods delivered may be influenced by the natural condition of goods, the packing patterns, and the actual handling of shipment, it is practical to allow the party in charge of shipment to decide the quantity.

How Should the More or Less Portion Be Priced?

Under a "more or less" clause, the payment for the over-load or under-load portion will be made either according to contract price or at the market price at the time of shipment. According to CISG, unless otherwise stipulated in the contract, the payment for the more or less portion of the goods shall be calculated at the contract rate.

However, if the parties are concerned about the possible great change in price at the time of delivery, they may include statements in the contract that settlement for this more or less part is based on the market price at the time when the goods are shipped on board the vessel. In doing so, the method of calculation should also be set forth in the contract in full detail (see Figure 4.9).

Examples

- 800 metric tons, 5% more or less at seller's option.
- 20 metric tons, the sellers are allowed to load 5% more or less, the price shall be calculated according to the unit price in the contract.
- 1,000 metric tons 5% more or less, at buyer's option with more or less portion priced at the market price at the time of shipment.

Figure 4.9 Examples of More or Less Clause

The quantity clause in a sales contract is the legal basis for the seller to effect shipment and for the buyer to take delivery and make payment later on. Considering the various measurement units and, it is crucial to stipulate the quantity clause in a manner, which is clear, specific, and reasonable at the same time, to make the fulfillment of the contract flexible and smooth.

4.4 Packing and Marking

Most internationally traded goods have to travel long distance. In the course of transit, goods are vulnerable to all sorts of unexpected events. To protect the safety of cargo to the fullest extent possible, usually exporters have to take every necessary step from the design of packaging to the choice of transport.

However, some exceptions do exist. For a special category of products, packing is neither necessary nor viable. These kinds of goods normally have a primitive and stable nature, hence not apt to be easily damaged, unless accidents out of the ordinary range happen. Raw material or industry products like rolled steel, lead ingots, rubber and timber fall into this category. According to the extent of encapsulation they are called nude cargo. Some finished products like complete automobiles are also considered nude cargo. There is another type of goods which shares the similar features of not easily being influenced by outside circumstances and hereby bearing no packing. These goods are addressed as bulk cargo. The differences are only that while nude cargo usually are in a solid state, bulk cargo are more of a liquid-like existence and normally of low value. Ore, coal, grain, and liquid chemicals not sold with individual container all belong to this category.

Appropriate packing nevertheless remains important for the majority of export goods. This part will discuss some basic knowledge and relevant operations related to the packing of export cargo.

Packing of commodity has become increasingly important in international trade. Increasingly fierce competition in the market space and the gradually diversified and particular demand of end consumers bring great challenges to the traders. New products, new ways of selling must be created promptly to meet the market needs. Packing in many cases has become an essential part of the product sold. More and more attention has been put on the design, use of material and technique for packing purpose. The functions of packing for export goods these days are various. The major ones can be briefly summarized as follows: protecting goods, facilitating delivery, reducing cost and promoting sales.

4. 4. 1   Transport Packing

Transport packing, also called shipping packing, outer packing or big packing, is mainly adopted to facilitate cargo transportation. There are various containers that can be used for packing; these include case, drums, bags, bales or crate, etc.

● Cases include wooden cases, crates, cartons, corrugated cartons, etc. This type packing gives complete protection and lessens pilferage plus an aid to handling. It is particularly prominent with surface transport an disused for goods that cannot be compressed tightly, such as machinery and other items of expensive equipment. However, it is becoming less popular as the cost of timber has risen sharply in recent years and containerization has lessened the need for such strong packing in certain trades.

● Drums/ Casks/Hogsheads/Barrels include wooden drum, iron drum and plastic cask, etc. They are used for the conveyance of liquid or greasy and powdered or granular goods. The main problems associated are the likelihood of leakage if the unit is properly sealed, and the ability of the drums becoming rusty during transit.

● Bags include gunny bags cloth bags and paper bags, etc. They are ideal for a wide variety of powdered granular or bulk goods such as cement, fertilizer flour, oil cakes up on pallets to facilitate handling but they subject to damage by water, sweat, leakage or breakage.

● Bales or bundles are suitable for wool, cotton, feather, silk and piece goods, etc. These goods are to be compressed into bales first and then packed with cotton cloth and gunny cloth strengthened by metal or plastic straps outside.

● Crates or skeleton cases are a form of container halfway between a bale and a case. They are of wooden construction. Lightweight goods of larger cubic capacity, such as machinery, domestic appliances, refrigerators, cycles, and certain foodstuffs, for instance, oranges are suitable for this form of packing.

According to the method of packing, transport packing can be categorized into unit packing and collective packing. Unit packing is used on the smallest shippable unit of cargo. It can be in different forms of containers used for packing such as cases, cartons, drums, bags, bales, bundles, etc. Collective packing is also called "group shipping packing ". By means of collective packing, a certain number of units of cargo are grouped together to form a large collection. Pallet, flexible container are the commonly used equipments for collective packing in international trade .

Pallet is a large tray or platform allowing a number of units of cargoes to be grouped together. Pallets can move cargoes in loads instead of single pieces from a lorry into a train or onto a ship.

Flexible container is a big bag of different sizes facilitating the carrying of large quantities of cargo.

Container is a large metal case, of standard shape and size, for carrying goods by specially built road vehicles, railway wagons and ships. Some views argue that a container is only part of the transport vehicle rather than away of collective packing.

Apart from protecting the cargoes, collective packing also facilitates and speeds up the loading and unloading operation. At present, in order to improve the speed of operation and utility of ports and docks, some countries enforce regulations stipulating that cargoes shall be transported in collective packing.

4. 4. 2　Marking

Making refers to different diagrams, words and figures which are written, printed, or brushed on the outside of the shipping packing. When goods are well enwrapped by the packing, there is no clue to the identification of them. People need certain kind of indicator to guide their handling of the cargo along the flow of logistics. Marking plays an important role in the use of transport packing. According to its function, marking can be classified into four types: shipping marks, indicative marks, warning marks and supplementary marks.

Shipping Marks

Shipping marks are a type of marking on the shipping packing. It quickens the identification and transportation of the goods and helps avoid shipping errors.

Therefore, the design of shipping marks should be simple, clear and easily identifiable. For instance, the position of the shipping marks should be proper; the color of the marks should be durable. Besides, no advertising words and pictures are allowed to insert into the shipping marks .

International standard shipping marks recommended by The International Standards Organization are made up of the following four parts :

- Consignee's Code: usually the initials or abbreviation of a consignee
- Destination: the name of the place of destination
- Reference No: the number of the relevant contract, order, invoice, etc.
- Number of Packages: the consecutive number of each package

Adopting standard shipping marks can not only make it easy to identify and transport the goods, but also simplify the process of checking the relative documents and certificates.

Indicative Marks

Indicative marks are diagrams and simple words used to call the attention of the relevant people when they handle the goods in the process of loading, unloading, carving and storing. Indicative marks, when necessary, are also painted or stenciled on the outer packing of the commodities.

Warning Marks

To ensure safe carriage of dangerous goods in international transportation, warning marks, also called dangerous marks are used to indicate dangerous cargoes such as explosive, corrosive, inflammable and radioactive products.

Supplementary Marks

Sometimes, in accordance with the rules and regulations laid down by the importing and exporting countries, or the agreement entered into by the parties concerned, some supplementary marks are inserted such as the mark of weight or volume and the mark of origin.

- Mark of Weight or Volume

Weight or volume marks are marks indicating the volume or the gross/net weight of the package to facilitate loading/ unloading or booking shipping space.

- Mark of Origin

Marks of origin are required by many exporting or importing countries for customs statistics and taxation. In some countries, mark of origin avoids confusion over the real origin of the goods.

The mark of origin must be legible, indelible and clearly visible. In addition, each country has relevant regulations governing the way of marking for imported goods, including language and marks to be used and even the size of the letters of making. Therefore, the way of marking depends on the requirements of the particular country.

Usually, exporters, who put on the mark of origin, should look into the importing country's marking regulations before packing and marking their goods for export shipment.

4. 4. 3　Sales Packaging

Sales packaging is also called inner packaging, small packaging, immediate packaging or marketing packaging. The main purpose of such packaging is for sales promotion. To facilitate sales, different packaging are designed to match with various kinds of goods: carrying packaging, hang-up packaging, easy-open packaging, spraying packaging, pilling up packaging, gift packaging, and so on.

Instead of exploring the variety and novelty of the design of sales packaging, a topic probably more appropriate in the field of marketing, we will address several issues that are related to the sales packaging and relevant to the practice of international trade at the same time.

Neutral Packing

Neutral Packing is a special type of marking rather than a type of packing as its name may indicate. While neutral packing is required, no marking of origin or name of the manufacturer should appear on the product, on the shipping packing or sales packaging.

There are two cases for neutral packing: neutral packing with designated brand and neutral packing without designated brand. In the case of neutral packing without designated brand, neither brand or mark nor the country of origin is indicated on the packing. There are cases when the exporter is required to use a specified brand or trade designated by their importer, it becomes neutral packing with designated brand.

Neutral packing is generally adopted as a means to promote export sales. By means of neutral packing, exporters may break down the high customs duties or unreasonable import quota levied on the imports by the importing countries. However, in recent years, neutral packing has been restricted by many countries. For instance, as mentioned in the previous section, mark of origin is compulsory for imported goods in some countries. Goods packed in neutral packing, in absence of mark of origin, are not allowed to enter into these countries. Therefore, cautions must be taken while exporters agree to adopt neutral packing.

OEM

OEM is the short form of Original Equipment Manufacturer. It refers to a kind of international trade practice that sellers use the brand name or trademark designated by buyers on their manufactured goods.

OEM is popular internationally. Supermarkets, large department stores and monopolized-commodity stores in many countries adopt OEM for the goods sold to secure a rise both in their reputation and price. To large and regular orders, exporters are also willing to accept OEM. By means of OEM, sellers can make good use of buyers' well established brand names or trade marks, good reputations and status to expand sales and competitiveness in the world market.

In case of neutral marking with trade marks or brands designated by the buyer, or OEM, close attention should be paid to issues concerning infringement on a third party. Usually, the buyers are required to produce certificate evidencing their right in using the brand name or trade mark they designate. Still it is advisable for sellers to insert a safeguard clause or similar stipulations in the contract to that effect (see Figure 4. 10).

Example

" As to the trade marks designated by the buyer, if the seller is charged with the infringement by any third party, the buyer shall take up the matter with plaintiff and it has nothing to do with the seller. Any losses thus sustained shall be compensated by the buyer. "

Figure 4. 10   Example of Protecting Clause

4. 4. 4   Packing Clause in the Sales Contract

A sales contract should expressly indicate the packing method (e g. the material, dimensions, weight of every single piece, filling material used and reinforcement conditions, etc.) the packing cost and the shipping marks.

Details of the packing methods must be clear in the packing clause and shall include such details as shape, size, material used, etc. For example, 30 cm × 40 cm × 50 cm cartons, new iron drums containing 190 kg net weight, or 4 ply kraft paper bags containing 25 kg. In addition, reinforcement conditions or filling material, if applicable, must also be properly stipulated. For instance, "packed in wooden drums reinforced with iron band " "packed in cartons, 10 cartons to a plastic pallet " "packed in iron drums, 80 drums to a container" . Ambiguous expressions should be avoided in the contract. When the buyer is required to supply the packing material, wholly or partly, the seller should also stipulate in the contract the time limit within which the packing material shall arrive (see Figure 4. 11) .

Examples

- Packing: In new iron drums of 175 kg each, net.
- Packing: Each piece in a polybag, half dozen in a box and 10 dozen in a carton.
- Packing: In new single jute bags , 100 kg net each.
- Shipping Marks : At seller's option.
- Shipping Marks : At buyer's option, the relevant shipping marks should reach the seller × × days before the time of shipment.

Figure 4. 11   Examples of Packing Clause

Shipping cost division must be specified in the packing clause. Packing expense can be included in the price of the commodity; in this case, it is the seller who bears the packing cost. However, it can also be excluded from the price with packing cost borne wholly or partly by the buyer. In the latter case, the seller should specify in the contract the expense to be paid by the buyer and the method of payment as well.

Shipping marks may be supplied by the buyer or the seller. When the buyer designates the shipping marks, it is advisable for the seller to stipulate the arrival time of the marks in the contract to ensure smooth delivery .

In trade practice, unless otherwise stipulated in the contract, the cost of packing are included in the contract price and the design of the shipping marks are at a seller's option.

Chapter 5

International Cargo Transport and Shipment Clause

★ LEARNING OBJECTIVES

- To understand the role the ocean transport plays in international cargo transport.
- To learn the operation of liner shipping.
- To learn other important modes of transport, such as air cargo transport and containers.
- To understand the functions the B/L plays in ocean transport.
- To learn how to process a B/L and an air waybill.
- To be familiar with some common clauses of shipment.

International cargo transport means moving cargoes from an exporting country to an importing country or to any other regions. In other simple words, it aims at how to transfer goods from the seller's premise to the buyer's warehouse, which is usually very far away from the seller's place. It plays a key part in both a safe payment and a successful international business transaction.

5. 1　Ocean Transport

5. 1. 1　Features of Ocean Transport

Ocean transport is the most popular and most widely-used mode of transport in the international trade, which means moving cargoes by shipping vessels among different nations and areas. It is estimated that more than two thirds of the world total trade volume and up to 90% of China's exports and imports are transported by sea. Ocean transport has many obvious advantages as follows:

- Ocean transport possesses a large capacity. The capacity of vessels is far much larger than ones of trains, airplanes and trucks. For example, the dead weight of the largest oil tank may be up to 600,000 tons, while the one of some cargo vessels can be 500,000 tons, and some ships are capable of being loaded with over 4,000 TEUs.

- Ocean transport enjoys a low cost. Owing to the large capacity as mentioned above, economics of scale can be achieved and thus the unit distribution cost is reduced to the degree equal to one twentieth of railway transport, and one thirtieth by air.

- Ocean transport has good adaptability to cargoes of different size, weight, shape, etc., without being restricted to railways, roads and highways. The earth is naturally an easy passage which needs no investment and maintenance, for the reason about its 70% is covered by water.

Nevertheless, ocean transport also has disadvantages. It is vulnerable to seasonal wind, heavy fog, iceberg and other natural disasters which may very possibly result in bad punctuality, high riskiness and slow passage.

5.1.2   Major Types of Cargo Vessels

There are different types of cargo vessels designed and built to suit the needs of shipping different cargoes.

- Container Vessel—These ships are mainly designed to carry large steel containers that are usually 20 feet or 40 feet in length. There are some vessels which barges with containers loaded on can be pulled in directly to this kind of vessel through its open bow. These ships have to be loaded and off loaded by large cranes to and from trucks, and they hence are limited to ports that have handling equipment, which is called as container terminals.

- Bulk Vessels—This kind of ship is mainly used for the carriage of bulk commodities like wheat, sulfur, iron ore, coal etc. They can be called as a grain ship, a collier and an ore ship.

- General cargo ships (sometimes called Breakbulk Carriers) —The average general cargo ship is about 500 feet long. These ships will usually have four or five holds (a hold is the cargo space in a ship), with one or, in a few cases, two holds aft the engine room, and four to five holds generally in the front of the engine room. They have long protruding rigging for winches by each hold. These winches are used to load and unload the cargo, which is usually packaged and moved as single parcels, or assembled together on pallet boards. Some general cargo ships may also have refrigerated spaces for perishable cargo.

- Oil Tanker Vessels—It is used to carry crude oil or fuel oil. Oil tanks account for about 50% of the world merchant fleet in deadweight terms. They can be as large as over 500,000 deadweight tons.

- Reefer Vessel (refrigerated vessels) —It is used for the carriage of frozen cargoes or temperature-controlled cargoes like fruits, meat, fish etc.

- Ro-Ro Vessels—It is used for the carriage of wheeled cargo like cars, buses, trucks, excavators, etc. Loaded trailers or any other wheeled modes can be driven onto these vessels to enable faster loading and unloading.

- Multi-purpose Vessels—It is used for the carriage of a combination of above cargoes. They are very versatile, popular and useful vessels especially along certain routes which require self-geared vessels and do not have shore handling facilities.

- A LASH (Lighter Aboard Ship) —This is a vessel designed to carry lighters on which cargoes are loaded. Barges (lighter) are loaded with goods at inland river or shallow ports, and then, are towed to ocean ports to meet the LASH, the mother vessel, which on arrival lifts the barges on board with its own crane. It offers high efficiency and is ideal on shallow inland waterways.

5. 1. 3　Operation Modes of Ocean Transport

According to the different operating mode, the ocean transport can be divided into liner shipping and charter shipping.

Liner Shipping

Liner shipping involves the transport of cargo to ports on a regular basis, operating on a particular geographic route according to advertised timetables. A liner works as buses, trains and other public transportation, regardless of volume of commodities, seasons.

(I) *Features of Liner Shipping*

Liner shipping has the characteristics of fixed route, fixed calling ports, fixed liner freight, and fixed sailing schedules. The four "fixeds" enable a shipper to make shipment on time, to make a right budget, and thus make successful transactions. In liner shipping, the shipping company is responsible for loading and unloading of the goods, which is included in freight, without any responsibilities on shippers' or consignees' part. Demurrage and dispatch are taken into no account.

(II) *The Freight Rate of Liner Shipping*

The freight rate is collected according to the freight rate tariff advertised by shipping companies. Transport policies and regulations, commodities classification, shipping lines, surcharge and special freight for refrigerated products and live animals and other issues are dealt with in most freight rate tariffs, which, published by different shipping companies, may have slight difference in both contents and form. Freight rate tariff can be divided into commodity rate freight tariff and class rate freight tariff. The former one lists the basic rate for every single commodity, and the latter, also called freight class list, classifies commodities into classes (usually 20 classes), with distinctly designated freight rate for each class, and the first class is charged with lowest price, while the 20th class is charged with the highest. In practice, freight class list is widely applied. The procedures of calculating freight are as follows:

- Look up your commodity in the freight class list which is provided by your carrier, and identify the class that the commodity belongs to, and freight ton (weight ton, measurement ton, or ad valorem) that applies to the commodity.

- Look up, in the freight class list, the applicable shipping line, the port where the goods are shipped and the port of destination, and then figure out the basic rate.

- Look up, in the freight class list, the surcharges which are imposed on your goods, its applicable rate and basis (often percentage of basic freight, or amount of containers), and the currency to be used.

- Add total basic rate to surcharge. The sum is the total freight.

It can be inferred from the above that the freight of shipping liner consists of basic rate and surcharge.

(Ⅲ) *Basic Rate*

The basis of basic rate calculation includes six types as follows:

- weight ton （WT） —weight ton is mainly applicable to the cargo of which the weight per cubic meter is more than one metric ton. It marked with "W", one weight ton equals to one metric ton （1,000 kgs） . The basic rate will be calculated according to the cargo's weight.

- measurement ton （MT） —measurement ton is mainly applicable to light cargo of which the weight per cubic meter is less than one metric ton. It marked with "M", one measurement ton equals to one cubic meters. The basic rate will be calculated according to the cargo's measurement. Both weight ton and measurement ton are named as freight ton as a basic unit to calculate freight.

- weight ton /measurement ton （W/M） —the shipping company will choose the larger freight ton between weight ton and measurement ton.

- Ad valorem freight （A. V. ） —For valuable cargoes such as jewelry and precious metals, sophisticated instruments and expensive art products, the freight will be collected in proportion to the value of the cargo. Usually it is 3% –5% of the FOB price.

- Weight ton/ measurement ton or Ad valorem （W/M or A. V. ） —The freight will be calculated on the basis of weight ton, weight ton or the value of the goods, subject to the highest rate.

- freight per—For cargoes like truck and animals, freight is charged for each unit, such as per head, per container, per bill of lading.

- open rate—for bulk primary cargoes of low value, the freight is open to negotiation between the shipper and the carrier.

(Ⅳ) *Surcharge*

Surcharge is part of freight. Literally, it means the additional charge in addition to the basic rate, used to compensate the liner shipping company for extra expenses or losses resulting from the vessel, cargo, port or other factors. There is a variety of surcharges. Some of them can be canceled, and new ones can be created in accordance with new situations.

Main surcharges includes: Bunker Surcharge or Bunker Adjustment （BAF ）, Devaluation Surcharge or Currency Adjustment Factor （CAF）, Deviation Surcharge, Suez Canal Surcharge, Transshipment Surcharge, Direct Additional, Port Surcharge, Port Congestion Surcharge, Heavy-lift Additional, Long Length Additional, Cleaning Charge, Optional Fees or Optional Additional, Alteration Charge, Terminal Handling Charge （THC） .

Total freight can be calculated using the following formula:

Total freight = Total Quantity × Basic Freight Rate × （1 + Surcharges/Additional Rate）

Eg. A batch of shipment exporting from Guangdong Whampoa port to another port abroad, its freight rate basis is W / M. the total number is 300 cartons, the gross weight per carton is 25 kgs, the volume per carton is 50 cm × 30 cm × 20 cm, the basic freight rate is USD60 per freight ton, with 5% special fuel surcharge, 10% port congestion surcharge. How much is the total freight?

Calculation:

Total weight: 25 kgs × 300 cartons = 7,500 kgs = 7.5 M/T

Total measurement = 50 cm × 30 cm × 20 cm × 300 cartons = 9 CBM

Measurement > Weight, "M" is the freight calculation basis

Total freight = basic freight rate × (1 + surcharges rate) × total measurement

$$= 60 × (1 + 5\% + 10\%) × 9 = USD\ 621$$

(V) *Freight Forwarder*

A large number of exporters and importers do not have a large enough traffic to justify the use of a tramp. Liners, ideal for cargo of small quantity, can provide good service but with simple procedures, and have proved to be a very economical means of international cargo distribution. Nevertheless, international cargo transport is a complex business by nature. Small exporters are unable or unwilling to afford their own export shipping staff, and thus they, to handle international shipments, turn to a freight forwarder, a professional shipping agent. A freight forwarder, known as forwarder, or forwarding agent, sometimes also known as a non-vessel operating common carrier (NVOCC), is a person or company that acts as a bridge between shipping company and the shipper. They are experts on the availability of different modes of transportation for different kinds of markets, on the cost of transport and on the suitability of each mode. A forwarder himself does not move the goods but acts as an expert in the logistics network, organizing shipments for shippers to get goods from the manufacturer producer to a market, customer or final point of distribution. He contracts with a carrier or often multiple carriers to move the goods, who can use a variety of shipping modes, including ships, airplanes, trucks, and railroads, for a single shipment. For example, the freight forwarder may arrange to have cargo moved from a plant to a port of shipment by truck, flown to the destination city, then moved from the port of destination to a buyer's building by another truck. A shipping agent's job includes booking space, processing documentation, arranging for customs clearances, inspection, insurance, payment, collecting cargo, grouping consignments and performing other activities pertaining to international shipments.

Charter Shipping

The other type of merchant vessels is the charter shipping, also called as tramps. Charter shipping means that a cargo ship is hired from a shipping operator. It is a tramp service, which ideal or bulk cargoes, referring to a cargo ship that does not operates on regular routes and schedules. A charter party is needed to stipulate the sailing date, shipping route and calling ports as the shipper requires, as well as shipper's and ship operator's responsibilities and rights. Tramps can be classified into voyage charter, time charter and bareboat charter.

(I) *Voyage Charter*

A voyage charter is the hiring of a vessel, or just its some space and crew for a voyage which may involve more than one calling port. Under voyage charter, the ship operator takes all the responsibilities for ship operation, including the payment for the port costs, fuel costs and crew costs, while the charterer are free out of responsibilities as soon as goods have been loaded on board. He or his customer waits for the delivery of goods at the port of destination indicated in the charter party. A voyage charter can be further divided into single voyage charter, round voyage charter and consecutive voyage charter. Obviously, the ship operator has fulfilled his obligations when voyages are finished within a time limit. Therefore, the ship operator will make effort to shorten the journey time to achieve faster circulation of the ship. A voyage charter party should include the following main clauses:

- shipper, i. e. charterer, carrier, name of vessel, flag of vessel
- cargo type, packaging, quantity
- ports and alternatives
- laydays date—Laydays date refers to the earliest and the latest date the charterer can accept for loading the cargo, and when the ship must have the vessel ready. If the ship arrives before the laydays specified, the charterer does not have to take control or start loading. If the ship arrives at the designated ports or places after the laydays, then the contract can be cancelled and hence laydays are often presented as the term Laydays and Cancelling, shortened to Laycan.
- how (by freight ton or a lump-sum), when and where the freight is paid (prepaid at the port of shipment or to collect/ payable at the destination)
- loading and unloading fees—loading and unloading fees should be clearly stipulated in the charter party. Five terms might be involved：

a. Gross Term or liner terms indicate that the carrier is responsible for the cost of loading and unloading of the goods;

b. F. O. indicates that the charterer of a vessel is responsible for the cost of unloading goods from the vessel;

c. F. I. indicates that the charterer of a vessel is responsible for the cost of loading goods onto the vessel;

d. F. I. O. indicates that the charterer of a vessel is responsible for the costs of loading goods onto the vessel and unloading goods from the vessel;

e. F. I. O. S. T. indicates that the charterer of a vessel is responsible for the costs of loading goods onto the vessel and unloading goods from the vessel, as well as the costs of stowing and trimming goods.

- Laytime, demurrage and dispatch—laytime is a period, like 10 – 15 days, allowed for loading and unloading the cargo. There are different ways to calculate it. They are running or consecutive days, working days and weather working days. If laytime is exceeded, the charterer must pay demurrage. It can encourage the charterer to finish loading and unloading within the specified period. Demurrage is understood as the compensation charterer has made for the possible losses sustained from failing finishing loading and unloading within the specified period. If laytime is saved, the charter party may require the ship owner to pay dispatch to the charterer. Dispatch is generally as half much as demurrage.

(II) *Time Charter*

A time charter is the hiring of a vessel for a specific period of time, within of which the charterer is in the position of deciding applicable goods for shipments, selecting the ports and directing the vessel where to go. The charter hire must be paid to the ship on the basis of day, or month within specified time, even when the vessel is out of business; the ship owner still manages the vessel and pays for ship technical maintenance, ship insurance and the crew while the charterer pays for all fuel the vessel consumes, port charges, commissions, besides a daily hire to the owner of the vessel. A time charter party should include the main clauses as follows:

- shipper, i. e. charterer, carrier, i. e. the ship owner
- description of ship: its deadweight, flag of vessel
- trade limit: cargo applicable to the vessel, areas allowed to travel
- charter period
- delivery of vessel
- charter hire: Charter hire is decided by the capacity of the vessel and time

(III) *bareboat charter*

A bareboat charter or demise charter is a special form of time charter. It means the hiring of a vessel only for a long period of time, without any crew and ship's captain, where by no administration or technical maintenance is included as part of the agreement. The charterer pays for all operating expenses, including fuel, crew, port expenses and hull insurance in addition to fixed costs. The charterer obtains possession and full control of the vessel along with the legal and financial responsibility for it. A bareboat charter is seldom used in ocean cargo transport.

5.2   Others Modes of Transportation

5.2.1   Air Transport

Air cargo industry has seen a rapid growth since World War II, and has gained a increasing large part of international cargo transport. Today, there are over 1000 airliners, over 30000 airports more than 1000 of which are suitable for international flight. According to the statistics issued by IA-TA (International Air Transport Association), over 500 million cargoes were shipped by air in 2014, bringing total value worth US $ 62. 5 billion for air cargo industry. From 2004 to 2014, the industry increased at the rate of 3. 9% per year.

Advantages of Air Transport

- high speed and quick transit. Air transport is ideal for perishables, seasonal goods, and goods in urgent need.
- low risk of damage and pilferage. Air transport offers tighter security than other modes of transport for the goods carried. As a result, loss and damage can be reduced, and packing can be

simplified and insurance underwriters offer favorable insurance rates.

- fast capital flow and small inventory. If the transportation of goods is completed in a short time and in a good condition, financial settlement will be usually faster. In this way the high speed brings smaller amount of capital tied up in transit and interest savings.

- bigger trade and new markets. Goods like tulips from Holland cannot possibly be transported to China without quick air transport.

- simple procedures and good service. IATA members up to 262, accounting for 83% transport volume, must compete on service quality while they adhere to IATA guidelines in the same price setting.

Air Cargo Company and Air Freight Forwarder

- Air cargo companies are a kind of carrier. They own airplanes, and have the responsibilities of moving cargoes to destinations places. Air China Cargo in Beijing, China Cargo Airlines LTD in Shanghai and China Southern Cargo Airlines LTD in Guangzhou are the biggest air cargo companies in China.

- Air cargo companies focus on flight, but pay little attention to sales. In practice, air freight forwarders are the experts of making arrangements for air cargo before and after shipments. An air freight forwarder functions the same as a shipping agent mentioned above, jobs of whose includes booking space, collecting and consolidating cargoes, issuing air waybills, collecting freight charges, processing documents, arranging for customs clearances and inspection, making and taking delivery of goods, arranging for transshipment and performing other activities. She act as a bridge between a shipper and an air cargo company, sending a shipping letter to and taking delivery of goods from the latter on behalf of his customer (shipper), while issuing a master air waybill or house air waybill to the shipper on behalf of an air cargo company. Sinotrans Limited in China is a big air freight forwarder which cooperates with many international airliners.

Operating Modes of Air Cargo Transport

- An airline, exactly like a shipping liner, has fixed flying schedule, fixed airline and fixed destination airports. It is usually a passenger plane, ideal for a variety of high value, small volume perishable goods or goods in bad need. It is estimated that 50% of all air freight is moved in this way.

- Air charter is the business of renting an entire aircraft by an air chartering company (airline) to a charter or charters (usually a air freight forwarder). The chartered aircrafts are used to transport goods to designated places. Air chartering mainly applies to individual private itineraries, urgent or time-sensitive cargoes, and other forms of ad hoc air transportation.

- Consolidation means that an air freight forwarder consolidates several lots of cargoes from different shippers into one whole consignment. He presents only one shipper's letter of instruction for cargoes to and receive only one copy of a master air waybill from the air cargo company, and then issues house air waybills to every actual shipper. Consolidation can reduce freight charge, and is the main service provided by an air freight forwarder, thus the freight charge can be reduced.

Air Freight Rate

Air freight rate can be collected on the basis open rate and IATA rate. The former indicates that the freight rate to be collected is specified in a carriage contract signed by air cargo companies and the shipper (usually the air freight forwarder), while the latter, published by IATA, means the air cargo tariff which includes published through rate and unpublished through rates.

(I) *Published Through Rate*

includes the following five types of rate:

- General Cargo Rates (GCR) are the basic rate, which can be classified into two types

a. "N" standing for normal rate, which is applied to general cargoes under 45 kgs

b. "Q" standing for quantity rate. Q45 is used to indicate cargoes over 45kgs under 100kgs, while Q100 indicates cargoes over 100kgs under 300kgs, and Q300 indicates cargoes over 300kgs.

It is noted that air freight charge is also collected according to weight ton or as volume ton, as the ocean freight of liners, that is the freight charge is calculated on the basis of the larger one between weight ton and volume ton. However, the measurement ton referred in the part, is 6,000 cubic centimeters as one single unit, rather than 1 cubic meter, and weight ton is 1 kilogram as one unit, rather than 1 metric ton.

- Commodity Class Rate (CCR) means an additional charge or special discount on the basis of general commodity rate (GCR). "S" stands for surcharge, and "R" stands for reduction. Urgent parcels, medical products, valuable plant, live animals, coffin, perishable goods, valuable commodities, guns and bullets and other special goods are charged with 150% or 200% of normal GCR, while reading materials like magazines, newspaper and personal luggage will be charged with 50% of normal GCR.

- Specific Commodity Rates (SCR) are reduced rates applicable to a wide range of commodities specified in TACT (The Air Cargo Tariff). They are restricted to specific places of dispatch and places of delivery, shipping time and the minimum weight (usually 300kgs). It is usually applied to bulk cargoes, or seasonal goods in low value.

- Minimum Charge is the possibly least freight collected for a single consignment. If all possible rates calculate according to GCR, or CCR or SCR is less than minimum charge, the latter is the air freight.

- Unit Load Device rate is air freight charge collected according to the numbers of containers, or pallets.

(II) *Unpublished Through Rates*

If there is no Published Through Rate to look up for a special journey, Unpublished Through Rates can be referred to. It consists of construction rate and combination of rate. Construction rate can be calculated by adding the freight rate specified in TACT to add-on amounts, which can also be found in TACT but cannot be used separately. It is worth pointing out that construction rate calculation must be consistent. If the known freight rate specified in TACT is collected according to GCR, then the add-on amounts must be collected in accordance with GCR. It can be inferred that if the known freight rate specified in TACT is collected according to ULD, then the add-on amounts must be collected in accordance with ULD. Mixing up different types of rate is not allowed. Combination of rate can be used when there is no published through rate and no add-on amounts published in TACT for reference. An airport between the shipping airport and the destination airport can be chosen as the middle point. The rate from the shipping airport to the middle airport plus the rate from the middle airport to the destination airport equals the total air charge, also called combination rate. Choosing a different airport as a middle port may very possibly result in different total freights. As this is the case, the lowest total freight can be the option as long as it is based on the actually existing airlines.

5.2.2　Railway Transport

Railway transport is an important land mode of transport. It used to play an essential role in the beginning of China's opening up. It can be closely comparable to air cargo transport in speed, and to ocean cargo transport in capacity. Most importantly, it is secure, punctual, low cost and not vulnerable to vile weather. It contains two main modes.

International Railway Transport

International Through railway transport means connection of the railways which are built across more than two countries for moving cargoes under only one international through railway bill. The railway company takes the responsibilities of arranging customs clearance, and thus no shipper , no consignees are involved in the whole process of making delivery.

The freight charge of international through railway transport depends on the weight and volume of goods. Goods can be delivered in full car load or less than car load by railway. Full car load indicates that under one copy of railway bill goods need to be delivered in a whole car without any other consignments from other consignors. China Railway stipulates that delivery in full car load is necessary provided that a 300,000 tonnage car is needed to accommodate the weight, the measurement, shape or nature of the goods. FLC must be used for the following 7 types of goods:

- Goods in need for keeping cool or warm;
- Harmful and hazardous materials;
- Contaminants;
- Live bees;
- Bulk goods without unit numbers;
- Live animals without containers;
- A single item over 2 tons in weight, or 3 cubic meters in volume, or over 9 meters in

length. LCL indicates that under one copy of railway bill goods to be delivered is less than 5,000 kilograms in weight, and needs no a whole car to be loaded in terms of volume.

Freight Rate of International Railway Transport

In China, the Freight rate of international railway transport are mainly based on Agreement of

International Commodity Railway Through Transportation (CMIC) signed by Soviet and other 7 eastern European countries in 1951, and Railway Cargo Transport Tariff, published by Ministry of Railways of the People's Republic of China. Freight charge consists of three parts. They are rate for railways in the importing country, rate for transitional countries, and rate for the exporting country.

- The first kind of rate is collected according to the tariff regulations issued by the government of an importing country.

- The transit freight is charged by the transitional countries. It will be collected at the prevailing rate of the date when cargoes are crossing these countries' territories, which can be prepaid by the shipper, or can be paid by the consignee at the destination. If a country is not a member of CMIC, it can refer to any other applicable agreement to calculate the freight. In our country, if delivery happens at the seller's train, then a Chinese exporter is only responsible for the distance to the domestic border station, while a Chinese importer must take all responsibilities from the transitional country to China.

- Procedures for calculating transit freight:

a. Find out in the railroad brochures what the respective distances are in kilogram that the goods have to go through in transitional countries;

b. Identify in commodity class rate the applicable class and chargeable weight (weight ton or measurement ton) in accordance with the cargo name;

c. In the railroad brochures, find out the applicable rate for slow trains in accordance with the total distances in kilogram and the class whereby the goods are classified. The formula is as follow

Basic rate = freight rate × chargeable weight

Total rate = basic rate × mark-up rate

Mark-up rate here means the percentage added to the basic rate. Express trains for the same goods as mentioned in slow journey will be charged on the basis of slow trains, with another 100 %; if in LCL, with another 50% and then another 100%; with another 200% if goods are shipped in a passenger train in FCL.

- Procedures for calculating domestic freight:

a. Find out, in Railway Cargo Transport Tariff, the distance from the starting station to the border;

b. Identify, in Railway Cargo Transport Tariff, commodity class and chargeable weight (weight ton or measurement ton) in accordance with the cargo name;

c. In Railway Cargo Transport Tariff, find out the applicable rate and the class whereby the goods are classified.

d. Chargeable weight times applicable rate equals domestic freight, as the following formula:

Freight = rate × chargeable weight

An Overview of International Railway Transport in China

- The Trans-Asian Railway Network Agreement is an agreement signed on 10 November 2006, by seventeen Asian nations in China. The network will consist of four main railway routes. They are the northern corridor, the southern corridor, a southeast Asian network, the north-south corridor. The Northern Corridor will link Europe and the Pacific, via Germany, Poland, Belarus, Russia, Kazakhstan, Mongolia, China, and the Korea. The existing Trans-Siberian railway, 9,250 kms long, connecting Moscow to Vladivostok covers much of this route. The Southern Corridor will go from Europe to Southeast Asia, connecting Turkey, Iran, Pakistan, India, Bangladesh, Myanmar and Thailand, with links to China's Yunnan Province and, via Malaysia, to Singapore. A Southeast Asian network; this primarily consists of the Kunming-Singapore Railway. The North-South Corridor will link Northern Europe to the Persian Gulf. The main route starts in Helsinki, Finland, and continues through Russia to the Caspian Sea. The plan has sometimes been called the "Iron Silk Road" in reference to the historical Silk Road trade routes.

 • The New Eurasian Land Bridge, also called the Second or New Eurasian Continental Bridge, is the southern branch of the Eurasian Land Bridge rail links. The Eurasian Land Bridge is the overland rail link between East Asia and Europe extending from Lianyungang to Rotterdam, across Kazakhstan, Russia, Belarus, and Poland, a distance of 11,870 kilometers. As of 2013 express freight trains were being used by manufacturers such as Hewlett-Packard to ship products from factories in the interior of China through Kazakhstan, Russia, Belarus, and Poland to Europe. The Customs Union of Belarus, Kazakhstan and Russia reduces inspections, delays and theft. Shipment from a factory in a city in central or western China such as Chongqing or Chengdu to distribution centers in Europe takes about 3 weeks, that is, less than the 5 weeks taken by ship transport but 25% more expensive while it takes about a week by air, but costs 7 times as much as rail and adds 30 times more carbon to the atmosphere.

 • Hong Kong is a free trade zone, where a lot of shipments from China can be made without any taxes imposed. Train services to Hong Kong terminate at the Hung Hom Station in Kowloon. Within Hong Kong the cross-boundary services use the tracks of the East Rail Line. There are three through-train routes, Beijing line (to/from Beijing), Shanghai line (to/from Shanghai) and Guangdong line (to/from Zhaoqing and Guangzhou East). Another express train service linking Hong Kong and Guangzhou with intermediary stop in Shenzhen has been approved and construction in the China section has commenced. This new express rail line will reduce the train travel time between Hong Kong and Guangzhou from 2 hours to 1 hour. Hong Kong cross-boundary services consist of two parts, one of which is charged with in China Mainland, and the other in Hong Kong. The procedure is as follows: firstly, shipments arrive at Shenzhen Northern Station, and Sinotrans Limited, Shenzhen Branch takes the delivery, and secondly, Sinotrans Limited, Shenzhen Branch will arrange customs formalities, and thirdly, China Travel Service (Hong Kong) limited, the agent for Sinotrans Limited, Shenzhen Branch in Hong Kong, will be responsible for delivering the goods to the destination station in Kowloon.

5.2.3　Containerization

It is estimated that goods shipped in containers account for 60% of the world's seaborne trade.

The containers are rectangular, closed box models, known under a number of names, such as inter-modal container indicating it can be used by different modes of transport, cargo or freight container, ISO container, shipping, sea or ocean container. Containerization is a system of intermodal freight transport using intermodal containers to bundle cargo and goods into larger, unitized loads. That is to say, they can be loaded and unloaded, stacked, transported efficiently over long distances, and transferred from one mode of transport to another—container ships, rail transport flatcars, and semi-trailer trucks in the same container—without being opened. The handling system is completely mechanized and done with cranes and special forklift trucks. So containerization is a modern, highly efficient and effective mode of transport. It has brought majors changes as follows:

- Packing cost reduction. A long distance of transport asks for strong packing to protect goods from possible damages. Containers serve the purpose because they are closed, made of corrugated weathering steel used for the sides and roof and thus contributes significantly to the container's rigidity and stacking strength. Goods need no packing for rough transport.

- Better service. Goods can arrive in better condition because containers can provide strong packing for transportation. The containers with goods loaded on may have to be transferred from different modes of transport—from ship to rail to truck—without unloading and reloading their cargo. In this way, containers can reduce risks of theft, dampness, contamination that goods might undergo in a long distance of transport, which in turn bring a more favorable premium offered by insurance companies.

- Both handling cost and operating cost are reduced. Prior to highly mechanized container transfers, crews of 20 – 22 longshoremen would pack individual cargoes into the hold of a ship. After containerization, large crews of longshoremen are no longer necessary at port facilities. Containers terminals, without vulnerability to vile weather, offer higher efficiency in loading and unloading, thus enables quick transit and reduces the operating cost for the carrier.

Types of Containers

There are some criteria to classify containers, but these following ones are often used.

Dry containers also known as general container, are used for general dry goods which needs no controlled temperature. They are closed on all sides. The type accounts for 80% of the world's containers.

Bulk containers are either closed models with roof-lids, or open-top units for top loading, they are designed for the transport and storage of bulk liquid and granulated substances, such as chemicals, food ingredients, solvents, pharmaceuticals, etc.

Refrigerated containers has an integral refrigeration unit for controlling the temperature inside the container or the container is supplied with cold air via ship's central cooling plant. Refrigerated containers are used for goods which must be transported at a constant temperature above or below freezing point, like fruit, vegetables, meat and dairy products.

Standard containers are often used to measure the capacity of container ships and container terminals, also known twenty foot equivalent unit (TEU). A standard container is 20 feet long (about

6. 096m）, 8 feet（about 2. 44m）wide and 8. 6 feet（about 2. 491m）high, the capacity of 20-foot container is about 17, 500kgs or 25cbms. The container with the same width and height as a standard container but a doubled length of forty feet（12. 192m）is called a 40-foot container, and the container with the same width and length as 40-foot container is called as high cube container which is 40 feet long and 9. 6（about 2. 896m）feet high. For 40-foot container, its capacity is about 24, 500kgs or 55 cbms, and for 40-foot high cube container, its capacity is 26, 000 kilograms or 67cbms.

Flat rack containers consist of a floor structure with a high loading capacity composed of a steel frame and s softwood floor and two end walls, which may either be fixed or collapsible. The end wall are stable enough to allow cargo securing means to be attached . Flat racks can be 20-foot containers and 40-foot ones, any heavy or bulky out-of-gauge cargo, like machinery, semi-finished goods or processed timber.

Container Service Terms

- FCL vs LCL

A full container load（FCL）is an standard container that is loaded and unloaded under the risk and account of one shipper and only one consignee. In practice, it means that the whole container is intended for one consignee. FCL shipment tends to have lower freight rates than an equivalent weight of cargo in LCL. FCL is intended to designate a container loaded to its allowable maximum weight or volume, but FCL in practice on ocean freight does not always mean a full payload or capacity. In fact many companies will prefer to keep a "mostly" full container as a single container load to simplify logistics procedures and increase security compared to sharing a container with other goods.

Less-than-container load（LCL）literally means that a shipment which is not large enough to fill a standard cargo container. The abbreviation LCL formerly applied to "less than（railway）car load" for quantities of material from different shippers or for delivery to different destinations carried in a single railway car for efficiency. In practice, LCL not only means the goods is less than a container, but also means that the shipment of cargo that cannot fill a container full has to be merged with cargoes from other shippers into an full container for more than one consignee.

- CY vs CFS

Container yard（CY）is the manner in which the cargo is delivered and received in CY. CY is also the location where the shipper sends goods before loading on board, or where the goods are sent directly after unloading, with the same full container intact. Container freight station（CFS）is the manner of the cargo in which it is delivered and received in LCL. CFS is the place where the goods are bundled into a whole container before goods are loaded on board, or where the goods are separated after the goods are unloaded at the port of destination .

- Containers Service Mode

a. Door to door: The carrier comes to the seller's factory to pick up the container after it is loaded by the shipper. The carrier delivers the same container to the consignor's warehouse after the container is unloaded at the port of destination. The service is used for FCL. Naturally this is the most expensive, but least hassle service. It is used for FCL.

b. Door to CY: The carrier comes to the seller's factory to pick up the container after it is loaded by the shipper, and takes the responsibility after delivering the container to the container yard at the port of destination where the container shall be kept intact. It is used for FCL.

c. Door to CFS: The carrier comes to the seller's factory to pick up the container after it is loaded by the shipper, and deliver the container to the container freight station the port of destination, where the container will be unpacked to separate different cargoes. It is the consignee's responsibility to take this loose cargo and move it from the destination port to his/her final location, which may be a warehouse, distribution center, retail location, etc.

d. CY to door: The shipper loads the goods onto the container and delivers it to the container yard at the port of shipment, and then the carrier will be responsible for transporting the goods to the buyer's warehouse. It is used for FCL.

e. CY to CY: The shipper loads container and delivers it to the container yard at the port of shipment, and then the carrier moves the container to the container yard at the port of destination. It is used for FCL.

f. CY to CFS: The shipper loads container and delivers it to the container yard at the port of shipment, and then the carrier moves the container to the container freight station at the port destination. However, at the destination side, the container will be unpacked at the container freight station. It is the consignee's responsibility to take this loose cargo and move it from the destination port to his/her final location, which may be a warehouse, distribution center, retail location, etc.

g. CFS to CY: The shipper delivers loose cargo the CFS at the port of shipment, where his/her freight forwarder will then pack the goods from different shippers into one shipping container. At the destination side, the cargo will be delivered in the same container to the CY at the port of destination.

h. CFS to CFS: In this case, the shipper delivers his loose cargo to CFS at the shipping port, where his/her freight forwarder will then pack the goods from different shippers into one shipping container. At the destination side, the container will be sent to CFS where it will be unpacked to separate different cargoes for different consignees. The consignee is responsible for arranging the pickup of the cargo at the destination port and moving it to his or her final location.

Container Service Charge

Container service charge consists of inland haulage charge, LCL charge for grouping and unpacking goods, CY service, container chartering charge and terminal handling charge as well as ocean freight charge. The calculation of container freight depends on the size of the cargo, there are two ways:

- Full Container Load (FCL). Charge at box rate on each unit of container. It can be adopted FCL when the cargo's measurement ton or weight ton reaches 85% or 95% of the maximum load of the container respectively. Then the container will be delivered directly to container yard (CY).

● Less than Container Load (LCL). When the cargo's measurement ton or weight ton is less than a container's maximum load, it should be transported to the container freight station (CFS) designated by the shipping company, and then packed into a full container load (FCL) together with other shipper's cargo. LCL freight always being calculated on measurement tons or weight tons, and the shipping company will charge the higher one. It shows M/W or R/T on the freight tariff.

Eg. 1,000 Children bicycles to be exported to France, the port of destination is the Marseille port. The cargo will be packed in cartons, each containing six bicycles. The volume of each carton is 0.0576cbm, and gross weight of each carton is 21kgs. Please calculate the container freight for this batch of bicycles.

Step 1: select the right size of container:

The total number of cartons = 1,000 ÷ 6 = 166.6, that is 167 cartons (CTNS)

The total volume = 167 × 0.0576 = 9.6cbms

Total gross weight = 1,000 ÷ 6 × 21 = 3,500kgs = 3.5CTNS

According the specification of different types of container, it is known that both the measurement tons and weight tons of this batch of goods can not reach the maximum load of a 20-foot container, therefore, it should be packed by LCL.

Step 2: Upon inquiry, the LCL freight to port of Marseille is as follows:

Basic freight: USD151.00 per measurement ton (MTQ), USD216.00 per weight tons (WTS)

Step3: According to the above information, the sea freight is calculated (suppose the US dollar and the RMB exchange rate is 6.3):

Shipment by volume = 9.6 × 151 = 1,449.6 (USD)

Shipment by weight = 3.5 × 216 = 756 (USD)

The shipping company will charge the higher one, so the ocean freight is USD1,449.6.

Total freight = 1,449.6 × 6.3 = 9,132.48 (RMB)

5.2.4　Multimodal Transport

Multimodal transport, which developed in connection with the "container revolution", is (also known as combined transport or international freight transport) the transportation of goods under a single carriage contract, but performed with at least two different means of transport; the carrier who signed the carriage contract with the shipper is liable (in a legal sense) for the entire carriage, even though it is performed by several different modes of transport (by rail, sea and road, for example).

The carrier does not have to possess all the means of transport, and in practice usually does not; the carriage is often performed by sub-carriers. The carrier responsible for the entire carriage is referred to as a multimodal transport operator, or MTO. Multimodal transport operators provide customers with so-called door-to-door service, with a combination of ocean vessels with trucks, trains, and planes. A MTO may be a large sea carrier, also known as vessel operating common carrier (VOCC), who performs the carriage of goods by sea, and sub-contracts the rest of carriage to other actual carriers to perform, but must be responsible for the whole journey of the goods, including loading and unloading fee, warehousing fee during transit. A MTO can also be a forwarder, and in practice, he usually is the MTO, who is referred to as a non vessel operating common carrier (NVOCC) because he does not possess any sea vessel. He has moved away from his traditional role as an agent for the sender, accepting a greater liability as a carrier. The NVOCC contracts with multiple carriers to move the goods, who can use a variety of shipping modes, including ships, airplanes, trucks, and railroads, and takes all the responsibilities for the whole journey from the shipper's factory to the consignee's warehouse.

5. 2. 5　Land Bridge Transport

A intensive network of land transportation, as well as intermodal containers, has also made great contributions to the development and prosperity of intermodal freight transport. Land bridge means the land portion of the trip which a containerized ocean freight shipment must travel across. The land works as a bridge which connects the port of shipment in an exporting country and the port of destination in an importing country. The land, a significant part of the trip, enroute to the final destination of goods, is referred to as the "landbridge", and its mode of transport used is rail transport or truck transport. There are three applications for the term.

● Land bridge—An intermodal container shipped by ocean vessel from the port of shipment in a country to the port of destination in another country, must travel across an entire body of land, enroute. To be simple, the mode is port-land-port. For example, a container shipment from China to Germany, is loaded onto a ship in China, unloads at a Los Angeles (California) port and travels via rail transport to a New York, and loads on a ship for Hamburg.

● Mini Land bridge—An intermodal container shipped by ocean vessel from a country to another country, passes across a large portion of land in the exporting country. The mode is land—port—port. For example, a container shipment from Wuhan, China to Los Angeles (California), travels via trail transport to Shanghai, and gets loaded there, unloaded at a Los Angles, the final destination.

● Micro Land bridge—An intermodal container shipped by ocean vessel from a country to another country, passes across a large portion of land to reach an interior inland destination. The mode is port—port—land. For example, a container shipment from China to Denver (Colorado), is loaded onto a ship in China, unloaded at a Los Angeles (California) port and travels via rail transport to Denver (Colorado), the final destination.

5. 3　Shipping Documents

Shipping documents must be required to specify the responsibilities and rights of each party involved. These documents are closely relevant, but serve different functions to secure the goods through the whole journey.

5. 3. 1　Ocean Shipping Documents

Main relevant Documents include as follows:

- Booking note（B/N）: This is an application form which a shipper needs to complete according to the sales contract and terms in a letter of credit if necessary, and presents to a carrier or a freighter forwarder when booking shipping space. It shall be furnished with name, weight, volume of goods, and loading port, discharge port, and the date of shipment.

- Shipping order is the shipping document which is furnished with vessel name, voyage number, issued by a carrier when he accepts the space booking. Shipping order must be shown both for loading goods on board and for clearing the goods from customs. The shipping order will be signed by the customs after the goods are cleared. Then, it has become a signed shipping order. (signed S/O).

- Container load plan is the document which is furnished with the name, number, measurement, weight and marks of the commodity to be loaded onto the container as well as stowage plan. It is issued by the CSF if the goods are delivered in LCL; and it is completed by the shipper if the goods are loaded, stowed and sealed by the shipper and shipped in FCL. The document can be used for customs clearance, cargo receipt, container's loading and unloading and lodging claims.

- Loading list is the document that the carrier furnishes with discharge ports of all the cargoes on the ship. It lists the numbers of shipping orders, names of commodities, number of packages, weight and measurement. If the goods are dangerous, hazardous or special, it shall clearly state the requirements and handling instructions. Therefore, this is a an important document for planning stowage.

- Mate's receipt is the document evidencing that the ship has received goods on board. It is issued by the mate, in accordance with the shipping order and the loading list.

- Ocean bill of lading is a shipping document that serves as a receipt of goods, issued by a carrier to a shipper.

- Delivery order is the document authorizing the release of imported cargo from the carrier's unloading office to the consignee who presents a original bill of lading and other document. It is required for import customs clearance.

Ocean Bill of Lading

（Ⅰ）Functions of Bill of Lading

Ocean bill of lading, the most important shipping document, is issued by a carrier or his agent to a shipper which details a shipment of merchandise. It can be served as follows:

- A receipt that the carrier has received goods from the shipper.

- Document of title to goods that gives title of that shipment to a specified party who thus must present an original B/L when take delivery of goods. The B/L is a financial instrument for payment, which is used as a mortgage in a bank just because it is a document of title to goods.

• Contract for carriage. The carrier is responsible to send the goods to the destination place/port specified in the B/L. If he fails, some disputes can be settled down with reference to the terms printed on the reverse of B/L.

In addition to the above, the B/L is also an important document used for claims and negotiation.

(Ⅱ) Types of Ocean B/L

B/L can be classified according to different criteria

• According to the criterion whether the goods have been board or not, a shipped B/L and a received for shipment B/L can be classified. A shipped B/L is issued only when the cargo has been shipped or loaded on board the carrying vessel. Such a B/L is obligatory for Group C contracts. It is also safe for buyers as it proves that the goods have been loaded or shipped. When the goods have been received by the shipping company but not yet loaded on board vessel, a received for shipment B/L, received B/L for short, is issued. It proves only the receipt not loading or shipment of the cargo. It therefore bears no date of shipment and generally no name of a vessel. A received for shipment B/L becomes a shipped B/L with an "on board" notation. This notation should be duly signed and dated.

• According to the criterion whether there are some clauses expressly stating the dubious condition the goods are in, a clean B/L and a dirty B/L can be classified. A clean B/L which bears no superimposed clauses clearly declaring a defective condition of the goods or packaging such as " insufficiently packed", "on case leaking" etc. This kind of B/L indicates that the cargo has been delivered to the carrier in apparent good condition. A dirty B/L, also known as claused B/L and unclean B/L, contains a clause of disagreement with the printed statement of the B/L such as " in apparent good order and condition". It shows that the cargo has been delivered to the carrier in doubtful condition, like " used cases" or " one carton short". This kind of B/L may very possibly not be accepted by the negotiating bank.

• According to who the B/L can be consigned to, a straight B/L and an order B/L can be classified. A straight B/L has a clearly designated consignee, filled in with his/her name, address, telephone and fax numbers. This kind of B/L is not transferable to other parties by endorsement, i. e. it is not negotiable. An order B/L makes the consignee "to order" or "to the order of shipper" or " to the order of ******bank". This kind of B/L can be conferred, by endorsement, to another party which is usually designated by a bank, or shipper or sometimes a freight forwarder. That is, it is negotiable.

• According to the criterion of the transport mode, there are four types of B/L classified, i. e. a direct B/L, a transshipment B/L, a through B/L and a multimodal B/L. A direct B/L means that the cargo involved in the B/L must be shipped from the port of loading directly to the port of discharge on the same vessel. A transshipment B/L means the cargo involved in the B/L has to be transferred from one ship to another one during transit. This kind of B/L must be furnished with "transshipped at ****" A through B/L means that the cargo involved in the B/L has to be transferred from one mode of transport like ships to another different means of transportation like trucks during transit. The principal carrier or the freight forwarder who issued the through B/L is liable under a contract of carriage only for its own phase of the journey, and acts as an agent for the carriers executing the other phases. In this sense a through B/L is the same as a transshipped B/L. A multimodal B/L means that the cargo involved in the B/L has to be transferred from one mode of transport like ships to another different means of transportation like trucks during transit.

Unlike in case of a through bill of lading, the principal carrier or the freight forwarder who issued the multimodal B/L takes on full liability under a contract of carriage for the entire journey and over all modes of transportation. It is also called combined bill of lading, combined transport bill of lading, or intermodal bill of lading.

• According to the vessel operating mode, a liner B/L and a charter B/L can be classified. A liner B/L is issued by a shipping liner to a shipper, while a charter B/L is issued also by the ship but must be based on and be in accordance with the charter party. The latter is often used with the charter party against a bank.

• According to the issuer of the B/L, a master B/L and a house B/L can be classified. A master B/L is issued by the shipping company, which the consignee must present when taking delivery of the goods from the shipping company. A house B/L is issued by a freight forwarder to a shipper, as a receipt of goods which must be bundled with other cargoes from different shippers into one whole container.

• Other types. A stale B/L is the one that is presented to a bank for negotiation more than 21 days after its issuing date , or after the L/C has expired. The bank will usually dishonor this kind of B/L, unless otherwise stated in the L/C. An ante-dated B/L is issued , as requested by the shipper, before the goods are actually loaded on board. An advanced B/L looks exactly the same as a shipped B/L. When the cargo has been received by the shipping company but not yet loaded on board, the kind of B/L is issued by a carrier upon the letter of guarantee the shipper submits. Both an ante-date B/L and an advanced B/L are illegal because they are not issued at the actual date of loading.

5.3.2   Other Type of Transportation Documents

Air Waybill

Air waybill is a document or consignment note used for the carriage of goods by air. It is basically a receipt of goods for dispatch and evident of the contract of carriage between the carrier and consignor. An air waybill is only the evidence of a transportation contract between the consignor/shipper and the carrier/airline, therefore it is not a negotiable document because it is only a notice of arrival to release cargo, not a document of title to the consignment.

It should be pointed out that the airline industry has adopted a standard format for AWB which is used throughout the world for both domestic and international traffic. A set of AWB usually has three originals and at least six copies. The first original is held by the carrier after signed by the shipper; the second is sent with consignment and delivered to the consignee at the destination after signed by both the carrier and the shipper; the third is held by the shipper after signed by the carrier.

Sea Waybill

Sea Waybill is issued by the carrier or the agent evidencing that the goods have been received (taken over or loaded on board the vessel) and that the goods are to be delivered to the designated consignee at the port of destination. Sea waybill is evident of a transport contract, not a document of title to the goods, therefore it is not transferable and not used for taking delivery of goods.

Like a sea waybill, documents such as rail waybill, parcel post receipt and express mail receipt are only evident of a transport contract, not document of title and is not transferable of negotiable.

5. 4   Shipping Clause in the Sales Contract

The shipping clause in a sales contract mainly refers to the details of the shipment , such as the time of shipment, port of shipment, port of destination, shipping notice, partial shipment, transshipment and other issues.

5. 4. 1   Time of Shipment

Time of shipment is the time period or the deadline by which the seller must effect the shipment of the contract goods. In negotiating this clause, some factors in relation to the time shall be taken into consideration as follows:

- Availability of goods, quantity needed, inventory, and shipping space, shipping lines, rain /freezing seasons .
- Clarity of stipulation and absence of ambiguity such as " immediate/prompt shipment" .
- Some flexibility to allow reasonable length of time for loading the cargo. For example, more days should be allowed if shipment is to be made in raining season.
- Nature of the cargo because some goods like tobacco should be loaded in good weather and more days should be loaded in good weather.

There are basically two ways to indicate the time of shipment. One is to clearly specify a period time, like "shipment before June 30th" , or " shipment during March" or " shipment during July/ August in two equal lots" . The other one is to create a link between the time of shipment and the deadline, like " shipment with 30 days after receipt of L/C" . In this way, the buyer's corresponding responsibility should be also stipulated, like relevant " L/C must reach the seller not later than ****" .

The first methods is straightforward and easy to understand, while the second provides more flexibility and security for the transaction as shipment will not be made until payment is guaranteed. Generally speaking, the L/C should arrive at least 15 days before the time of shipment to allow sufficient time for the seller to check and amend, if necessary, the L/C and to arrange shipment. The

L/C shall expire 7 – 15 days after the deadline for shipment to allow sufficient time to prepare documents for negotiation.

5.4.2   Port of Shipment and Port of Destination

Port of shipment and port of destination must be specified when the contract is signed. In the case whereby a decision cannot be made at the time when the contract is signed, several optional ports can be listed and a time limit must be set by which the seller or the buyer must notify the other party which port is to be the port of shipment or destination. There are several ways to specify the port of shipment or destination as follows:

- Only one port of shipment is specified, like CIF Shanghai.
- Two optional ports of shipment are specified, like FOB Shanghai/Tianjin.
- No specific port is indicated, like China PORT, or European Main Port.

In choosing port of shipment and port of destination, some considerations shall be given as follows:

Firstly, the port of shipment or destination must be clearly stated. Some ports share the same name, like Perth in Britain and in Australia, Cadiz in Spain and in Philippine. In this case, the name of port shall be followed by its country, like Kingston in Canada, thus to avoid misunderstanding.

Secondly, flexibility shall be provided by allowing optional ports. Sometimes when the sales contract is concluded, it might not be possible to decide precisely on the exact port where the delivery is made. It is then usually stipulated that the buyer has the right to name precisely the port later on, like the following clauses:

One port out of *** / *** / ***is to be declared by the buyer on or before the opening of the L/C;

One port out of *** / *** / ***at the buyer's option, the buyer must declare the definite port of destination to the carrier 2 days before the vessel's expected time of arrival at the first discharging port and bear the optional fees thus incurred.

If the buyer fails to do so, he must bear the risks and additional costs sustained from.

Thirdly, political situation shall be taken into consideration. For instance, when a country is under an international sanction or when a country is an enemy of some other countries, sellers might have to avoid going to the ports of that country.

Fourthly, port regulations, facilities and charges should be considered. Ports operating under unfamiliar regulations, with poor facilities and high charges should be avoid. For example, some ports do not allow certain dangerous cargoes to enter, to be loaded or unloaded; some ports need special permit or documents if this kind of cargo is to be load or unloaded there; some ports require special type of packing when dangerous cargoes are loaded or unloaded; and some some ports allow unloading alongside the carrying vessel which must leave as soon as loading is finished. The party involved will be fined, sued, or even imprisoned if he has violated the regulations.

5.4.3   Partial Shipment and Transshipment

Partial shipment means shipping the commodity under one contract by more than one shipment. Partial shipment may possibly result in inconvenience for the seller and the buyer, thus shall be clearly stipulated in the contract, like "Partial shipments (not/ to be) allowed. Transshipment (to be) prohibited", or "Ship 200 M/T during September and 100 M/T during Oct. " or "Shipment during November and December in two equal lots".

Transshipment means the cargo being shipped will be transferred from one ship to another one, or truck, train, before reaching the port of destination. The clause must also specify whether transshipment is allowed by using phrases such as "transshipment (not/to be) prohibited" or "to be transshipped at ***".

If transshipment is carried out in order to reach the agreed destination, the seller would have to pay the cost of transshipment. If, however, the carrier exercises his right under transshipment clause in order to avoid unexpected hindrance such as ice, congestion and labor disturbances, then any additional expense would be on the buyer's account.

As partial shipments and transshipment concerned, UCP 600 stipulates as follows:

● When the L/C does not clearly stipulate that partial shipments or transshipment is prohibited, then they are construed to be allowed.

● A presentation consisting of more than one set of transport documents evidencing shipment commencing on the same means of conveyance and for the same journey, provided that they indicate the same destination, will not be regarded as covering a partial shipment, even if they indicate different dates of shipment or different ports of loading, places of taking in charge or dispatch. If the presentation consists of more than one set of transport documents, the latest date of shipment as evidenced on any of the sets of transport documents will be regarded as the date of shipment.

● Presentation consisting of one or more sets of transport documents evidencing shipment on more than one means of conveyance within the same mode of transport will be regarded as covering a partial shipment, even if the means of conveyance leave on the same day for the same destination.

● A presentation consisting of more than one courier receipt, post receipt or certificate of posting will not be regarded as a partial shipment if the courier receipts, post receipts or certificates of posting appear to have been stamped or signed by the same courier or postal service at the same place and date and for the same destination.

● If a drawing or shipment by installment within given periods is stipulated in the credit and any installment is not drawn or shipped within the period allowed for that installment, the credit ceases to be available for that and any subsequent installment.

5.4.4   Shipping Line and Vessels

Choosing a shipping line contributes to the cost and time in transit. Sometimes, a sales contract is agreed to stipulate that shipping vessels shall be named by the buyer, for some different reasons.

They can be that the buyer has entered into a business relationship with some shipping company, or the buyer trusts the shipping company who owns modern equipment or a quick transit for an early delivery, or the importing governments encourage the buyer to use the named shipping company for the development of its shipping industry. Thus, the sales contract stipulates as " shipment to be effected per American President Line steamer" or "shipment should be made per RIL vessel".

Under CFR and CIF, it is the seller who is to book shipping space. If the buyer needs the goods in a short time or the goods cannot undergo a long time of transport, it is wise for the buyer to require the seller to use a certain shipping line which can provide a shortest time of journey, with the stipulations like "Shipment via Suez canal" or "Shipment during October via Panama canal".

5. 4. 5   Advice of Shipment

Advice of shipment is required to coordinate the responsibilities of the exporter and the importer. It is not as important when the transaction is conducted under CIF as done under CFR or FOB. Under CFR, the seller is responsible for arranging carriage and delivering goods on board, and it is the buyer who takes all risk after the goods are loaded on board. The goods shall be covered as soon as they are on board, in case that the accident happens before insurance is procured. Advice of shipment is required for insurance procurement. In practice, advice of shipment must be sent to the buyer no later than 3 days after goods are shipped for arrangement of insurance and taking delivery. The INCOTERMS 2010 stipulate that under CFR or FOB the seller has the responsibility of providing sufficient notice to the buyer that the goods have been delivered on board. "Sufficient" means exactly "immediate" in time, and "clear" in statements.

Chapter 6

Cargo Transportation Insurance
and Insurance Clause

- Understand the fundamental principles of international cargo transportation insurance.
- Know about the risks and losses in marine transportation.
- Master the scope of Marine Cargo Clauses 2009 of China Insurance Clause.
- Know how to stipulate insurance clauses in international trade contracts.

In an international trade, goods usually travel a long distance to another country. At any stage of the transit, it is possible for the goods to suffer from all kinds of material losses, expenses or liabilities caused by risks such as natural disasters or fortuitous accidents. Therefore, it is extremely important for the traders to buy cargo transportation insurance in advance to hedge against the risk of a contingent, uncertain loss with the limited cost.

The basic categories of international cargo transportation insurance mainly include marine insurance, land cargo transportation insurance, air transportation cargo insurance and parcel cargo transportation insurance. Ocean Marine Insurance is the oldest form of insurance, probably dating to the Middle Ages. And it is the most extensive use of insurance in human history. Other cargo insurance is developed based on marine insurance. Therefore, this chapter will focus on the marine cargo insurance.

6. 1 Fundamental Principles of Cargo Insurance

Insurance is a contract whereby one party, in consideration of a premium paid, undertakes to indemnify the other party against loss from certain perils or risks to which the subject matter insured may be exposed to. There are two basic parties in involved in a cargo insurance contract: the insurer and the insured. The amount of money to be charged for a certain amount of insurance coverage is called the premium.

An insurer, or insurance carrier, is selling the insurance, which usually refers to the insurance company. Another name for the insurer is the underwriter. The insured, also referred to as the insurant or policy holder, is the person or entity buying the insurance policy. In cargo insurance practice, the insurant can be either the seller or the buyer, depending largely on the incoterms adopted for transaction and on which party is subject to main risks in transit. For example, under a CIF contract, the seller must obtain insurance contract even though he is not the party subject to marine risks in transit. Once damage or loss within the insurance coverage occurs, the party suffers will lodge a claim against the insurer. This party is referred to as the claimant. In cargo insurance, the insured and the claimant may not necessary be the same party. This is mostly decided by who has the insurable interest in the goods damaged or lost. There may also be other parties involving in the insurance such as insurance agents, insurance brokers, insurance notary, etc.

In the law, the insured and the insurer must follow the following general principles in signing and performing the contract.

6. 1. 1   Principle of Insurable Interest

Can everything be insured? Must be capable of financial measurement. There must be large number of similar risks. The person applying for insurance must be having insurable interest in the insurance contract.

Insurable interest refers to the legal right to insure arising out of a financial relationship (recognized under law), between the insured and the subject matter of insurance. There must be certain property, right, interest or life capable of being insured. The property, right, interest etc, must be the subject matter of insurance. The insured must be having benefits from the safety or well being of the subject matter and would be suffering by its loss or damage. The relationship between the insured and subject matter of insurance must be recognized at law.

In accordance with article 12 of China's Insurance Law, policy holder or the insured shall own insurable interests when the insurance accidents happen, otherwise, the insurance contract shall be invalid. This principle can make the insured be unable to gain additional benefits from the insurance contract if he has no insurable interest, which will avoid the change of insurance contract into a gambling contract.

6. 1. 2   Principle of Utmost Good Faith

According to the Principle of Utmost Good Faith or the Bona Fide Principle, both parties have a duty to disclose, accurately and fully, all facts material to the risk being proposed, whether requested or not, especially applicable to insurance contracts because in case of insurance, the product is intangible. When an insurer decide whether to insure, usually far away from the cargo location, it is difficult to survey the subject-matter insured and only rely on the statement of insurance. Only the proposer knows the details of risk being proposed.

As the underwriter knows nothing, and the man who comes to ask an insurance knows everything, it is the duty of the assured to make a full disclosure to the underwriter of all material circumstances, without being asked.

Of course, the principle of good faith is reciprocal duty on both parties. Proposer to disclose all material facts relating to the subject matter of the insurance. The insurer fully discloses the product features and benefits. The insurer does not make untrue statement at the time of negotiating the contract.

Breaching of good faith in case of—misrepresentation—may be innocent or fraudulent. If related to material facts, the contract may be avoided by the aggrieved party. Non-disclosure—which may be innocent or fraudulent. If fraudulent, it is called concealment. If substantially false and material, then contract may be avoided.

6.1.3　Principle of Proximate Cause

Principle of Proximate Cause means the most direct cause. When loss is a result of two or more causes, simultaneously or successively, the proximate cause becomes relevant. No matter how many causes there are, only the most direct one can be considered. It is the immediate cause, not the remote or distant one, that should be understood in terms of predominance in certain consequential losses. The losses may be suffered by the insured subject as a result of a peril which is not the proximate cause, are customarily paid. For example, the damage of property is caused by water from fire fighter. Other example, a house caught fire due to the air bombing of an enemy aircraft. The loss was due to fire, but the proximate cause is war (which in this case is excluded). Hence no claim is payable.

Although there is no provisions for the principle of proximate cause in China's Insurance Law and Maritime Code of PRC, in the practice of the international cargo insurance, the principle of proximate cause is an important principle which is commonly used to determine whether the insurer is responsible for the loss of the subject matter insured, and is the appropriate insured liability.

6.1.4　Principle of Indemnity

Principle of indemnity is a mechanism to provide financial compensation to place insured in the same pecuniary position as enjoyed immediately before the loss. It may be at the option of the insurer in the form of cash payment, repair or replacement. It is applicable to property, liability and other non-life insurance.

Insurance contract for international carriage of goods is a compensatory contract of property insurance. Therefore, in the case of excess insurance and double insurance, the insurer only indemnify the claimant to the actual loss suffered by the insured. The total indemnification shall be limited to the insurance amount, for the purpose of insurance is to compensate rather than benefit from the insurance.

6. 1. 5   Principle of Subrogation

Insurance Subrogation is the derivative product of the principle of indemnity for insured losses. As indemnity is only applicable to property insurance, the principle is not applicable to life and personal accident insurance.

The principle of subrogation is applicable to contracts of indemnity. The right of a person, having indemnified another under a legal obligation, stands in the place of that other and avail himself of all the rights and remedies of that other. That means when the loss of the property insured is caused by a third party, the insurer is automatically entitled to seek compensation from the third party after indemnification has been made to the insurant. Insured is entitled to indemnity, not more than that. Subrogation avoids the situation where an insured might profit from an insured event. For example, Mr. House Owner has his house worth Rs. 5 lacs insured with M/s. Fire Insurer. Assume that the house is totally destroyed by fire due to the faulty wiring by Mr. Contractor. The Fire Insurer has to pay to House Owner Rs. 5 lacs. The Fire Insurer can exercise the subrogation rights against Contractor, after fully indemnifying House Owner.

There are several explanations of the subrogation as follows: firstly, whether it is total loss or partial loss, the insurer shall make the indemnify first to the claimant before he can obtain the right of subrogation; secondly, under a valued policy, the compensation from the third party is less amount than that paid by the insurer, he may recover the difference, if more amount than the insured he must pass the surplus proceeds to the insured; thirdly, the insured can obtain interest on the losses prior to the insurer's indemnification, the insurer can obtain interest on the losses after his compensation, but also can comply with the provisions of the contract.

6. 1. 6   Principle of Contribution

Principle of Contribution is also the derivative product of the principle of indemnity for insured losses. It is applicable in case of two or more insurers on one risk. It means that under the circumstance of double insurance, all the insurers shall share the cost of indemnity. The insurer who has paid the claim has the right to recover a proportionate amount from other insurers. Insurers word their contribution conditions to state that they are only liable for their ratable proportion of the loss. Principle of Contribution avoids the situation where an insured might get more than the actual loss.

6. 2   Marine Risks and Losses

Risks in cargo transportation are of many kinds. Different risks could result in the loss of or the damage to the cargo in transit. The different risks are covered by different insurance clauses and different insurance clauses mean different premiums. So we need to have a good understanding of the different risks and losses before we know how to effect insurance. This section will explain the major risks involved in the ocean transport, the possible losses resulted in and the expenses incurred.

In marine cargo transportation insurance, risks fall into the perils of the sea and the extraneous risks.

6.2.1　Marine Risks

The Marine risks consist of the perils of the sea and extraneous risks.

Perils of the sea

In marine insurance, the perils of the sea include both natural calamities and fortuitous or unexpected accidents.

- Natural calamities refer to disasters caused by force natural events such as heavy weather, lightening, tsunami, earthquake, volcanic eruption, thunderbolt, floods and so forth. It should be noted that the natural calamities hereby referred to do not actually include all disasters due to natural forces.

- Fortuitous or unexpected accidents neither do not actually include all accidents due to haphazard causes. It generally include the accidents due to unexpected causes such as fire, explosion, vessel stranding, grounding, sinking or capsizing, collision, missing and so on.

Extraneous risks

Extraneous risks refer to risks caused by other reasons other than perils of the sea, and can be divided into two categories: general extraneous risks and special extraneous risks.

- General extraneous risks refer to the common risks such as theft, breakage, leakage, contamination, sweat and heating, taint of odor, hook damage, shortage, rust, etc.

- Special extraneous risks refer to risks resulting from military affairs, political factors and government rules such as war, striking, confiscation, failure to deliver, rejection, and so forth.

6.2.2　Marine Losses

When used for insurance, the word "average" has not the usual meaning. Average refers to any loss or damage due to natural calamities and fortuitous accidents and the related costs incurred in the process of transit. Marine insurance defines its coverage in terms of the nature of the loss or damage, according to the extent of the loss, it can be classified into total loss and partial loss.

Total loss

Total loss refers to the loss of the entire value of the subject matter to the insured, normally involving the maximum amount for which a policy is liable. Most insurance policies provide for the payment of total loss up to the insurance amount. Total loss is divided into actual total loss and constructive total loss according to the situation of losses.

- Actual Total Loss means the loss that the insured property is totally destroyed or is damaged in such a way that it can be neither recovered nor repaired for further use, or the insured is irretrievably deprived of it. For example, that a batch of tea under a contract was immersed in seawater and could not be used any more even if the tea arrived at the final destination. When an actual total loss occurs, the insured is not required to give the insurer notice of abandonment (i. e. , the surrender of all rights, title and interest) in the insured property in return for the sum insured.

● Constructive Total Loss refers to the loss where an actual total loss appears to be unavoidable or the cost to be incurred in recovering or reconditioning the goods together with the forwarding cost to the destination named in the policy that would exceed their value on arrival. For example, when a ship or the consignment has to be abandoned because the cost of salvage or recovery would exceed the value of the ship and the consignment in sound condition upon the arrival of the port of destination. They are treated as totally lost. In constructive total loss cases, the insured may abandon the property by giving a "notice of abandonment" to the insurer who then assumes all rights to the property.

Partial loss

A partial loss is any loss other than a total loss. It is a partial damage to or the total loss of part of the insured cargo. A partial loss may include a particular average loss, a general average loss and particular charges. Thus there are two distinct types of partial loss.

● General average. A loss may be by accidental or intentional. A general average loss is caused intentionally, and such a loss is beneficial to others, then those who have benefited should share that loss. Thus a general average loss is a loss resulting from a voluntary sacrifice or expenditure in time of peril, for the safety of hull, cargo and freight.

A partial loss can be treated as general average if it is formed upon the following conditions: firstly, there must be an event which is beyond the ship owner's control, which imperils the entire adventure; secondly, there must be a voluntary sacrifice; thirdly, the action of the ship's master shall be successful, there must be something saved. The voluntary sacrifice might be the jettison of certain cargo, the use of tugs, or salvors, or damage to the ship, voluntary grounding, knowingly working the engines that will result in damages.

Where there is a general average loss, all parties concerned in the maritime venture (Hull/Cargo/Freight/Bunkers) have to contribute to make good the voluntary sacrifice. They share the expense in proportion to their respective interested value in the adventure. This is called a General Average Contribution. Internationally general average shall be adjusted at any port or place at the carrier's option according to the York-Antwerp Rules 1974, as amended in 1990, and any other amendments thereto. In our country, the general average shall apply to Provisional Rules of the General Average Adjustment of China Council for the promotion of international trade (Beijing G. A. Settlement Rules for short) .

● Particular Average is the accidental damage to part of the cargo. The average is not caused by deliberate act of a person for the common benefit and that loss must be borne exclusively by the owner of the property suffering the loss and is termed a particular average. If a loss or damage suffered by the claimant is covered by an insurance policy, it shall be reimbursed by the insurance company.

Although both general average and particular average belong to the category of partial loss, there are still some differences between them:

a. The causes of loss and expenses are different. General average is caused by intentional and reasonable measures taken for the safety of the ship and the consignments.

b. Loss of composition is different. Particular average only include the loss of the goods caused by the perils of the sea itself, and general average includes both the loss of the goods, and fees paid for the lifting of the common dangerous cargo.

c. Different ways to bear the loss. The party, whose cargo is damaged, while general average should be proportionally contributed among all parties benefited from the intentional measures, assumes particular average.

6.2.3　Expenses Incurred

Expenses are often covered by marine cargo insurance which consist of two types of expenses:

Sue or labor expenses

When the subject matter suffers the risks within insurance coverage, the insured takes measures to avert or minimize the loss or damage to the subject matter insured. Expenses incurred to limit physical damages, or to take legal action to protect the ship and its cargoes from further loss once a loss has occurred, usually reimbursed by the insurer to the extent who reduce the loss otherwise being payable by the insurer, according to policy terms.

Salvage charge

The term "salvage" refers to the practice of rendering aid to a vessel in distress. A third party other than the insured or his agents, or any person employed for hire by them takes measures to save the subject matter. Such expenses, where properly and effectively incurred, can be recovered as particular charge or as a general average loss, according to the circumstances under which they were incurred. Salvage charges often take the "No-Cure, No-Pay" principle.

6.3　Cargo Clauses of Marine Cargo Insurance of CIC

Ocean Marine Cargo Clauses of the People's Insurance Company of China was constituted in 1972 and revised many times, the latest version of C. I. C. is 2009 version. It is the main basis for Chinese import and export company to apply for the insurance of marine cargo transportation. The main provisions of C. I. C include the insurance coverage, exclusions, duration and obligation of the insured and the time of validity of a claim etc.

6.3.1　Basic Ocean Cargo Coverage of China Insurance Clauses (CIC)

There are mainly two types of insurance coverage: basic coverage and additional coverage. Basic coverage can be taken out independently. But additional coverage must be insured together with a basic coverage. Moreover, basic coverage mainly includes FPA, WPA and All Risks. Additional coverage includes general additional coverage and special additional coverage.

Basic Coverage

This insurance is classified into the following three conditions—Free From Particular Average (F. P. A), With Average (W. A.) and All Risks.

(Ⅰ) Free From Particular Average (F. P. A.)

FPA is a limited form of cargo insurance cover under which partial loss or damage resulted from natural calamities is not recoverable, unless these natural calamities occur before or after fortuitous accidents. It provides coverage covering the following losses:

- Total or Constructive Total Loss of the whole consignment hereby insured caused in the course of transit by natural calamities: heavy weather, lightning, tsunami, earthquake and flood. In case a constructive total loss is claimed for, the insured shall abandon to the Company the damaged goods and all his rights and title pertaining thereto.

- Total or Partial Loss caused by accidents such as the carrying conveyance being grounded, stranded, sunk or in collision with floating ice or other objects as fire or explosion.

- Partial loss of the insured goods attributable to heavy weather, lightning and/or tsunami, where the conveyance has been grounded, stranded, sunk or burnt. Irrespective of whether the event or events took place after or before such accidents.

- Partial or total loss consequent on falling of entire package or packages into the sea during loading, unloading and transshipment.

- Reasonable cost incurred by the insured on salvaging the goods or averting or minimizing a loss recoverable under the policy, provided that such cost shall not exceed the sum insured of the consignment so saved.

- Losses attributable to discharge of the insured goods at a port of distress following a sea peril as well as special charges arising from loading, warehousing and forwarding of the goods at an intermediate port of call or refuge.

- Sacrifice in and contribution to general average and salvage charges.

- Such proportion of losses sustained by the ship owners as is to be reimbursed by the Cargo Owner under the Contract of Affreightment "Both to Blame Collision" clause.

(Ⅱ) With Particular Average (W. P. A) /With Average (W. A.)

WPA is a wider cover than FPA. It provides a more extensive cover against all loss or damage due to marine perils throughout the duration of the policy, including partial loss or damage which may be attributed to natural calamities like heavy weather, lightning, tsunami, earthquake and flood. WPA provides coverage covering the following two parts:

- Risks covered under F. P. A. condition.

- Partial losses of the insured goods caused directly by nature calamities such as heavy weather, lightning, tsunami , earthquake and/or flood.

(Ⅲ) All Risks (A. R.)

All Risks is the most comprehensive among the three basic types of coverage. However, even though the name of this coverage implies All Risks, this coverage shall not cover All Risks of loss or damage. It does not cover loss, damage or expense caused by delay, inherent vice or nature of the goods, or special external risks of war, strike, etc. The coverage of All Risks is shown as following:

- Risks covered under the F. P. A. and W. A.
- Total or partial loss of or damage to the insured goods caused by general extraneous risks during the course of transit.

Additional coverage

(Ⅰ) General Additional Coverage

General additional risks cannot be covered independently and should go with FPA or WPA. They are included in All Risks coverage.

- *Theft*, *Pilferage and Non-delivery* (*T. P. N. D. for short*) . It covers the loss of or damage to the insured goods on the insured value caused by: a. Theft and/or pilferage, the risk against the possibility of the cargo being stolen in transit during the voyage; b. Non-Delivery of entire package, the risk against loss of complete packages due to improper unloading or for some other unknown reasons.

- *Fresh Water and/or Rain Damage.* This insurance refers to the risk of loss or damage to the goods directly caused by fresh water and/or rain, not seawater.

- *Risk of Shortage in Weight.* It covers the shortage of weight for the goods during the course of transit due to the breakage of external package. But, normal shortage of weight in transit is not covered.

- *Risk of Intermixture and Contamination.* It covers risk of intermixture and contamination occurring during the course of transit.

- *Risk of Leakage.* It covers the risk of leakage and seepage of the liquid goods caused by the damaged container during the course of transit.

- *Risk of Clash and Breakage.* It covers the breakage of the fragile goods by vibration, clash or pressing during the course of transit.

- *Risk of Odor.* It covers risk of taint of odor of the insured edibles, herbal medicine, cosmetics or raw materials etc.

- *Sweating and Heating Risk.* It covers risk of loss of or damage to the goods by sweating, heating and wetting occurring during the course of transit arising from sudden change of temperature or breakdown of ventilation of the carrying vessel.

- *Hook Damage.* It covers the damage to the goods caused by hooks in the process of loading and unloading, such as the loss of cereals due to the damage of packing bags by hooks.

- *Risk of Rust.* It covers the risk of damage to the goods by rust caused by contamination with seawater during the course of transit. However, goods getting rusty by itself or due to its own flaws are not covered.

- *Breakage of Packing Damage.* It covers the damage to the goods resulting from rough handling.

All the general extraneous risks are included in the coverage of All Risks, so the general additional coverage would better not be insured together with All Risks.

(Ⅱ) Special additional coverage

Special additional risks cover the damage or losses arising from special additional reasons such as political events, military affairs, national policies and acts, and administrative measures. The special additional risks include:

- *War Risks.* It covers loss of or damage to the insured goods caused directly by or in consequence of war, warlike operations, conventional weapons, etc. But it does not cover loss, damage or expenses arising from any hostile use of atomic or nuclear weapons of war.

- *Strike, Riots & Commotions (SRCC)* . It covers loss of or damage to the insured goods directly caused by acts of strikes, locked-out workmen, etc.

- *Failure to Delivery Risk.* It refers to the insurer shall pay a total loss of the insured goods in case the goods once loaded on board the seagoing ship, fail to be delivered at the destination within 6 months of scheduled date for arrival due to whatever cause it might be.

- *Import Duty Risk.* It covers the loss caused by Import Customs duty at the port of destination on the portion of the goods damaged by a peril insured against.

- *On Deck Risk.* It covers the loss of or damage to the goods caused by the special risks when they have been shipped on deck. Usually they are large volume goods, goods of toxic, flammable and explosive or pollution, and must be stored on the deck customarily.

- *Rejection Risk.* This insurance is to indemnify the insured for rejection and/or condemnation at the port of entry by the government of the importing country.

- *Aflatoxin Risk.* It covers the loss of the insured goods when the cargo is rejected or confiscated by reason of the existence of aflatoxin, to an extent exceeding the limit sanctioned by the importing country.

- *Fire Risk Extension Clause for Storage of Cargo of Destination Hongkong Including Kowloon, or Macao.* After being discharged at the final destination at Hongkong, including Kowloon, or Macao from the carrying conveyance, if the insured cargo is directed to be stored in a warehouse specifically designated by the bank to whom the interests in the cargo are assigned, this insurance shall extend to cover fire risk at such warehouse from the time the marine coverage eases until the termination of the said bank's interests in the cargo or the expiration of thirty days counted from the day the marine coverage ceases, whichever shall first occur.

These types of additional coverage also cannot be covered independently and are usually taken out together with FPA, WPA and All Risks.

6. 3. 2   Exclusions

Exclusions refer to the risks and losses not covered under either the basic coverage or the additional coverage by the insurer in marine cargo insurance practice. The exclusions of CIC include:

- Loss or damage caused by the intentional act or fault of the insured.
- Loss or damage falling under the liability of the consignor.
- Loss or damage arising from the inferior quality or shortage of the insured goods prior to the attachment of this insurance.
- Loss or damage arising from normal loss, inherent vice or nature of the insured goods, loss of market and/or delay in transit and any expenses arising there from.
- Risks and liabilities covered and excluded by the ocean marine (cargo) war risks clauses and strike, riot and civil commotion clauses of this Company.

6.3.3 Duration of Marine Cargo Clause of C. I. C.

Just as the usual practice in international insurance market, with regard to the duration of liability of the insurer in ocean marine cargo insurance, CIC adopts "warehouse to warehouse Clause" (W/W clause for short).

W/W clause, which means insurance coverage of risks to a shipment of goods from the time the goods leave the warehouse for commencement of transit and continue during ordinary course of transit until delivered to final warehouse at destination, or until the expiration of 60 days as of the moment of the insured goods are unloaded (if the shipment fails to reach the aforesaid warehouse), but with exception that the goods are transported to other place of destination not indicated on the insurance documents.

If the insured goods are to be transshipped to other place of destination not indicated on the insurance policy, the insurance liability shall be terminated on the commencement of transit to such other destination.

If the insured goods are to be used for allocation or distribution on delivery to any other warehouse or place of storage, whether prior to or at the destination named herein, the insurance shall terminate on delivery to the warehouse.

The period of cover under the War Risk is more limited and not complies with the W/W clause, being from loading on to the oversea vessel until discharge at the final port or place of discharge. If the goods remain unloaded, the duration of insurance terminates on the expiry of 15 days after the midnight of the day when the vessel arrives at the port of destination.

6.3.4 The Validity of A Claim

The time of validity of a claim under this insurance shall not exceed a period of two years counting from the time of completion of discharge of the insured goods from the seagoing vessel at the final port of discharge.

When presenting a claim to the insurer the insured shall submit the documents such as original insurance policy, invoices, bills of lading, packing list, tally sheet, weight memo, certificate of loss or damage, survey report and/ or statement of claim. If the third party liability is involved, the claimant must also provide the documents, correspondence letter and supporting documents relating to the third party.

6. 4   Coverage of Marine Cargo Insurance of ICC

In the international insurance market, the national insurance organizations have developed their own insurance policy. However, the most widely used is the London Insurers Association developed the "terms of Institute Cargo" (Institute Cargo Clause; abbreviated ICC) .

China's enterprises on CIF or CIP terms of exports, generally based on "China Insurance Clause", but if the foreign customer requires to cover insurance according to "Institute Cargo clauses", that is generally acceptable.

The Institute Cargo Clause (ICC) has undergone many revisions since its first publication in 1912. A more recent revision was completed in January 1, 1982. And this revision in 1982 brought revolutionary change to the international marine insurance market and replaced the old S. G policy which had been regarded as the Bible of marine insurance for a long time. It also changed its status as the affixed terms. However, with the development of global economic integration, international trade becomes increasingly frequent and the modes of international transport are more various than ever. Thus there is an urgent need of corresponding provisions to adapt to this kind of advancement. Joint Cargo Committee took on comprehensive investigation and consultation all over the world and collected suggestions from a large number of specialists and published the new Institute Cargo Clause on November 24th, 2008. This clause was come into force on January 1st, 2009. The existing provisions of ICC are mainly composed of the following 5 clauses:

- Institute Cargo Clause (A) [referred to as ICC (A)]
- Institute Cargo Clause (B) [referred to as ICC (B)]
- Institute Cargo Clause (C) [referred to as ICC (C)]
- Institute War Clause-Cargo
- Institute Strike Clauses- Cargo

ICC (A) , ICC (B) and ICC (C) have independent and complete structure and can be covered independently. Institute War Clause and Institute Strike Clause also have independent and complete structure and can be covered independently only after agreement from the insurance company.

6. 4. 1   Institute Cargo Clause (A)

Risks Covered

- It covers all risks of loss of or damage to the subject-matter insured except as excluded.
- It covers general average and salvage charges, adjusted or determined according to the contract of affreightment and/or the governing law and practice, incurred to avoid or in connection with the avoidance of loss from any cause except those excluded.
- It indemnifies the Assured against such proportion of liability under the contract of affreight-

ment "Both to Blame Collision" Clause as is in respect of a loss recoverable hereunder. In the event of any claim by ship owners under the said Clause the Assured agree to notify the Underwriters who shall have the right, at their own cost and expense, to defend the Assured against such claim.

Exclusions

This insurance covers all risks of loss of or damage to the subject matter insured except the following exclusions:

(I) General Exclusion Clause

- Loss, damage or expense attributable to willful misconduct of the Assured.

- Ordinary leakage, ordinary loss in weight or volume, or ordinary wear and tear of the subject-matter insured.

- Loss damage or expense caused by insufficiency or unsuitability of packing or preparation of the subject-matter insured to withstand the ordinary incidents of the insured transit where such packing or preparation is carried out by the Assured or their employees or prior to the attachment of this insurance (for the purpose of this Clause "packing" shall be deemed to include stowage in a container and "employees" shall not include independent contractors) .

- Loss damage or expense caused by inherent vice or nature of the subject-matter insured.

- Loss damage or expense proximately caused by delay, even though the delay is caused by a risk insured against (except expenses payable under its covered general average and salvage charges.) .

- Loss damage or expense caused by insolvency or financial default of the owners managers charterers or operators of the vessel where, at the time of loading the subject matter insured on board the vessel, the Assured are aware, that such insolvency or financial default could prevent the normal prosecution of the voyage. This exclusion shall not apply where the contract of insurance has been assigned to the party claiming hereunder who has bought or agreed to buy the subject matter insured in good faith under a binding contract.

- Loss damage or expense arising from the use of any weapon of or device employing atomic or nuclear fission and/or fusion or other likes reaction or radioactive force or matter.

(II) Unseaworthiness and Unfitness Exclusion Clause

- Loss damage or expense arising from unseaworthiness of vessel or craft, unfitness of vessel craft conveyance container or lift van for the safe carriage of the subject-matter insured, where the Assured or their servants are privy to such unseaworthiness or unfitness, at the time the subject-matter insured is loaded therein.

- The Underwriters waive any breach of the implied warranties of seaworthiness of the ship and fitness of the ship to carry the subject-matter insured to destination, unless the Assured or their servants are privy to such unseaworthiness or unfitness.

(III) War Exclusion Clause

- Loss damage or expense caused by war civil war revolution rebellion insurrection, or civil strife arising therefrom, or any hostile act by or against a belligerent power.

- Loss damage or expense caused by capture seizure arrest restraint or detainment (piracy excepted), and the consequences thereof or any attempt thereat.

- Loss damage or expense caused by derelict mines torpedoes bombs or other derelict weapons of war.

(Ⅳ) Strikes Exclusion Clause

- Loss damage or expense caused by strikers, locked-out workmen, or persons taking part in labor disturbances, riots or civil commotions

- Loss damage or expense resulting from strikes, lock-outs, labor disturbances, riots or civil commotions

- Loss damage or expense caused by any terrorist or any person acting from a political motive.

ICC (A) risk is similar to that of all risks under China's insurance clause, a wider scope of its responsibilities, so use of insurance "excluded liability" means outside of all the risks that their insurance coverage.

6.4.2   Institute Cargo Clause (B)

Risks Covered

(Ⅰ) Loss of or damage to the subject-matter insured reasonably attributable to

- Fire or explosion
- Vessel or craft being stranded, grounded, sunk or capsized
- Overturning or derailment of land conveyance
- Collision or contact of vessel, craft or conveyance with any external object other than water
- Discharge of cargo at a port of distress
- Earthquake, volcano eruption or lightning

(Ⅱ) Loss of or damage to the subject-matter insured caused by

- General average sacrifice
- Jettison or washing overboard
- Entry of sea, lake or river water into vessel, craft hold, conveyance container, lift van or place of storage

(Ⅲ) Total loss of any package lost overboard or dropped whilst loading on to, or unloading from vessel or craft

Exclusion

Except for the exclusions of ICC (A), ICC (B) shall not cover:

- Deliberate damage to or deliberate destruction of the subject matter insured or any part thereof by the wrongful act of any person or persons (while ICC (A) only excludes loss, damage or expense attributable to willful misconduct of the Assured)
- Actions of pirates
- Any risks that not in the list of the covered risks of this coverage.

The scope of ICC (B) insurance is similar to that of W. P. A under CIC.

6.4.3 Institute Cargo Clause (C)

Risks Covered

(Ⅰ) Loss of or damage to the subject-matter insured reasonably attributable to

- Fire or explosion
- Vessel or craft being stranded, grounded, sunk or capsized
- Overturning or derailment of land conveyance
- Collision or contact of vessel , craft or conveyance with any external object other than water
- Discharge of cargo at a port of distress

(Ⅱ) Loss of or damage to the subject-matter insured caused by

- General average sacrifice
- Jettison

The exclusions of ICC (C) are the same as that of ICC (B)

ICC (B) insurance and ICC (C) insurance have adopted the manner set out in the risk that their insurance coverage.

The coverage of ICC (C) is similar to that of FPA under CIC, but the coverage is smaller.

6.4.4 Institute Strikes Clause (Cargo)

There are the risks covered under Institute Strikes Clauses (Cargo) .

Loss of or damage to the subject-matter insured caused by:

- Strikes, locked-out workmen, or persons taking part in labor disturbances, riles or civil commotions;

- Any act of terrorism being an act of any person acting on behalf of, or in connection with, any organization which carries out activities directed towards the overthrowing or influencing, by force or violence, of any government whether or not legally constituted;

- Any person acting from a political, ideological or religious motive.

The coverage of Institute Strikes Clauses (Cargo) is similar to that of Strike Risk under CIC.

6.4.5 Institute War Clause (Cargo)

There are the risks covered under Institute War Clauses (Cargo) .

Loss of or damage to the subject-matter insured caused by:

- War, civil war, revolution, rebellion, insurrection, or civil strife arising therefrom, or any hostile act by or against a belligerent power;

- Capture, seizure, arrest, restraint or detainment, arising from risks covered under the above, and the consequence thereof or any attempt threat;

- Derelict mines, torpedoes, bombs, or other derelict weapons of war.

The coverage of Institute War Clauses (Cargo) is similar to that of War Risk under CIC.

6. 5   Insurance Terms in the International Sales Contract

6. 5. 1   Insurance Clause of the Contract

There aren't unified contents or unified form of insurance clause in the contract of international sales of goods. And it should be clearly stipulated the provisions such as who is the insured, the insurance coverage, the insured amount, the insurance clauses and which party bears the the insurance premiums etc.

There are the examples of insurance clause:

• In the contracts under FOB, CFR, FCA and CPT trade terms, the buyer is to arrange insurance and pay premium. In the contract regarding insurance, it can only stipulate:

" To be effected by the Buyers" .

• In the contracts under DAP, DAT, and DDP trade terms, the seller is to arrange insurance and pay premium. In the contract regarding insurance, it can only stipulate:

" To be effected by the Sellers" .

• In the CIF or CIP contract, the insurance clause can stipulate:

Insurance to be covered by the sellers for 110% of the total invoice value against All Risks and War Risks, as per and subject to the Ocean Marine Cargo Clause 2009 and Ocean Marine Cargo War Risks Clause 2009 of the People's Insurance Company of China.

6. 5. 2   Procedures of Cargo Insurance

In export trade, under CIF or CIP contract, in accordance with international trade practices, it is the seller who arranges insurance with an insurance company. In our country, usually the insured applies to the insurer one transaction by transaction. While Filling in the insurance policy, the contents such as the insurer's name, the name of the subject matter insured, quantity, packing and insurance coverage, the amount of insurance policy, name of transport and transport routes should be stated strictly in accordance with the import and export contracts or letters of credit provisions in detail, which can ensure that the insurance policies obtained after the payment of the premium can be successfully used for settlement of export foreign exchange.

The date of insurance policy should not be later than the time of shipment stipulated in the contract or letter of credit. The recommended minimum insured amount is the total CIF or CIP value plus 10% for other fees and normal marginal of profit under letter of credit, according to the Uniform Customs and Practice for Documentary Credits, the International Chamber of Commerce, No. 600 publication (UCP600 for short) .

In import trade, under the terms as FCA, FOB, CFR, CPT, the buyer effects insurance. In our country, sometimes in order to simplify the insurance procedures to avoid missing of insurance or failing to apply for the insurance in time, generally open cover is recommended for the buyer, that is the general contract between the insured (normally the importer) and the insurer, often used in import transactions in China. Usually the import & export companies enter into a long-term pre-agreement with the insurance companies. It will no longer need to fill the policy in subsequent each batch of imported goods after shipment, only if the shipping advice will be forwarded to the insurance company in time, which can be regarded as completing insurance procedures, the insurance company will automatically cover the goods.

6.5.3　To Calculate the Insurance Amount and the Insurance Premium

Generally speaking, the Insurance amount, also known as the Insured Amount or the Sum Insured, stands for the amount of insurance indicated in the insurance policy and agreed upon by the insurant and the insurer. It is the highest amount of indemnity given by the insurer when the subject matter is subject to a loss or damage and serves as the basis for the calculation of insurance premium. The value to be insured is based on the value of the commercial invoice where the insurance premium is included as that in a CIF or CIP contract.

Usually, the parties use the following calculation formula to determine the insured amount.

$$\text{Insured amount} = \text{CIF (CIP)} \times (1 + \text{markup rate})$$

$$\text{Insurance Premium (I)} = \text{insurance amount} \times \text{premium rate (R)}$$

$$= \text{CIF (CIP)} \times (1 + \text{markup rate}) \times \text{premium rate (R)}$$

6.5.4　Insurance Documents

In international transaction practices, an insurance policy is actually a contract, serving as evidence of the agreement between the insurer and the person taking out insurance. It is not only usual for a certificate of insurance, but also the written basis for the claimant to lodge an insurance claims and for the insurance companies to handle the settlement of claims. The most commonly used insurance documents are insurance policy, insurance certificate and open policy, etc.

Insurance policy

Insurance policy is the most commonly used document that contains all the details concerning the goods, coverage, premium and insured amount on the face and the detailed contract terms at the back. (See Specimen 6.1) All this information must be sufficient enough for the insured party to assess the risk and make insurance decision.

In addition as the written basis of counterclaim and claim settlement, Insurance policy can also be one of documents presenting to the bank for negotiation. Under CIF contract, the insurance policy is one of the documents submitted by the seller to the buyer.

Insurance certificate

Insurance certificate, also known as a small insurance policy, is a simplified version of insurance contract. The certificate carries the same contents as the insurance policy except for the detailed terms of insurance responsibility, rights and obligations on the back, serving the same functions in the transaction.

Open policy

Also known as open cover, open policy is recommended for exporters and importers who do a large volume of business. There is a prior agreement between the insurer and the insured, including the insurance coverage, the premium, the maximum amount of each shipment, which shall be stated in an open policy, but the pre-agreement does not specify the total amount of insurance. An open policy provides coverage for all goods shipped by the insured while the policy is in effect. The duration of this policy may run from six to twelve months or may, sometimes, last indefinitely.

Combined certificate

Combined certificate is the combination of the invoice and insurance policy, much simpler than the insurance certificate. It is not often used now.

Settlement for International Trade

- Be familiar with the paying tools and methods in international trade settlement.
- Be familiar with the kinds of Bill of Exchange and grasp the content.
- Grasp the nature, the functions and main content of L/C.
- Be familiar with the payment clause in international trade contract.

In international trade practice, the importer is obliged to pay the exporter the agreed amount, in the agreed currency, within the agreed period of time and by the agreed method of payment. How and when an exporter receives payment for the goods he sends abroad are problems that concern him the most. Payment in domestic trade is a fairly simple matter. It can be made either in advance or within a reasonably short period after delivery. However, these problems are magnified many times in international trade. Much time is unavoidably lost in correspondence, dispatch and delivery. Who is liable for this loss? Must the seller wait perhaps six months for his money or shall the buyer pay several months before he even sees his goods? What's more, in a case of non-payment, the seller will be involved in expensive legal action and possibly total loss. Because of these problems, different methods of payment have been adopted in international trade. Generally, in every contract for the sale of goods abroad, the clause dealing with the payment of the purchase price consists of four elements: time, mode, place and currency of payment. The various methods of financing exports represent the order and variations of these four elements.

In international trade, the most frequently used means of payment include currencies and bills. The former are used for account, settlement and payment; the latter for settlement and payment. In practice, sellers of goods, in general, almost never insist on their rights to demand cash, for payment, but readily take certain bills, such as bill of exchange (draft), promissory note and cheque (check) for substitutes. The latter two are only used occasionally in foreign trade. So, here more attention will be paid to bill of exchange.

7.1 International Trade Payment Tools

7.1.1 Bill of Exchange

Definition of Bill of Exchange (draft)

A bill of exchange writing, addressed by person to whom it is time, a sum certain in operation process of one also called draft, is defined as "an unconditional order in person to another, signed by the person giving it, requiring the addressed to pay on demand, or at a fixed or money, to, or to the order of a specified person determinable future or to bearer". The draft and includes: to draw, presentation, acceptance, payment, endorsement, dishonor and recourse. Drafts are negotiable instruments and may be sold.

Content of Bill of Exchange (draft)

The following specimens of bill of exchange can help us make clear the content of it.

(Specimen A)

No. 1602

$10,000
　　　　　　　　　　　　　　　　　　　　　　　　New York, 8th January, 2016

　　On demand pay to Tom Smith or bearer the sum of USD Ten Thousand only.

To: Mr. Felix Bryan

London
　　　　　　　　　　　　　　　　　　　　　　　　　　　(signed) David white

(Specimen B)

No. 677/96

　　Exchange for $7,500
　　　　　　　　　　　　　　　　　　　Shanghai, China, 8th August, 2015

　　At 60 days sight of this First of exchange (the SECOND of the same and date being unpaid) pay to or to the order of Shanghai A&G Import and Export , Corporation the sum of USD seven thousand five hundred only.

　　To: ABC Import and Export Co. , Ltd.

　　56 Linden Street

　　Miami, U. S. A.

　　　　　　　　　　　　　　　　　Shanghai A&G Import and Export Corporation

　　　　　　　　　　　　　　　　　　　　　　　Manager

　　　　　　　　　　　　　　　　　　　　　　　(signed)

According to the definition of bill of exchange, the two specimens may be decomposed into the following elements:

(Ⅰ) An unconditional order in writing

(Ⅱ) Addressed by one person/party (the drawer)

In A: David White, New York

In B: Shanghai A&G Import and Export Corporation

(Ⅲ) To another (the drawee)

In A: Felix Bryan, London

In B: ABC Import and Export Co. , Ltd, Miami

(Ⅳ) Signed by the person/party (the drawer) giving it

- Requiring the person/party to whom it is addressed (the drawee, or the payer)

- To pay

- On demand, or at a fixed or determinable future time

- √In A: on demand

- √In B: 60 days after sight (a determinable future time)

- A sum certain in money

- √InA: $ 10,000

- √In B: $7,500

- To, or to the order of, a specified person/party, or to bearer (the payee)

- √In A: Tom Smith (or bearer)

- √In B: Shanghai A&G Import and Export Corporation

The Parties

A bill of exchange involves three parties:

Drawer: the person who writes the order and gives directions to the person to make a specific payment of money. He is usually the exporter or his banker in import and export trade; usually, he is also a creditor of the drawee.

Drawee (or the payer): the person to whom the order is addressed and who is to pay the money. He is usually the importer or the appointed bank under a letter of credit in import and export trade. In addition, when a time bill has been accepted by the drawee, he becomes an acceptor who is the same person as the drawee. The drawer and the acceptor must be different persons.

Payee: the person (individual, firm, corporation, or bank) to whom the payment is ordered to be made. The drawer and payee may be often the same person. In this case, the bill may be worded "Pay to our order…". The payee is usually the exporter himself or his appointed bank in import and export trade. The payee may also be the bearer of the bill. The payee may be the original payee in the bill, or may be some party to whom the original payee has transferred the instrument. If a bill with such instruction "Pay…Co. or order" or "Pay to the order of … Co. ", it means to pay to the payee or to anyone to whom he in turn directs payment to be made. In this way, the bill should be endorsed by the payee, now the endorser, and can be passed on to a new payee, the endorsee, thus making it negotiable. A bill may have many number of endorsers.

The relationship among these parties in bill of exchange may be described as a triangle (see Figure 7. 1).

Figure 7. 1   The Parties in Bill of Exchange

Classification of Bills of Exchange

On the basis of different criteria, bills of exchange may be classified into severaltypes:

- Commercial Bill and Banker's Bill

According to different drawers, the bills of exchange can be classified into commercial bill and banker's bill. If the drawer is a commercial concern, the bill is called a commercial bill. It is often used in foreign trade finance. If the drawer is a bank, the bill is called a banker's bill. It is mainly used in remittance.

- Clean Bill and Documentary Bill

Clean bill is a draft to which no documents are attached. It is seldom used in business. It is usually adopted to collect commission, interest, sample fee and cash in advance. On the contrary, if a draft with some shipping documents, such as bill of lading, insurance policy, invoice etc, attached to it, the draft is called documentary bill. It is usually used to collect the payment of import and export goods.

- Sight (or Demand) Bill and Time (or Usance) Bill

According to the time when the bill falls due, bills of exchange may be divided into sight (or demand) bill or a time (or usance) bill. A sight bill requires the drawee to pay the bill on demand or at sight. As for a time bill, the drawee is required to accept it first and pay it at a fixed or determinable future time, in other words, it requires acceptance before payment. The fixed or determinable future time may be a certain number of days after acceptance:

√ At···days after sight, such as "30 days sight" or "60 days after sight";

√ At···days after date of draft, such as "90 days after date of this draft";

√ At fixed date in the future, such as "On July 18, 2016".

- Commercial Acceptance Bill and Banker's Acceptance Bill

In time or usance commercial bills, when the drawer is a commercial firm and the drawee is another commercial firm, the bill after acceptance by the commercial firm or the drawee is called a commercial acceptance bill; when the drawer is a commercial firm or bank and the drawee is a bank, the bill after acceptance by the bank or the drawee is called a banker's acceptance bill.

Use of Bill of Exchange in Foreign Trade

A bill of exchange (draft) is an order to pay. It is made out by an exporter and presented to an importer, usually through a bank. It may be payable immediately on presentation (a sight or demand draft), or so many days after presentation (a time draft). In the later case, the drawee writes "Accepted" across it and signs his name. The exporter can then get immediate payment by discounting the draft and supplying a letter of hypothecation. If a time draft is not honored at maturity, it will be noted and protested by a Notary Public, and represented to the drawee. Such a draft, and the corresponding payment terms, "Documents against Acceptance", obviously involve risk to the exporter or his bank.

Sight draft terms involve less risk, since title to a consignment is only released against payment; even so the shipper may be left with his consignment on his hands; or rather, afloat, if the draft is dishonored. When drafts are presented under a letter of credit, however, the shipper has full protection, and only the opening and confirming banks run any risk.

7.1.2　Promissory Note

A promissory note is an unconditional promise in writing made by one person to another signed by the maker, engaging to pay, on demand or at a fixed or determinable future time, a sum certain in money, to, or to the order, of a specified person or to the bearer.

The main difference between a promissory note and a draft lies in that there are three parties, namely drawer, drawee and payee involved in a draft but only two, drawer and payee in a promissory note. The payer of promissory note is the drawer himself.

Promissory notes can be made by commercial firms, called commercial promissory notes, or bankers, called bank promissory notes. Commercial promissory notes can be sight promissory notes or time promissory notes, while bank promissory notes can only be sight. In international trade, most promissory notes are drawn by bankers which are mostly not negotiable (see Figure 7.2).

Figure 7.2　Promissory Note Sample

7.1.3　Cheque（Check）

A cheque is an unconditional order in writing drawn on a banker signed by the drawer, requiring the banker to pay on demand a sum certain in money to or to the order of a specified person or to bearer.

In foreign trade, a cheque drawn on a bank overseas cannot be readily negotiated by the exporter. If the exporter's bank was prepared to negotiate it for him then he would receive payment immediately but at the cost of the discount. If his bank is not willing to negotiate it, the exporter would have to ask his bank to collect the cheque for him, which would be time-consuming as well as relatively expensive (see Figure 7.3).

Figure 7.3　Cheque Sample

7.2　Modes of International Payment

Trading with other countries is not the same as trading within one's own country. Both exporters and importers face risks in export or import transactions because they will inevitably experience the possibility that the other party may not fulfil the contract.

For the exporters, they are likely to take the risk of buyer default; the customers might not pay in full for the goods. There are several reasons for this: the importers might go bankrupt; a war might break out or the importers' government might ban trade with the exporting country; or they might ban imports of certain commodities. Moreover, the importers might run into difficulties getting the foreign exchange to pay for the goods, or they are even not reliable and simply refuse to pay the agreed amount of money.

For importers, they may face the risk that the goods will be delayed and they might only receive them a long time after paying for them. This may result from port congestion or strikes. Delays in fulfillment of orders by exporters and difficult customs clearance in the importing country can cause loss of business. There is also a risk that the wrong goods might be sent.

It is to prevent such risks that different methods of payment have been developed. The modes of payment in international trade can be generally divided into three categories, illustrated as follows (see Table 7.1).

Table 7.1   Modes of Payment in International Trade

Modes of Payment in International Trade	Remittance	M/T (mail transfer)	
		T/T (telegraphic transfer)	
		D/D (demand draft)	
	Collection	D/P (documents against payment)	D/P at sight
			D/P after sight
		D/A (documents against acceptance)	
	Letter of Credit		

Remittance and collection belong to commercial credit, letter of credit belongs to banker's credit. In international trade, "credit" stipulates who takes the responsibility of paying money and surrendering the shipping documents that represent the title to the goods. In remittance or collection transaction, the buyer is responsible for making payment, the seller handing over documents. While in letter of credit transaction, the banker is responsible for paying money and tendering documents on behalf of both parties.

7.2.1   Remittance

Under remittance, the payer instructs his bank or other institutions to have a payment made to the payee. Four parties are involved in the remittance business: the remitter, the payee, the remitting bank and the paying bank. In foreign trade, remittance is often adopted in those sales under the terms of cash in advance, cash with order, cash on delivery or open account. There are three kinds of remittance.

Types of Remittance

Mail Transfer (M/T)

Mail transfer is the most common method of remittance. The buyer gives money to the remitting bank (his local bank) which, then, issues a trust deed for payment and sends it to the paying bank (his branch or correspondent bank) in the seller's place by mail instructing him to pay the specific amount to the seller. This method costs less, but slower.

Telegraphic Transfer (T/T)

The procedure under T/T is similar to that of M/T except that the instructions from the buyer's bank to its branch or correspondent bank at the seller's end are made by cable instead of by mail.

This means that the payment can be effected more quickly and the seller can receive the money promptly. But the buyer has to be charged more for it.

Demand Draft (D/D)

Under demand draft, the remitting bank, at the request of the buyer, draws a demand draft on its branch or correspondent bank instructing it to make a certain amount of payment to the seller on behalf of the buyer.

Advantages and Disadvantages of Remittance

In international trade, most of transactions are paid through M/T and T/T if remittance is used. T/T is beneficial to the seller because it enables him to obtain money promptly, accelerate the turnover of funds, increase the income of interests and avoid the risk of fluctuation in exchange rate. But it is disadvantageous to the buyer in that he has to bear more cable expenses and bank charges. In practice, if T/T is not definitely stipulated in transaction, the buyer had better make payment by M/T. When the amount of payment is comparatively large, or the money market fluctuates greatly, or the currency of settlement being used is likely to devaluate, it is wise for the buyer to use T/T. In a word, the choice of T/T or M/T should be clearly stipulated in the contract according to specific situation. As far as D/D is concerned, it is transferable, which is different from M/T and T/T.

7.2.2　Collection

When funds are not required immediately, or where bills are not sufficiently attractive to a banker for negotiation, they may be handed by an exporter to his bank for collection. The exporter asks his bank to arrange for the acceptance or payment of the bill overseas, and the bank will carry out his task through its own branch office abroad or a corresponding bank.

Under the collection, the exporter issues a draft with shipping documents attached, then forwards the draft to a bank in his place (i. e. the remitting bank) to make an application for collection and entrusts the remitting bank to collect the purchase price from the buyer through its correspondent bank abroad (i. e. the collecting bank). In the course of collection, banks only provide the service of collecting and remitting and are not liable for non-payment of the importer.

The procedure of collecting payment is illustrated as follows (see Figure 7.4).

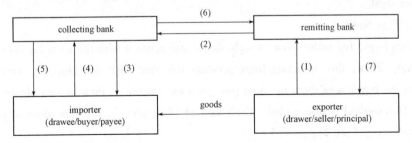

Figure 7.4　Procedure of Collecting Payment

Remarks:

(1) The exporter despatches the goods and draws a draft, then sends the draft together with shipping documents to the remitting bank to make an application for collecting money on his behalf.

(2) The remitting bank sends the draft and shipping documents to a correspondent bank overseas—the collecting bank.

(3) The collecting bank presents the draft and documents to the importer for acceptance (D/A) and/or payment (D/P).

(4) The importer makes payment (D/P) or endorses the bill for acceptance.

(5) The collecting bank hands over the documents to the importer.

(6) The collecting bank notifies the remitting bank of crediting the money to their account.

(7) The remitting bank makes payment to the exporter.

Collection means the handing by banks, on instructions received, of documents (financial documents and/or commercial documents), in order to: a. obtain acceptance and/or, as the case may be, payment, or b. deliver commercial documents against acceptance and/or as the case may be, against payment, or c. deliver documents on other terms and conditions. Collection is of two types: clean collection and documentary collection. The parties to a collection financial documents not accompanied by commercial documents (financial documents means bill of exchange, promissory notes, checks, payment receipts or other similar instruments used for obtaining the payment of money; commercial documents means invoice, shiping documents, documents of title or other similar documents whatsoever, not being financial documents).

Documentary collection means collection of: a. financial documents accompanied by commercial documents; b. commercial documents not accompanied by financial documents. The documentary collection is widely used in international trade. Documentary collection falls into two kinds: documents against payment (D/P) and documents against acceptance (D/A).

Documents Against Payment (D/P)

Under this payment term, the exporter is to ship the goods ordered and deliver the relevant shipping documents to the buyer abroad through the remitting bank and the collecting bank with instructions not to release the documents to the buyer until the full payment is effected. According to the different time of payment, document against payment can be further divided into D/P at sight and D/P after sight.

(I) D/P at Sight

Under this term, the seller draws a sight draft, and sends it with the shipping documents to the collecting bank. Then, the collecting bank presents the sight draft and shipping documents to the buyer. When the buyer sees them he must pay the money at once, then he can obtain the shipping documents. This method is also called "Cash against Documents", the procedure of which can be seen in the following (see Figure 7. 5)。

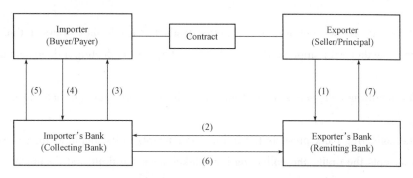

Figure 7. 5   Procedure of D/P at Sight

Note:

(1) According to the contract, the exporter loads the goods and draws a sight draft, then sends the draft together with shipping documents to his bank for collecting a documentary bill on his behalf.

(2) The remitting bank sends the documentary bill to a correspondent bank overseas—the collecting bank for collecting money.

(3) The collecting bank presents the bill and documents to the importer for payment.

(4) The importer makes payment.

(5) The collecting bank hands over the documents to the importer.

(6) The collecting bank notifies the remitting bank of crediting the money to their account.

(7) The remitting bank makes payment to the exporter.

(II) D/P After Sight

Under this term, the seller draws a time (or usance) draft. The collecting bank presents the time draft and shipping documents to the buyer. When the buyer sees them, he just accepts the time bill and then effects payment at maturity of the draft. When receiving the money from the buyer, the collecting bank hands over the shipping documents to him. The procedure of D/P after sight is shown as follows (see Figure 7.6).

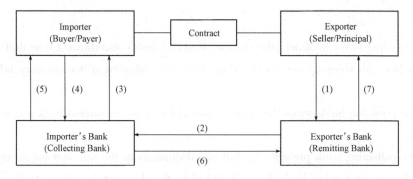

Figure 7. 6   Procedure of D/P after Sight

Note:

(1) According to the contract, the exporter loads the goods and draws a time draft, then sends the draft together with shipping documents to his bank for collecting a documentary bill on his behalf.

(2) The remitting bank sends the documentary bill to a correspondent bank overseas—the collecting bank.

(3) The collecting bank presents the bill and documents to the importer for acceptance. After the importer accepts the draft, the collecting bank takes back the draft and documents.

(4) The importer makes payment when the time falls due.

(5) The collecting bank hands over the documents to the importer.

(6) The collecting bank notifies the remitting bank of crediting the money to their account.

(7) The remitting bank makes payment to the exporter.

Document Against Acceptance

This term of payment is applicable only to a time bill that is used in documentary collection. In which the collecting bank will release the shipping documents to the buyer without any payment but merely against the acceptance of the bill by the buyer to honor the draft at a certain future date agreed upon between the seller and the buyer. D/A is always after sight (see Figure 7.7).

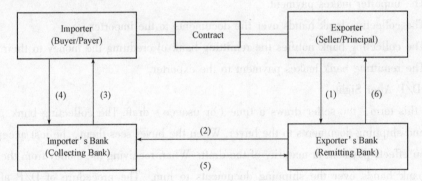

Figure 7.7   Procedure of D/A

Note:

(1) According to the contract, the exporter loads the goods and draws a time draft, then sends the draft together with shipping documents to his bank for collecting a documentary bill on his behalf.

(2) The remitting bank sends the documentary bill to a correspondent bank overseas—the collecting bank.

(3) The collecting bank presents the bill and documents to the importer for acceptance, after that, the collecting bank takes back the draft and gives the shipping documents to the importer.

(4) The importer makes payment when time falls due.

(5) The collecting bank notifies the remitting bank of crediting the money to their account.

(6) The remitting bank makes payment to the exporter.

Advantages and Disadvantages of Collection

There are some advantages for the exporter under the mode of collection. Since the remitting bank instructs the collecting bank not to hand over the shipping documents to the buyer until the draft is accepted or paid, thus, a lot of risks in "shipment earlier, settlement later" can be avoided. What's more, the buyer's lack of commercial integrity can also be prevented. For example, under D/P, without making payment, the buyer can not obtain the documents of title to the goods and take delivery of the goods. The ownership of the goods still belongs to the seller. If the buyer dishonors the draft, the seller can sell the goods to others. When facing a market with fierce competition, the exporter can use collection to win customers and promote its sales. On the other hand, the importer may also get advantages from collection. Firstly, it can facilitate the importer to obtain financing. Under D/P after sight, capital tie-up of the importer can be avoided or reduced. Secondly, the expenses are low. Under collection, the importer does not need to pay the service fee as required for the opening of L/C.

However, problems still exist. Under D/P, there is the possibility of the buyer or his bank refusing to honor the draft and take up the shipping documents, especially at a time when the market is falling. In such a case, the seller may face the risk of non-payment or late payment by the buyer, although he is still the owner of the goods.

As for D/A, the most striking disadvantage is that after the buyer accepts the draft, the documents of title will be surrendered to him. So, if the buyer goes bankrupt or become insolvent before the payment of the draft, the seller will bear the losses of all the payment. Therefore, D/A is more risky than D/P for the exporter.

In international trade, payment through collection is accepted only when the financial standing of the importer is sound or where a previous course of business has vinspired the exporter with confidence that the importer will be good for payment. As far as the seller's benefit is concerned, and D/P at sight is better than D/P after sight, D/P is better than D/A.

In order to prevent the various disadvantages of collection, a much better mode of payment in international trade has been developed, that is, letter of credit.

7.2.3   Letter of Credit (L/C)

The methods of payment just described above are used if a buyer and a seller have gained a degree of confidence in each other. But trading partners have other alternatives. If an importer's bank is satisfied with its customer's credit rating, it will open a letter of credit in favor of and addressed to the exporter. This, letter pledges to pay the exporter if the merchandise is shipped in accordance with the conditions in the letter of credit conditions that are based on the contract between the buyer and seller. Thus, the bank backs up the business transaction between the importer and exporter. Letter of credit has become a safer and quicker method of obtaining payment in international trade.

Definition of Letter of Credit (L/C)

The letter of credit, sometimes just called credit or L/C for short, is a letter addressed to the seller, written and signed by a bank acting on behalf of the buyer. In which, the bank promises that it will pay or accept drafts drawn on itself if the seller conforms exactly to the conditions set forth in the letter of credit. And only the documentation submitted is precisely specified will the bank pay or accept the exporter's draft. Through the letter of credit, a bank substitutes its own commitment to pay for that of its customer, that is, the buyer. Although a letter of credit is not really a guarantee, it serves almost the same purpose in that the bank assures payment to the seller as long as he complies with stated conditions.

Circulation of Letter of Credit

After signing, with the exporter, a contract agreeing to make payment by L/C, the importer requests his bank to issue a letter of credit in favor of the exporter. If it accepts the importer's application, the opening bank issues a letter of credit and then informs its foreign branch or correspondent to advise the beneficiary (the exporter), who then examine the letter of credit. If it does not conform to the conditions set in the sales contract, the exporter may request an amendment. If it is an irrevocable letter of credit, and they usually are, it can not be changed unless all parties agree to amend.

After confirming the letter of credit, the exporter delivers the goods to the shipper who then issues a bill of lading. Other documents, such as invoices and insurance documents, are prepared by the exporter. The next step occurs when the exporter draws a draft on the opening bank and present it, with the letter of credit plus documents to his or her own bank. Usually this bank will investigate the documents and, if they are in order, it will pay the draft. The letter of credit and documents are sent to the opening bank. It is the bank's responsibility to examine the documents in relation to the letter of credit issued. If discrepancies exist, they will have to be corrected, either by a new letter of credit, by new documents, or by amendments. Discrepancies include any one of the following: the letter of credit expired; the draft was not properly drawn; there was no indication on the bill of lading that goods were received on board; there was insufficient insurance; or an invoice description did not match that of the letter of credit. If no discrepancies are found after careful checking, the opening bank will reimburse the money to the exporter's bank (negotiating bank) in accordance with the terms of the credit. The opening bank then presents the documents to the buyer for payment or acceptance. Documents will be released to the buyer upon his payment of amount due or acceptance of the draft. With the documents, the buyer can take delivery of the goods.

The circulation of letter of credit can be illustrated as follows (see Figure 7. 8) .

Note:

(1) The buyer makes application for a letter of credit with his bank and signs opening bank's agreement form. The opening bank approves the application issues the actual letter of credit document.

(2) The opening bank forwards the letter of credit to the advising bank.

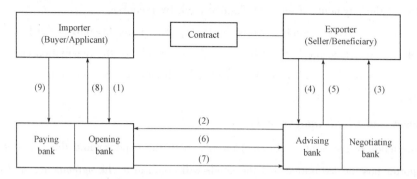

Figure 7.8   Circulation of Letter of Credit

(3) The advising bank delivers the letter of credit to the beneficiary.

(4) Having examined the letter of credit, the beneficiary (seller) ships the goods to the buyer. After that, the beneficiary prepares documents, draws a draft and presents them to his bank.

(5) The beneficiary's bank negotiates the documents and pays funds to the beneficiary in accordance with the letter of credit.

(6) The negotiating bank forwards the documents to the opening bank.

(7) The opening bank receives the documents and checks them. If the documents are in order and comply with the letter of credit, the opening bank credits the negotiating bank's account.

(8) The opening bank notifies the buyer to make payment for documents.

(9) After making payment, the buyer receives the documents and takes delivery of the goods.

Major Parties Involved in a Letter of Credit

● The applicant, who is usually the importer who applies to the bank for the letter of credit.

● The opening bank, or issuing bank, which opens the letter of credit upon the request of the importer. It holds itself responsible for the payment of the goods.

● The advising bank, or notifying bank, which is authorized by the issuing bank to transfer the letter of credit to the exporter's bank. It is in the exporter's country and usually the correspondent bank of the issuing hank. It is only responsible for the authenticity of the letter of credit.

● The beneficiary, who is usually the exporter and is entitled to use the letter of credit for the payment of the goods.

● The negotiating bank, which is willing to buy on discount the documentary draft drawn by the beneficiary. It can either or not be designated in the letter of credit. In the transfer of the bill of exchange, the negotiating bank can be taken as the bona fide bearer of the bill of exchange. The negotiating bank and the advising bank can be the same bank, as is to be decided by the letter of credit.

● The paying bank, which is designated by the letter of credit to pay the draft. In most cases, it is the issuing bank. it might also be some other bank, as when the currency used in the letter of credit is that of a third country, the paying bank can be a bank of that country. Once the paying bank has effected the payment, it can not claim it back by recourse.

● The confirming bank, which is asked by the opening bank to confirm the L/C.

If a bank has confirmed the L/C, it holds itself responsible for the negotiation or payment of the L/C. Mostly, the confirming bank is advising bank, but it can also be some other bank in the country of the exporter.

Advantages and Disadvantages of Letter of Credit

Under the mode of payment by letter of credit, both the exporter and the importer are protected. The importer has the assurance that the goods will conform to the agreement, and the exporter is assured that the goods will be paid for, The importer might be able to get better trade terms by using the letter of credit system, though this will be offset somewhat by the bank's charges. when he wants to open a letter of credit to cover his purchase from abroad, the importer may apply to his banker for a L/C for any amount.

The banker, usually, will note ask for payment of the full amount of the L/C to be opened but will only collect a deposit, thus, the importer's capital will not be tied up. In other words, the bank finances or guarantees the balance of the purchase price. As far as the exporter is concerned, he knows that the goods will be paid for by the importer's bank, even if the exporter is unfamiliar with foreign bank's credit standing. The confirming bank pays the drafts drawn under the letter of credit and has no recourse to collect from the exporter, even if the opening bank issuing the letter of credit does not reimburse the confirming bank. Therefore, from what have been mentioned above, one will readily see that a banker's letter of credit for the payment of the purchase price is most desirable.

However, a letter of credit cannot absolutely eliminate the risks—the businessmen are likely to encounter. Since the opening bank makes payment according to the submission of relevant documents rather than the goods, it is likely that importers may suffer from the fraudulent acts. The beneficiary may get the payment from the bank with mock documents that do not conform to physical goods, or even with documents of no goods. In addition to the fraudulent acts, the wrong practice of the related personnel and improper procedures, ambiguous expressions or negligence, may also bring loss to the related party.

Illustration of Letter of Credit

The chief contents of the L/C can be seen as a combination of the chief contents of the sales contract, the required documents and the bank assurance. It might be further broken into the following provisions :

● The parties involved, including the applicant, the issuing bank, the negotiating bank, the paying bank.

● Remarks about the L/C: such as the No. of the L/C, its type, the issuing date, the amount of the L/C , the expiry date and place, etc.

● The clauses of the bill of exchange, such as the amount of the bill, the drawer and drawee, the paying date, etc.

- The clauses about shiping, such as the port of shipment, the port of destination, partial shipment and transshipment, the latest date of shipment.

- The clauses about the description of the goods, such as the name of goods, specifications, quantity, packing, unit price, total amount, and so on.

- The clauses about the documents, include what kinds of documents are required, the required number of copies of the documents, the content of the specific document, etc.

- The period for presentation. The documents should be presented within 21 days after the date of shipment indicated on transport document but within the credit validity. Otherwise the banks will not accept the presentation.

- Guarantee clauses of the opening bank, which testifies that the opening bank will hold itself responsible for the payment to the beneficiary or the holder of the draft.

- Particular clause, such as the special provisions about the deal in accordance with the particular business or political situations of the importing country.

L/C is of different forms. Thought several standard forms have been recommended by the International Chamber of Commerce, they are rarely used by the issuing banks. Most of them use their own forms with reference to the standard forms. The legal document concerning the use of L/C is Uniform Customs and Practice for Commercial Documentary Credits.

Types of Letter of Credits

In international trade, different kinds of letters of credit are used to meet various requirements of trading and payment. According to different criteria, letters of credit may be mainly classified into the following types:

- Irrevocable and Revocable Letter of Credit

An irrevocable letter cannot be modified, amended or withdrawn by either the opening bank or the buyer before the expiry date of credit validity without the agreement of the beneficiary. As long as the shipping documents surrendered by the exporter are in conformity with the stipulations of the L/C, the opening bank will surely take the responsibility of making payment. So, it is very commonly use in foreign trade.

A revocable letter of credit, as its name implies, can be cancelled or amended at any moment during the term of credit validity without notice to the beneficiary. Such kind of L/C is unlikely to be acceptable to exporters and seldom used in trading with other countries because it doesn't provide them with any guarantee.

- Confirmed and Unconfirmed Letters of Credit

A confirmed letter of credit is confirmed by another bank (usually the advising bank) on behalf of the opening bank. The exporter, then, has a confirmed irrevocable L/C and he is double guaranteed against payment by both confirming and issuing bank. The confirming bank is responsible to the beneficiary at first.

An unconfirmed letter of credit is one which advised by the advising bank, but it is unwilling to take the responsibility for the documents, merely notifies the beneficiary that the credit has been opened.

- Transferable Letters of Credits

In the case of transferable letter of credit, the beneficiary authorizes the advising bank or negotiating bank to transfer the right of issuing a draft in whole to another person (the second beneficiary). It is the second beneficiary who draws the bill of exchange, presents the shipping documents and obtains the payment. But the first beneficiary still bears the responsibility of the seller stipulated in the sales contract. A transfer L/C can only be transferred once. The second beneficiary can not transfer it again.

- Sight and Usance Letters of Credit

Sight letter of credit also means "L/C available by draft at sight" or "L/C by sight Draft". Under a sight letter of credit, the negotiating bank makes payment immediately upon the presentation of the sight draft and shipping documents as stipulated in the L/C by the seller. Similarly, when the negotiating bank delivers the sight draft and the documents to the opening bank, the latter also makes reimbursement at once.

When the seller presents a time draft together with the shipping documents stipulated in the L/C to the negotiating bank, the latter doesn't make payment immediately. He hands over the draft and the documents to the opening bank. The opening bank doesn't pay the money immediately either. He just accepts the draft and returns it to the seller. He makes payment when the time draft falls due. This kind of letter of credit is called usance letter of credit.

- Revolving Letter of Credit

If a contract is concluded on long term and allows partial shipments, or the parties wish the contract to proceed without interruption, the buyer often requests to establish a revolving L/C. It notifies the seller that when a shipment has been made and documents presented and paid, the credit, automatically becomes re-available in its original form and another shipment can be made and so on. The L/C can be used again and again under the same terms and without issuance of another L/C until the stipulated times of use and the stipulated total amount have been reached. It also simplifies formalities and reduces expenses.

- Back-to-Back Letter of Credit

In a back-to-back letters of credit, two letters of credit are involved. One is in favor of the exporter who is not the actual supplier of goods; the other is opened by the exporter in favor of the actual supplier. It is usually used if the supplier does not want his identity known. The back-to-back credit will show a lesser amount for the value of the goods, the difference being the profit the exporter makes. The tenor is often reduced by a few days to arrange for the substitution of in voices.

- Red Clause Credit

A red clause credit is similar to a normal letter of credit except that it contains a clause (originally typed or printed in red) authorizing the negotiating bank to make clean advance to the exporter. By it, the beneficiary may take the advantage of the importer's credit standing. Nowadays red clause credits are mainly used in situations where the buying company (importer) has, in the exporting country an agent whose role is to purchase merchandise to the exported. To finance these purchases, the importer may arrange for opening a red clause letter of credit.

7.3   Other Methods of Payment

L/C offers a safe way for the settlement of goods in international trade. But under certain occasions L/C can not be used, as when the deal covers a long period of time or when the business is done in some other area rather than trade. Under such case, letter of guarantee and standby L/C can be used.

Along with the development of international trade, some other methods have been developed for the settlement of payment, among them factoring is now more and more often used.

Also, various methods of payment can be combined together for the settlement of payment for various reasons and under various occasions, among them are the combination of L/C with collection, combination of L/C with remittance, the documentary collection with down payment, and the combination of documentary collection with standby L/C.

7.3.1   Letter of Guarantee

A letter of guarantee is a subordinate document issued by a bank which the issuing bank promises the beneficiary that it will hold itself responsible for the debt, losses caused by the fault or infringement of contract of a third person. It can be used under various occasions. Here we focus on two groups: tender guarantee and performance guarantee.

Tender guarantee is issued by a bank upon the application of the bidder. By which the bank promises to the beneficiary (the tender inviter) that it will answer for the losses caused by the bidder if he withdraws the bid or make amendments to his bidding documents before the bidding begins, or refuses to sign or pay the margin after he has been awarded the bidding.

By a performance guarantee, the issuing bank promises to the beneficiary it will pay him a certain sum of money or adopt certain remedial measures if the principal has not duly fulfilled his liabilities in time. In international trade practice, the performance guarantee can be either supplied by the importer or the exporter. Under the former case, the issuing bank will promise to the seller that if the importer can not duly effect payment of the goods, the bank will answer for it. Under the latter case, the issuing bank will promise to the buyer that if the exporter can not delivery the goods in time, the bank will answer for the losses sustained by the buyer.

A letter of guarantee is different from an L/C. Under the L/C, the opening bank holds itself for the payment of the goods, but under a letter of guarantee, the issuing bank holds itself responsible only after the principal has not fulfilled its obligations. Under an L/C, the seller can get the payment through negotiation, but he can not do so under a letter of guarantee. Also, under an L/C, the issuing bank handles only the relevant documents, it has nothing to do with the sales contract, but under an L/G, when the beneficiary has presented a written documents declaring that the principal has not fulfilled his obligations and asks the issuing bank for recompense, the bank will have to make clear how and why his principal has not fulfilled his obligations. In this way, he may get involved into the disputes. At some countries, like USA and Japan, do not allow their banks to get involved in commercial disputes, so the banks of these countries do not issue the L/C, instead they use stand by L/C.

7.3.2 Standby L/C

Standby L/C also called commercial paper L/C, or guarantee L/C, or performance L/C. It is a kind of clean L/C. It is a document by which the opening bank promises to the beneficiary that it will bear some liabilities on behalf of the applicant if the later has not duly fulfilled his obligations as required by the contract. Should the applicant have duly fulfilled his obligations, the standby L/C will not be used while if he has not, the beneficiary has to render a written document stating that the applicant has not duly performed the contract and ask the opening bank to make recompense.

Standby L/C is similar to the letter of guarantee in that they both use bank credit and are both a kind of promise to the beneficiary that it holds itself responsible should the principal have not duly performed a contract or effected a payment. They are different in that under the letter of guarantee, the opening bank holds itself responsible only if the applicant has not duly performed his liabilities, and in so doing it might possibly get involved in the disputes between the applicant and the beneficiary, while under a standby L/C the opening bank will effect the payment if the beneficiary has presented to opening bank the required statement, and the opening bank stands aloof from the contract between its applicant and the beneficiary.

With letter of guarantee or standby letter of credit, progress payment or deferred payment can be used. Progress payment is used when the sale is made by installments. Under such cases, the buyer will first pay a certain sum of front money, the balance of the payment will be made by installments along with the installments deliveries of the goods. Under deferred payment, the buyer first pays a certain sum of earnest money, then the balance of the payment will be made by installments during a rather long time afterwards. This is in fact a kind of credit sale, by which the buyer can make the purchase by using foreign funds. Deferred payment involves interest, and an interest clause should be given in the contract. Also, the two methods can be used in a combined way, that is, each installment payment under progress payment will not be effected immediately after each delivery, instead, it will be made some time after each delivery.

7.3.3 Combined Use of Different Method of Payment

In international trade, usually only one method is used for a deal. But sometimes more than one method is used for the payment of the goods.

The usual combined methods are combination of L/C and collection, combination of L/C and remittance, combination of documentary collection and down payment, combination of documentary collection and standby L/C, etc.

Combination of L/C and Collection

Payment effected in this way will partly be made by L/C and partly by collection. This will require the buyer to open an L/C for a certain percentage of the whole payment of the goods, the balance is to be collected. Also in the sales contract, it should be declared clearly that the amount of payment under L/C will be available against a clean draft, while the balance of the payment will be available against documentary draft on collection basis. The shipping documents will not be released to the buyer unless he has effected all the payment.

Combination of L/C and Remittance

Payment made in this way is partly done by L/C and partly by remittance. If the remittance is made before the shipment of the goods, the money is taken as front money or advanced payment. If it is done after the shipment of the goods, it is often used for the balance of the payment for the consignment which can be varied in amount.

Combination of Documentary Collection and Down Payment

To make sure that the exporter will not sustain much loss because of collection, the seller can ask the buyer to make some down payment. When the shipment has been made, the down payment is to be deducted from the payment to be collected. In this way, if the draft has been dishonored, the exporter can ship the goods back, the loss thus sustained can be compensated by the down payment. But the exporter should make sure that the goods can be shipped back from the importing country.

Combination of Documentary Collection and Standby L/C

To avoid risks in documentary collection, the exporter can ask the buyer to open a standby L/C. Thus when the payment has been dishonored, the seller can ask the opening bank of the standby L/C for payment.

To use this method, the time of validity of the standby L/C must be sufficiently long, so that after the payment has been dishonored, the seller has enough time to ask for the issuing bank for the payment. In filling the form of application for export collection, the exporter should ask the remitting bank to tell the collecting bank to give immediate notice by electronic means should the draft has been dishonored.

7.4　Payment Terms in Contract

Payment can be in different ways, therefore, it is imperative to clearly stipulate the specific mode of payment and the detailed requirement. The following are some examples of terms of payment in contract.

7.4.1　Remittance

The Buyers shall pay 100% of the sales proceeds in advance by M/T to reach the Sellers not later than July 15.

All the payment shall be made in the US currency by the Buyer to the Seller, by telegraphic transfer to the Seller's designated account (s) with the bank (s) _ in (country) .

7. 4. 2 Collection

Document Against Payment at Sight

- Upon first presentation the Buyers shall pay against documentary draft drawn by the Sellers at sight. The shipping documents are to be delivered against payment only.

- Payment shall be made by net cash against sight draft with bill of lading attached showing the shipment of the goods.

- Payment shall be made by net cash against sight draft with bill of lading attached showing the shipment of the goods. Such payment shall be made through the (bank) of (place) . The bill of lading shall not be delivered to the Buyer until such draft is paid.

Document Against Payment After Sight

- The Buyers shall duly accept the documentary draft drawn by the Sellers at…days sight upon first presentation and make payment on its maturity. The shipping documents are to be delivered against payment only.

- The Buyers shall pay against documentary draft drawn by the Sellers at…days after date of B/L. The shipping documents are to be delivered against payment only.

- The Buyers shall pay against documentary draft drawn by the Sellers at…days after date of draft. The shipping documents are to be delivered against payment only.

Document Against Acceptance

The Buyers shall duly accept the documentary draft drawn by the Sellers at…days sight upon first presentation and make payment on its maturity. The shipping documents are to be delivered against acceptance.

7. 4. 3 Letter of Credit

Sight Letter of Credit

- The Buyers shall open through a bank acceptable to the Sellers an Irrevocable Sight Letter of Credit to reach the Sellers…days before the month of shipment, valid for negotiation in China until the 15th day after the month of shipment.

- By Irrevocable Letter of Credit, available by Sellers'documentary draft at sight, to be valid for negotiation in China. until 15 days after date of shipment. The L/C must reach the Sellers 30 days before the contracted month of shipment.

Time Letter of Credit

- The Buyers shall open through a bank acceptable to the Sellers an Irrevocable Letter of

Credit at…days Sight to reach the Sellers…days before the month of shipment, valid for negotiation in China until the 30th day after the month of shipment.

● By Irrevocable L/G available by Seller's documentary draft at…days sight, to be valid for negotiation in China until 15 days after date of shipment. The L/C must reach the Sellers 30 days before the contracted month of shipment.

7. 4. 4   Combination of Different Modes

The Buyers shall open through a bank acceptable to the Sellers an Irrevocable Sight Letter of Credit to reach the Seller. . . days before the month of shipment, stipulating that 50% of the invoice value available against clean draft at sight while the remaining 50% on Documents against Payment at sight (or D/P at. . . days after sight) . The full set of the shipping documents of 100% invoice value shall accompany the collection item and shall only be released after full payment of the invoice value. If the Buyers fail to pay full invoice value, the shipping, documents shall be held by the issuing bank at the Sellers' disposal.

Chapter 8

The General Terms and Conditions in the Contract

- Understand the conditions for breach of contract and settlement of claims.
- Knows ways of stipulating claim clauses in a sales contract.
- Knows ways of stipulating force majeure clauses in a sales contract.
- Be aware of the issues to be considered in arbitration.

Inspection is an indispensable clause of a contract, and its aim is to protect the interests of both the buyer and the seller. Therefore, it is of great significance in international trade. Laws and regulations in various countries as well as international conventions have imposed inspection liabilities on both the sellers and the buyers. The failure in inspection and unfavorable inspection results may result in disputes between the buyers and the sellers and claims lodged by one party against the other.

In export trade, disputes between the buyers and the sellers may arise at anytime despite the fact that both parties of the sales contract work very carefully in the performance of the contract. In such cases, with a view to maintaining good relationship, the two parties of a dispute are normally recommended to rely on friendly negotiation and mediation. However, in case these means are not workable, the disputes may be submitted for arbitration or even litigation.

8. 1　Commodity Inspection

In international trade, the quality and quantity of the goods delivered by the seller should be in conformity with the terms of the contract and should be packed in the manner required in the contract. Commodity inspection is the inspection conducted by an authorized third party to testify whether the quality, quantity, weight, package, etc. are the same as those specified in the contract. After inspection, certain kinds of certificates can be issued and serve as evidence in the performance of the contract.

8.1.1   Inspection Contents

Inspection certificates are issued only when they are proved to up to certain standards. Then what standards should followed as far as commodity inspection is concerned?

Inspection on import and export commodities performed by the commodity inspection authorities shall cover quality , specifications, quantity, weight, packing and the requirements for safety and hygiene. If the mandatory standards or other inspection standards which must complied with are specified the Law and the Regulations, the inspection. Of course, shall be performed according to the standards specified in the Law and the Regulations.

In the process of inspection, it is likely that standards specified in the Law and the Regulations are not consistent with those agreed upon by the seller and buyer in the sales contract. In such cases, if the compulsory standards or other inspection standards specified in the Law and the Regulations are higher than the standards agreed upon in the contract , the inspection shall be conducted subject to the higher standard . If the standards specified in the Law and the Regulations are lower than the standards agreed upon in the contract , the inspection shall be conducted according to the standards agreed upon in the contract. Similarly, if the trade is conducted against the sample , the inspection shall be performed according to the sample provided.

Finally, in the case where there lack the inspection standards specified in the Law and the Regulations, and where the inspection standards are either not agreed upon or unclear in the contract, the inspection shall be conducted according to the standards of the manufacturing country, or relevant international standards or the standards designated by the state inspection agency.

8.1.2   Time and Place of Inspection

The inspection of the goods can be conducted at various stages of the trade process. Before delivery, the manufacturers should make a precise and comprehensive inspection of the goods with regard to its quality, specifications, quantity or weight, and issue an inspection certificate certifying the technical data or the result of the inspection. Before shipment, the authorized department in the exporting country will conduct mandatory inspection as to the quality and quantity or weight of the goods and issue the relevant inspection as to the quality and quantity or weight of the goods and issue the relevant inspection certificates which serve as the basis for the negotiation of payments by the seller. After the arrival of the goods at the destination, the goods may be further inspected by relevant departments in the importing country, the unfavorable result of which serves as the basis for the claim made by the buyer.

Therefore, it is important to make clear when and where inspection should be conducted. CISG provides that unless the parties have agreed otherwise, the buyer has the right to inspect the goods and the time and place of inspection should be stipulated in detail in the contract.

There are generally three ways to stipulate the place and time of inspection in the sales contract; inspection at the factory or at the port of shipment, inspection at the port of destination, and inspection at the port of shipment and re-inspection at the port of destination.

Inspection at the Factory or at the Port of Shipment

Inspection at the factory or at the port of shipment refers to the inspection conducted by an authorized party agreed upon by both parties of the sales contract to testify the quality, quantity, weight, package, etc. of the goods at the factory or at the port of shipment before the delivery of the goods. Another name for this is known as shipping quality and weight. The inspection certificate (s) thus issued will be considered final for the delivery of the goods by the seller. In other words, the seller will not be obliged to any change to the goods in transit after the goods have been inspected before delivery. Although the buyer can re-inspect the goods at the port of destination, normally he cannot claim for compensation based on the result of the re-inspection. Therefore, this method is more favorable to the seller.

Inspection at the Port of Destination

Inspection at the port of destination is also referred to as landed quality and weight. It indicates the inspection conducted by an authorized party within the time period stipulated in the sales contract to testify the quality, quantity, weight, package, etc. of the goods at the port of destination. The inspection certificate (s) issued in the destination country is considered final for the delivery of the goods by the seller. Under this arrangement, if any discrepancies are found between the actual goods and terms described in the sales contract, the buyer can retain the right of claim for compensation. Therefore, the buyer often prefers this method of inspection.

Inspection at the Port of Shipment and Re-inspection at the Port of Destination

As discussed above, neither way in stipulating time and place for inspection is satisfactory to both parties. To avoid further disputes, a third method has been created, stating that the inspection certificate (s) issued by the authorized party at the port of shipment can be used as one of the documents for the seller to settle payment after the delivery of the goods. However, the buyer retains the right to re-inspection the goods at the port of destination. If he obtains from an authorized party designated in the contract at the port of destination the inspection certificate (s) showing that the goods received are not as described in the contract, the buyer has the right to claim for compensation from the seller as long as there are evidences to show that the seller should be responsible for the discrepancies. Obviously , this method benefits both sides to a certain degree and therefore is widely adopted in International trade practice.

8. 1. 3 Inspection Bodies

In international trade , inspection certificates should issued by authorized inspection bodies;

otherwise, they will not be accepted as effective documents. Therefore , before getting your goods inspected, you should know something about the inspection bodies and their respective roles played in the inspection practice.

Types of Inspection Body

There are normally two main types of inspection body: governmental and non-governmental inspection bodies. The former owned or supervised by governments and specializes in inspection of particular commodities or commodities subject to mandatory inspection by governments. The Food and Drugs Administration (FDA) in USA is one such organization. A non-governmental inspection body is a private body that enjoys the same legal status as notary organizations and undertakes commodity inspections internationally. The most notable bodies of this type are Societe Generable De Surveillance S. A. (SGS) in Geneva Swiss , Lloyds Surveyors Britain and Underwriters Laboratories (UL) in the USA.

In China, it is SACI (the State Administration for Commodity Inspection) and its designated provincial divisions which are in charge of the inspection of import and export commodities throughout the country. Inspections are carried out in accordance with the Law of the People's Republic of China on Import and Export Commodity Inspection (hereafter called the Law) and the Regulations for the Implementation of the Law of the People's Republic of China on Import and Export Inspection (hereafter called the Regulations)

Task of Inspection Body

The basic task of the inspection authorities is to inspect goods for import and export. Import and export commodity inspections in China fall into two categories : statutory inspection and non-statutory inspection. Statutory inspection refers to the mandatory inspection carried out by the governmental bodies imports and exports. While statutory inspection a must、non-statutory inspection, on the other hand , is usually made on a random basis. SACI , in the light of the needs in developing foreign trade , makes, adjusts and publishes a List of Import and Export Commodities Subject to Inspection by the Commodity Inspection Authorities. As an exporter, one should check the list of commodities in advance so as to ensure that he won't skip his obligation concerning commodity inspection.

Aside from inspecting the commodity for exports and imports, a second task falls upon SACI is to exercise supervision and control over the Inspection conducted by the inspection agencies and personnel designated or accredited by the consignees, consignors manufacturers, trading units, storage and transport units the import and exports commodities.

Finally, SACI and designated or approved inspection agencies may accept the entrustment of the trading or foreign inspection bodies for surveying services of import and export commodities within the specified scope and may issue relevant certificates of survey.

8. 1. 4   Inspection Certificate

After the commodities are inspected and proved up to the standard (s) recognized by the Inspection authorities, the inspection authorities then will issue inspection statements. Inspection statements issued by the commodity inspection authorities after the commodities have undergone inspection is called inspection certificates. The inspection certificates usually issued by the Chinese inspection authorities are as follows:

- Inspection Certificate of Quality. This certifies that at the quality and specifications of import and export commodities are in conformity with the contract stipulation.

- Inspection Certificate of Weight or Quantity. This certifies that the weight or quantity of import and export commodities is in conformity with the contract stipulation.

- Inspection Certificate of Value. This certifies that the price of import and export commodities in the invoice is a true reflection of the value of the goods transacted.

- Inspection Certificate of Origin. This certifies that the name of the place is where the export commodity has been produced or manufactured.

- Sanitary Inspection Certificate. This certifies that foods and animal products for eating are up to standards for export.

- Veterinary Inspection Certificate. This certifies that the animal to be exported in a condition good enough for export.

- Disinfection Inspection Certificate. This certifies that animal products to be exported have been disinfected.

- Inspection Certificate on Damaged Cargo. This certifies the degree to which the goods imported have been damaged and the causes Of the damage.

8.1.5　Inspection Clause in the Sales Contract

Inspection clauses in international trade contracts usually contain stipulations on the inspection right, the time and place of inspection or re-inspection, the inspection body, the inspection items and the inspection certificates. An example of inspection clause in the trade contract is as follows (see the example of in Figure 8.1):

"It is mutually agreed that the certificate of quality and quantity/Weight issued by the Manufacturer shall be part of the documents for payment under the relevant L/C. In case the quality, quantity or weight of the goods are found not conformity with those stipulated this contract after re-inspection by the China exit and entry inspection and quarantine bureau within...days after discharge of the goods at the port/place of destination, the buyers shall return the goods to or lodge a claim against the sellers for compensation of losses upon the strength of inspection certificate issued by the said bureau, with the exception of those claims for which the insurers or the carriers are liable. All expenses including inspection (s) and losses arising from the return of the goods or claims should be borne by the sellers. In such case, the buyers may, if so requested, send a sample of the goods in question to the sellers, provided that the sampling is feasible. "

Figure 8.1　Example of Inspection Clause in the Trade Contract

8. 2　Disputes and Claims

As has been shown in the first chapter , the last stage in the international trade procedure is settlement of disputes. Although it is not necessarily a stage for every trade transaction , as disputes from both the seller or the buyer are sometimes unavoidable , it is important as well to deal with it here.

8. 2. 1　Disputes

In international trade, disputes arise from time to time for various reasons, for instance, a buyer may breach a contract by refusing to accept goods with improper reasons or failing to pay for the goods when payment is due; a seller may violate a contract by failing to make an agreed delivery or delivering goods that do not conform to the contract. Breach of contract means the refusal or failure by one party of a contract to fulfill an obligation imposed on him under that contract. In international trade practice , the main reasons for disputes can be concluded into three categories : breach of contract by the seller , breach of contract by the buyer and breach of contract by both the seller and the buyer.

Breach of Contract by the Seller

A seller may be considered to have breached a contract if he

● fails to make delivery of the goods within the shipment time stipulated in the contract

● fails to deliver the goods

● fails to deliver the goods that are in conformity with the contract or the L/C in respect of quality, specifications, quantity and packing, etc. and

● fails to present the shipping documents within the stipulated time period or the documents presented are incomplete and inadequate

Breach of Contract by the Buyer

Considering the different obligations undertaken by the buyer , a buyer may breach a contract if he

● fails to the relevant L/C within the stipulated period under an L/C payment

● fails to accept the goods without sufficient reasons under an L/C payment, and

● fails to dispatch the vessel or notify the seller with sufficient information about the vessel according to the stipulations in the contract under FOB terms

Breach of Contract by Both Parties

Breach of contract by both parties is not rare to see in both export and import trade. For example, both parties may be held responsible due to misunderstanding or miscomprehension of a contract that is nor clearly stipulated. Different interpretation or understanding of the terms may arise if vague expressions like "immediate shipment", "quantity about 300M/T", "destination : China ports " etc. are used in the contract.

As the sales contract has a legal binding force upon both contracting parties , any party who has violated the contract shall be legally held responsible for the breach , and the injured party is entitled to remedies according to the stipulations of the contract or the relevant laws.

8. 2. 2　Claims

When disputes arise , the injured party usually will lodge a claim against the party concerned for compensation. Thus , claim can be defined as a demand made by one party upon another for a certain amount of payment on account of a loss sustained through its negligence , there are generally three types of claim related to international trade practice : claim regarding selling and buying , claim regarding transportation and claim concerning insurance . In this section , only the first type of claim will be covered as this is the only type of claim that is made between the two parties of the sales contract.

In import and export trade, claims regarding quality and quantity or weight are common , even though proper inspection of the goods has been conducted by designated surveyors or public inspection bureaus. The goods may have been damaged or lost during transit. Therefore , it is necessary to include a discrepancy and claim clause in a contract. In case the goods delivered are inconsistent with the contract stipulations , the buyer should make a claim against the seller within the time limit of re-inspection under the support of an inspection certificate or survey report.

8. 2. 3　Claim Clause in the Sales Contract

As is well known , disputes are very common in international trade and are detrimental to the business relationship between the buyer and the seller. To avoid or to properly handle future disputes, it is necessary to include a claim clause in a contract , normally , there are two ways to stipulate claim clause in the contract: discrepancy and claim clause and penalty clause.

Discrepancy and Claim Clause

In most trade contracts, it is only the discrepancy and claim clause that is stipulated. The clause in this respect normally stipulates the relevant evidences or proofs to be presented and the relevant authoritative body for issuing the certificate. The evidences or proofs provided should be complete and clear, and the authority should be in competent in issuing the relevant certificates. Otherwise , claims can be refused by the other party.

Besides the evidence to be presented , this clause shall also include a period within which a claim is lodged. Technically , the period for claim refers to the effective period in which the claimant can make a claim against a party for breach. Claims made after the agreed period can be refused by the party in breach. Generally speaking, a period that is too long may put the seller under a heavy responsibility and a period that is too short may make it impossible for the buyer to file a claim. Therefore , the period for a claim should be made as specific possible in the clause (See the example in Figure 8. 2) .

Example

Claim should be filed within × × × days

- after arrival of the goods at the port of destination
- after the discharging of the goods at the port of destination
- after the arrival of the goods at the business place of the buyer , or
- after inspection

Figure 8. 2   Example of the Way in Stipulating Period for a Claim

If a claim is justified, prompt and well – supported, it can be settled in the following ways: making are fund, compensating for direct losses or expenses, selling the goods at lower prices or replacing the defective goods with good ones. As the discrepancy and claim clause is closely related to the inspection clause, they are combined as the inspection claim clause in some contracts (see Figure 8. 3).

Example

"Claim: any claim by the buyers regarding the goods shipped shall be filed within 30 days after arrival of the goods at the port of the destination specified in the relative bill of lading and supported by a survey report issued by approved by the sellers. Claims in respect of matters within responsibility of insurance company, shipping company/other transportation organization will not be considered or entertained by the Sellers. "

Figure 8. 3   Example of Discrepancy and Claim Clause in the Sales Contract

Penalty Clause

For transactions where goods in substantial quantity or large mechanical equipment is concerned, a penalty clause will be stipulated in the contract together with a discrepancy and claim clause in case one party fails to implement the contract such as non-delivery, delayed delivery, delayed opening of L/C, etc. Under this clause , the party who has failed to out the contract must pay another party a fine, a certain percentage of total contract value. The provisions for the amount and the penalty ceiling should also be included in the clause (see Figure 8. 4).

Example

"Should the sellers fail to make delivery on time as stipulated in the contract; the buyers shall agree to postpone the delivery on condition that the sellers agree to pay a penalty which shall be deducted by the paying bank from the payment under negotiation, or by the buyers direct at the time of payment. The rate of penalty is charged at 0. 5% of the total value of the goods whose delivery has been delayed for every seven days, odd days less than seven days should be counted as s even days. But the to amount of penalty, however, shall not exceed 5% of the total value of the goods involved the late delivery in case the sellers fail to make delivery ten weeks later than the time of shipment stipulated in the contract, the buyers shall have the right to cancel the contract and the sellers, in spite of the cancellation, shall still pay the aforesaid penalty to the buyers without delay. "

Figure 8. 4   Example of penalty Clauses in the Sales Contract

8. 3　Force Majeure

After the conclusion of a contract, some events beyond the parties' control may occasionally take place, which makes it impossible for the parties concerned to fulfill the contract. In order to safeguard their interests, the parties of the contract usually stipulate a force majeure clause in the sales contract.

8. 3. 1　Conditions for Force Majeure

Force Majeure, also named Act of God, is a French word meaning "superior force", It refers to an event that can neither be anticipated nor be reduced to control. Force Majeure clause is a term in a sales contract , but at the same time , also remains an independent law itself. However , this law can only be observed when the following conditions are satisfied. First, the event occurred must be impossible to anticipate or beyond the control of both parties concerned ; second , the event must occur after the stipulation of the sales contract ; and third , the event occurred must not be caused by the negligence of either party of the contract. Should any such events occur , the stipulation of Force Majeure clause in the sales contract enables a seller to avoid his contractual obligations without paying a compensation or penalty.

8. 3. 2　Force Majeure Clause in the Sales Contract

A force majeure clause in the sales contract should include the scope of force majeure events , time limit of notice to the other contractor , certificates and the agencies who issue them , and the relevant consequences.

Scope of Force , Majeure Events

Since there is still no definite agreement as to which events should be regarded as force majeure, the seller and the buyer usually stipulate in their contract the scope of force majeure events. Force majeure events include certain natural disasters such as fire , flood , storm , heavy snow , earthquake and social disturbances like war , strike and sanctions. However , with a view to clarifying what is covered by the force majeure clause under a particular contract , different methods of stipulation have been adopted.

- Stipulation Made in a General Way

This means the force majeure events in the contract are not made specific but stated in a more general way by simply adopting the wording "force majeure events" (See the example in Figure 8. 5) .

As this way of stipulation doesn't list the specific force majeure events , it is ambiguous and is likely to give rise to different interpretations from the perspectives of different parties, it is not suggested that this be adopted in practice.

Example

"If the shipment of the contracted goods is prevented or delayed in whole or in part due to Force Majeure events, the seller shall not be liable for non-shipment or late shipment of the goods of this contract. However, the seller shall notify, the buyer by cable or telex and furnish the latter within 15 days by registered airmail with a certificate issued by the China Council for the Promotion of International Trade attesting to such event or events. "

Figure 8. 5   Example of Force Majeure Clause Made in a General Way

- Stipulation Made in a Specific Way

This way of stipulation lists the possible force majeure events in detail in the force majeure clause in the contract (See the example in Figure 8. 6) .

Example

"If the shipment of the contracted goods is prevented or delayed in whole or in part by reason of war, earthquake, flood, fire, storm, heavy snow, the seller shall not be liable for non-shipment of the goods of this contract. However , the seller shall notify the buyer by cable or telex and furnish the latter within 15 days by registered airmail with a certificate issued by the China Council for the Promotion of International Trade attesting to such events or events. "

Figure 8. 6   Example of Force Majeure Clause Made in a Specific Way

Although this way of stipulation appears to be specific and clear in-terms of force majeure events, the possibilities of leaving out details makes it far from satisfactory. (see the example in Figure 8. 7) .

Example 1.

"If the shipment of the contracted goods is prevented or delayed in whole or in part by reason of war, earthquake, flood, fire, storm, heavy snow or other cause of Force Majeure, the seller shall not be liable for non-shipment of the goods of this contract. However, the seller shall notify the buyer by cable or telex and furnish the latter within 15 days by registered airmail with a certificate issued by the China Council for the Promotion of International Trade attesting such event or events. "

Figure 8. 7   Examples of Force Majeure Clause Made in a Synthesized Way

- Stipulation Made in a Synthesized Way

Neither way being satisfactory, a synthesized way in stipulating force majeure clause in the contract is created (See the example in Figure 8. 8) .

Example 2.

"The seller shall not be held responsible for failure or delay to perform all or part of this contract due to war, earthquake, flood, fire, storm, heavy snow or other cause of force majeure. However, the seller shall immediately advise the buyer of such occurrence, and within 15 days thereafter, shall send by registered airmail to the buyer for their acceptance a certificate issued by the competent government authorities of the place where the accident occurs as evidence thereof, under such circumstance, the seller however, is still under the obligation to take necessary measures to hasten the delivery of the goods. In case the accident lasts for more than 3 weeks, the buyer shall have the right to cancel contract. "

Figure 8. 8   Examples of Force Majeure Clause Made in a Synthesized Way

This way of stipulation is specific and flexible. It avoids the problems in the first two ways of stipulation, and therefore, is widely adopted in Chinese international trade practice.

8.3.3 Consequences of Force Majeure Events

There are usually two consequences resulting from force majeure events: postponement of the contract and/or termination of the contract. Whether terminating the contract or postponing the performance of the contract depends on the detailed stipulations in the contract or on the degree to which the force majeure event has affected the performance of the contract.

If the performance of a contract is delayed by a force majeure ease on a temporary basis or for a short time, the contract may be postponed. After the force majeure accident finishes, the contract should resume. For example, if there is a delayed shipment because of industrial strike, the contract will be postponed. But when the strike is over, the contract should continue and the seller has to go on with the shipment.

However, generally speaking, the contract can be avoided if the occurrence of the force majeure event makes it impossible to perform the contract. For example, if goods ready for shipment under the contract is damaged so seriously in an earthquake that it is impossible for the seller to recover within a short period of time. It this case, the seller can ask for the termination of the contract.

8.3.4 Notification

As mentioned above, by stipulating a force majeure clause in a contract, a seller is able to avoid his contractual obligations without paying a compensation or penalty. However, this is possible only when he can promptly notify the buyer of the force majeure event and its effect on their ability of performing the contract right after the accident so that the latter is able to take necessary remedial measures. Otherwise, the seller will still be held responsible for the loss or extended loss caused.

Issuance of the Certificates

A force majeure event should be verified by a certificate that attests the occurrence of such an event, in other words , the party that has failed to perform the contract should provide effective documents describing the frustrating events. Thus , the issuer of the certificate should also be mentioned in the clause (See the example in Figure 8.9) .

Example

"Force Majeure: If the shipment of the contracted goods is prevented or delayed in whole or in part by reason of war, earthquake, flood, fire, storm, heavy snow or other causes of Force Majeure. the seller shall not be liable for non-shipment or late shipment of the goods or non-performance of this contract. However, the seller shall notify the buyer by cable or telex and furnish the latter by registered airmail with a certificate issued by the China Council for the promotion International Trade attesting to such event or events. "

Figure 8.9　Example of Force Majeure Clause in a Sales Contract

If the party who failed to perform the contract fails to provide such effective certificates or the facts identified are not in conformity with his description, the liability of his failure to perform the contract will not be exempted or exempted totally.

8. 4　Arbitration

In Section 2 of this chapter, we have discussed disputes and claims. If one party found to have breached the contract , the other part of the contract who suffers the damage or loss therefrom will lodge a claim against the former for refund or compensation. However, in most cases the claim is not able to be settled smoothly as expected due to various reasons. Then what are the usual ways to settle disputes in international trade ?

8. 4. 1　Meaning and Characteristics of Arbitration

In international trade practice, when disputes arise between exporters and importers, it can be settled through negotiation, mediation, arbitration or litigation.

Friendly negotiation is the best method of all and beneficial to both parties because the people concerned are most familiar with the problem and dealing with it personally can avoid damage to their friendship.

If the parties involved are not able to reach any agreement between themselves, they can then seek the help of a third party for mediation. This third party can any impartial party or international chamber of commerce or a government trade authority. As these experts are skilled and experienced mediators, having a competent third party to speak as an intermediary is often more effective than speaking for oneself. A large portion of the disputes in international trade practice is settled this way because it helps maintain friendly business relations between exporters and importers. However, mediation only facilitates negotiation, no award or opinion on the merits of the disputes is given.

If the disputes cannot settled through amicable , negotiation or mediation , arbitration is the next best alternative. Arbitration differs from mediation in that the third party to whom the dispute submitted decides the outcome. Arbitration is often provided for in a contract, parties who have not so provided can choose have their dispute arbitrated after it has arisen.

In case the three alternatives are not workable for the settlement, the parties involved can resort to the last one-litigation. However, litigation is usually costly and time-consuming , and what's more, detrimental to the relationship between the parties concerned. Therefore, it is not suggested to if other ways to settle the disputes are workable.

As arbitration is simpler, less costly, less time-consuming and more flexible in international trade practice, in case no settlement can be reached through amicable negotiations, the two parties would prefer to submit the dispute to arbitration rather than take it to the court for litigation. In international trade practice, an arbitration agreement be made by the parties concerned before they decide to submit it to arbitration.

8.4.2　Arbitration Procedure

Internationally, the main arbitration body is the International Court of Arbitration. In China, the main arbitration body is the China International Economic and Trade Arbitration Commission (CIETAC). Pursuant to the current arbitration rules of CIETAC which took effect on May 1st, 2005, the general arbitration procedure is: applying for arbitration, forming arbitration tribunal, hearing of an arbitration case, issuing an award, setting aside an award and enforcing an award.

8.4.3　Arbitration Clause in the Sales Contract

Different from other clauses in the sales contract, an arbitration clause shall be regarded as existing independently and separately from the other clauses of the contract. Similarly, an arbitration agreement attached to a contract shall be treated a part of the contract, but existing independently and separately from the other parts of the contract. An arbitration clause or agreement expresses the willingness of the parties to submit the disputes for arbitration.

An eligible, effective, complete and accurate arbitration agreement or arbitration clause should have the following elements, place of arbitration, arbitration body, applicable arbitration rules, arbitration award, etc. Figure 8.10 is an example of arbitration clause in a sales contract.

Example

"Any disputes arising from or in connection with this contract, shall be settled amicably through negotiation. In case no settlement can be reached through negotiation, the case shall then be submitted to the China International Economic and Trade Arbitration Commission of the China Council for the Promotion of International Trade, Beijing for arbitration which shall be conducted in accordance with the Commission's Arbitration Rules in effect at the time of applying for arbitration. The arbitral award is final and binding upon both parties."

Figure 8.10　Example of Arbitration Clause in a Sales Contract

Chapter 9

International Trade for the
Implementation of the Contract

★LEARNING OBJECTIVES

- Be familiar with the performance of export and import contract.
- Grasp the methods of auditing L/C, grasp the marks of amending L/C.
- Be familiar with main documents for settlement.

For most nations exports and imports are the most important international activities. Each country has to import the articles and commodities it does not itself produce, and it has to earn foreign exchanges to pay for them. It does this by exporting its own manufactured articles and surplus raw materials. Thus the import and export trades are two sides of the same coin, and both can have beneficial effects on the home market. Imports create competition for home-produced goods; exporting gives a manufacturer a larger market for his products, so helping to reduce the unit cost. In each case the effect is to keep prices in the home market down.

But there may be factors that compel governments to place restrictions on foreign trade. Imports may be controlled or subjected to a custom duty to protect a home industry, or because the available foreign exchange has to be channeled into buying more essential goods and exports, too, may be restricted, to conserve a particular raw material required by a developing home industry.

These factors mean that importing and exporting are subject to a lot of formalities, such as customs entry and exchange control approval, from which the home retail and wholesale trades are free. They also mean that the procedure of foreign trade is much more complicated than that of domestic trade, the latter involves specialized knowledge and highly trained personnel.

All or most of the following organizations are involved in an export and import transaction:

- The exporters
- The shipping agents at the port or airport of loading

- The railways (in some cases) in the exporters' country
- The road hauler (in some cases) in the exporters' country
- The port authority
- The shipping company (for sea freight)
- The airline (for air freight)
- The insurance company or brokers
- The exporters' bank
- The importers' bank
- The railways (in some cases) in the importers' country
- The road hauler (in some cases) in the importers' country
- The shipping agent at the port or airport of discharge
- The importers

Many specialists may be involved in export and import transaction, including:

- A shipping agent and/or freight forwarder (forwarding agent) will take responsibility for
- the documentation and arrange for the goods to be shipped by air, sea, rail or road. These services may be carried out by the supplier's own export department, if they have the expertise.
- Airlines, shipping lines, railway companies or haulage contractors will actually transport the goods.
- Both the importer's and exporter's banks will be involved in arranging payments if a letter of credit or bill of exchange is used.
- Customs and Excise officers may need to examine the goods, check import or export licences and charge duty and/or VAT.
- A Chamber of Commerce may need to issue a certificate of origin, if this is required by the importer's country.
- An insurance company insures goods in transit.
- A lawyer if a special contract has to be drawn up.

Many import or export deals are arranged through an exporter's agent or distributor abroad, in this case the importer buys from a company importing goods in his own country. Alternatively, the deal may be arranged through an importer's buying agents or a buying house acting for the importer, or through an export house based in the exporter's country. In these cases the exporter sells, directly to a company in his own country, the latter will then export the goods.

An export or import business is so complicated that it may take quite a long time to conclude a transaction. Varied and complicated procedures have to be gone through in the course of an export or import transaction. From the very beginning to the end of the transaction, the whole operation generally undergoes four stages: preparing for exporting or importing, business negotiation, implementing the contract, and settlement of disputes (if any). Each stage covers some specific steps. Since the export and import trades are two sides of the same coin, and one country's export is another country's import, hence, we will take the procedures of export transaction in the following diagram to illustrate the general procedures of export and import transaction. Before proceeding to the following units, it is advised that this general picture be kept in mind.

This unit tries to present a general picture and a brief introduction to export and import trades for the purpose of clarifying their complicated procedures.

9.1   General Procedures of Export Transaction

Different countries have different economic policies or systems. So before doing business with foreign countries, one has to understand the whole procedures of exporting. Take the contract on basis of CIF and payment by L/C as an example, you can see Figure 9.1 General Procedures of Export Transaction.

9.1.1   Preparation for Exporting

The most difficult part of exporting is taking the first step. Any exporter who wants to sell his products in a foreign country or countries must first conduct a lot of market research. Market research is a process of conducting research into a specified market for a particular product. Export market research, in particular, is a study of a given market abroad to determine the needs of that market and the methods by which the products can be supplied. The exporter needs to know which foreign companies are likely to use his products or might be interested in marketing and distributing the products in their country. He must think whether there is a potential for making a profit. He must examine the market structures and general economic conditions in those places. If the economy is in a recession, the demand for all products is usually decreased. So the exporter's products might not sell well at such times. Market research mainly covers:

Research on the Countries or Regions

Countries or regions with different political and economic systems hold quite different attitude toward foreign trade business. The exporter should investigate their political, financial and economic conditions; their policies, laws and regulations governing foreign trade, foreign exchange control, customs tariffs and commercial practices; their foreign trade situation (the structure, quantity, volume of exporting and importing commodities, trading partners and trade restrictions, etc.). All of these factors exert a great impact upon the choice of a new market.

Research on the Market

A research should also be conducted about the production, consumption, price and its trend, the major importing or exporting countries of a particular commodity in order to fix the right price of exporting commodities and properly handle other business terms.

Research on the Customer

In international trade, credit information is of greater importance than in home trade. The exporter should know what kind of reputation the buyer or importer has, the approximate size of his

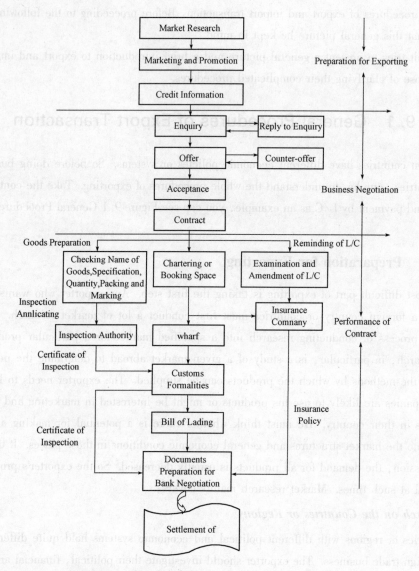

Figure 9. 1　General Procedures of Export Transaction

business, how he pays his accounts and information about his trade activities. Obviously, customers with sound reputation and good financial standing will facilitate the export trade. The exporter can facilitate the export trade. The exporter can obtain this information from various sources, such as references given by the buyer, his bank, various trade associations and enquire agencies. In this way, the potential customers can be identified.

　　In addition to conducting market research to collect information or data from external sources, the exporter can also take the initiative in marketing and promoting his products in the overseas market. The frequently adopted strategies are sales literature, point of sale advertising, packaging, sponsorship, showrooms, trade fair and exhibition, publicity, public relations, etc.

9.1.2   Business Negotiation

If a foreign company is interested in buying the exporter's products, negotiation should be organized. Business negotiation plays a very important role in the conclusion and implementation of a sales contract. It has a great bearing on the economic interest of the parties concerned.

No matter what way the negotiations are held, in general, they consist of the following links: enquiry, offer, counter-offer, acceptance and conclusion of sales contract. Among which, offer and acceptance are two indispensable links for reaching an agreement and concluding a contract.

Enquiry

An enquiry is a request for business information, such as price lists, catalogues, samples, and details about the goods or trade terms. It can be made by either the importer or the exporter. On receiving the enquiry, it is a regular practice that the exporter should reply to it without delay to start the negotiation.

Offer and Counter-offer

An offer is a proposal made by sellers to buyers in order to enter into a contract. In other words, it refers to trading terms put forward by offerors to offerees, on which the offerors are willing to conclude business with the offerees. There are two kinds of offers, one is firm offer, the other, non-firm offer. A reply to an offer which purports to be an acceptance but contains additions, limitations or other modifications is a rejection of the offer and constitutes a counter-offer.

Acceptance

Acceptance is a statement made by or other conduct of the offeree indicating unconditional assent to an offer. A contract is concluded once the offer is accepted.

Conclusion of sales Contract

As soon as an offer is accepted, a written sales contract or sales confirmation is usually required to be signed between the buyer and the seller to confirm the sale and stipulate their rights and obligations respectively. A sales contract or sales confirmation contains some general terms and conditions as well as the specific terms that vary with the commodity. But such terms as the names of seller and buyer, the description of the goods, quality and specification, quantity, packing, unit price, amount, payment, date of delivery, shipping, insurance, inspection, claim and arbitration are indispensable. The sales contract or sales confirmation is normally made out in two originals, one for buyer himself and the other for his seller.

9.1.3   Implementation of Contract

Under CIF contract with terms of payment by L/C, the implementation of export contract usually goes through the steps of goods preparation, inspection application, reminding of L/C, examination and modification of L/C, chartering and booking shipping space, shipment, insurance, documents preparation for bank negotiation and the settlement of claims, etc.

Preparing Goods for Shipment

After a contract is made, it is the main task for the exporter to prepare the goods for shipment and check them against the terms stipulated in the contract. The quality, specification, quantity, marking and the packing should be in line with the contract or the L/C, the date for the preparation should agree with the shipping schedule.

Inspection Application

If required by the stipulations of the states or the contract, the exporter should obtain a certificate of inspection from the institutions concerned where the goods are inspected. Usually, the commodity will be released only after the issuance of the inspection certificate by the inspection organization.

Reminding, Examining and Modifying, L/C

In international trade, a banker's letter of credit, is commonly used for the payment of the purchase price. In the course of the performance of contract, one of the necessary steps for the seller is to urge the buyer to establish L/C. According to the contract, the buyer should establish the L/C on time, but sometimes he may delay it for various reasons. For the safe collection of payment, the seller has to urge the buyer to expedite the opening of L/C. Upon receipt of a letter of credit, the seller must examine it very carefully to make sure that all terms and conditions are stipulated in accordance with the contract. If any discrepancies exist, the seller should contact the buyer immediately for necessary amendments so as to guarantee the smooth execution of the contract.

(I) L/C Reminding

If the payment is effected by L/C, the exporter is not to make the shipment until after he has received the L/C. Sometimes it is necessary to remind the buyer of the L/C. The buyer should also open the L/C ahead of the schedule. Because of financial difficulties or other reasons the buyer might not be able to open the L/C in time. The seller might feel it necessary to give him some time as allowance, and meanwhile retaining the right of claim against the buyer.

(II) Examine the L /C

Having received an L/C, the beneficiary should examine the L/C to see if it fully corresponds with the sales contract and the Uniform Customs and Practice for Documentary Credit. If any discrepancy happens, the seller should ask the importer to require the opening bank to make corrections. The seller cannot make the shipment until he has made an effective, complete, and acceptable L/C. The examination can be overall examination or only of the specified clause.

- Make Overall Examination

An L/C should be opened by a bank of a country or an area which keeps economic relations with our country. This principle is applicable even if the L/C has been opened by a branch bank of such nations or areas. The L/C must be in conformity with the trade agreements between the two trading nations, and finally the L/C should not carry deprecatory clauses or clauses which contain political bias.

The notifying bank should examine the seal or the test of the L/C to see if it is a false one, if the L/C has been delivered to the seller directly, he should present it to the notifying bank for such an examination. If the notifying bank doubts the credit status of the opening bank, certain measures should be adopted accordingly, such as asking the applicant to have the L/C confirmed by a bank of good reputation, having the paying bank to confirm the payment and add a payment insurance clause to the L/C, demand partial shipment and partial settlement if the consignment is of great quantity, and add a cable reimbursement clause to the L/C.

The beneficiary should also examine if the clauses on the L/C are self-contradictory, such as in the case when through B/L is required for negotiation but transhipment is not allowed; the price term is CFR, but insurance policy is required for negotiation, etc.

Finally, the beneficiary should check if there are any mis-spelt words, if he can fulfill all requirements (such as in the case when the L/C requires particular vessel for shipment) or if he can get all the documents ready within a given time, etc.

- Examine Individual Clauses

By this we mean to examine the individual clauses of the L/C one by one. The beneficiary of the L/C should examine the L/C to see the relevant clauses covering the name, specifications, quantity, packing, unit price, time for shipment, loading and unloading port, etc. correspond with those as listed in the sales contract. Particular attention should be devoted to the following:

(1) Expiry date of the L/C

Take "Expiry Date: June 15, 2016 in China for negotiation" for example, it gives the last date by which negotiation at a bank in the exporter's country should be done. Unless there are more secure means to avoid relevant risks, it is advisable for the beneficiary not to accept other clauses.

(2) The amount to be given

The amount given in the L/C should correspond with that as given in the sales contract, and in the same currency. If a "more or less" clause is given for the quantity of the goods in the sales contract, the L/C should bear similar quantity clause and the amount of the L/C should be flexible accordingly or the amount of the draft and the invoice should not be any more than that given in the sales contract.

(3) Time of loading, time of validity, and time for surrender of the shipping documents

The time of loading should be the same as that given in the sales contract. But it is acceptable if the time of loading given in the L/C is longer than given in the sales contract. The seller should ask for the extension of the time of loading if he cannot make the shipment in time because of various reasons, such as the late arrival of the L/C, the delay in the preparation of the goods, etc.

The L/C should have a time of validity, which is generally 10 to 15 days after the time of shipment so that the seller might have sufficient time to prepare the required shipping documents for payment settlement.

The time of surrendering the shipping documents is sometimes given in the L/C. This is given in the case when the time of validity of the L/C is rather long and the buyer wants to get the required shipping documents in time to take the delivery of the goods. It is usually given that the "surrender of the shipping documents should be made xxx days after the date of the bill of lading" . If no such time is given, the shipping documents should be made within 21 days after the shipment has been made. If the time for the surrender of the shipping documents is too short, such as two days, it is, usually unacceptable.

(4) Transshipment and partial shipment

If the sales contract states that transshipment and partial shipment are allowed, while the L/C does not bear such a clause, it should be taken that transshipment and partial shipment are allowed.

(5) Applicant and beneficiary

The exporter must make sure that the name and the address of the applicant and the beneficiary are correctly given and spelt in the L/C. Sometimes deliverer of the consignment is someone else other than the exporter, e. g. the branch agency or some other manufacturer, then the L/C should be transferable, in this way to avoid the trouble of different beneficiary and deliverer.

(6) Examination of terms

The exporter should also examine other terms item by item and think if they correspond with the sales contract or pose unnessary barriers to the deal, or if they conflict with each other.

(7) Time of payment

The time of payment must correspond with that given in the sales contract. If the L/C is a usance L/C and the sales contract states that the buyer should bear the interest thereupon, the L/C must bear similar clauses.

(Ⅲ) Amendment of the L/C

Once the exporter has found some flaws with the L/C, he should think whether it is necessary to make the amendment. If the flaw poses great barrier to the settlement of payment and hinders the performance of the sales contract, the exporter should ask for the importer to require the opening bank to make amendment in time. If the flaw is of minor significance and the deal can be performed smoothly accordingly, perhaps it will not be necessary to make the amendment, so as to save time and expenses.

If more than one clause should be amended, the exporter should ask for the amendment at once. The importer is obliged to accept or reject all of them, he cannot accept the amendment of one clause while rejecting the other (or others) . Once the amendment is made, it becomes part of the L/C and the deal is to be executed accordingly.

Chartering and Booking Shipping Space

After receiving the relevant L/C, the exporter should contact the ship's agents or the shipping company for the chartering and the booking of shipping space and prepare for the shipment in accordance with the importer's shipping instruction. Chartering is required for goods of large quantity which needs full shipload; and for goods in small quantities, space booking would be enough.

Customs Formalities

Before the goods are loaded, certain procedures in customs formalities have to be completed. As required, completed forms giving particulars of the goods exported together with the copy of the contract of sale, invoice, packing list, weight. Memo, commodity inspection certificate and other relevant documents, have to be lodged with the Customs. After the goods are on board, the shipping company or the ship's agent will issue a bill of lading which is a receipt evidencing the loading of the goods on board the ship.

Insurance

The export trade is subject to many risks. For example, ships may sink or consignments may be damaged in transit, exchange rates may alter, buyers default or governments suddenly impose an embargo, etc. It is customary to insure goods sold for export against the perils of the journey. The cover paid for will vary according to the type of goods and the circumstances. If the exporter has bought insurance for the goods, he will be reimbursed for the losses.

Documents Preparation for Bank Negotiation

After the shipment, all kinds of documents required by the L/C shall be prepared by the exporter and the importer and presented, within the validity of the L/C to the bank for negotiation. As to the shipping documents, they include commercial invoice, bill of lading, insurance policy, packing list, weight memo, certificate of inspection, and, in some cases, consular invoice, certificate of origin, etc. Documents should be correct, complete, concise and clean. Only after the documents are checked to be fully in conformity with the L/C, the opening bank makes the payment. Payment shall be disregarded by the bank for any discrepancies in the documents.

9. 1. 4　Settlement of Disputes

Sometimes complaints or claim inevitably arise in spite of the careful performance of a contract by the exporter and importer. They are likely to be caused by various reasons such as more or less quantity delivered, wrong goods delivered, poor packing, inferior quality, discrepancy between the samples and the goods which actually arrived, delay in shipment, etc. In accordance with specific conditions, complaints and claims may be made to the exporter, importer, insurance company or shipping company. Once disputes arise, it is advised that arbitration is better than litigation.

9. 2　General Procedures of Import Transaction

So far we have studied the general procedures of export transaction and dealt with different stages and steps, from the point of view of an exporter. Having been familiar with the process of the export business, we find it much easier to understand how an importer handles his import business. After all, the export and import trades are two sides of the same coin. When handling an import trade, the trade conditions and terms you are striving for are sometimes just the opposite to those you do in an export trade. The terms of delivery remain the same meaning regardless of whether you work as an importer or an exporter. A bill of lading is a bill of lading no matter who uses it for some practical purposes. The knowledge we have acquired from the previous sections is also applicable to import procedures. With the fundamental knowledge of export procedures we can grasp the essential points of import procedures easily and manage import trade well and smoothly.

The General procedures of import transaction can be summarized as follows:

- to conduct market investigation;
- to formulate import plan for a certain commodity;
- to send inquiries to the prospective sellers overseas;
- to compare and analyze the offers or quotations received;
- to make counter-offers and decide on which offer is most beneficial;
- to sign a purchase contract;
- to apply to a bank for opening a letter of credit;
- to book shipping space or charter a carrying vessel for taking over the cargoes, if the contract is in terms of FOB;
- to effect insurance with the insurance company upon receipt of shipping advice;
- to apply for inspection if necessary;
- to attend to customs formalities to clear the goods through the customs;
- to entrust forwarding agents with all the transport arrangements from the port to the end—user's warehouse;
- to settle disputes (if any).

Figure 9. 2 shows Procedures of Import Transaction.

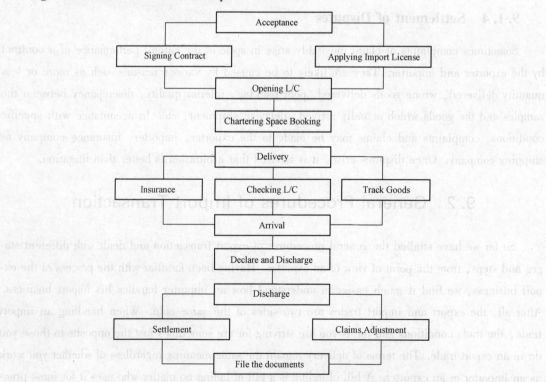

Figure 9. 2　Procedures of Import Transaction

9.3   Documents Needed in Export and Import Transaction

International trade attaches so great importance to shipping documents that, to a certain degree, it can be called trade of documents, or "symbolic" trade. This is because shipping documents represent the title to the goods. For example, under letter of credit, the buyer cannot take the delivery of the goods until he obtains the shipping documents; on the other hand, only if the seller releases the shipping documents can he receive the payment. What documents to be used and how to carefully and accurately complete them deserve our adequate attention. As a rule, every contract of sale stipulates the kinds of shipping documents required. Any slightest negligence in these documents might result in serious problems, which is not infrequent in practice. It is, therefore, imperative for both an exporter and an importer to abide by such stipulations. Generally, commercial invoice, bill of lading, insurance policy or certificate, packing list, and weight memo etc. , are called shipping documents. In addition, other documents required by the buyers and related to the matter of duty to be paid on the imported goods, sometimes, are also included in shipping documents, they are the proforma, consular invoice, certificate of origin, certificate of value, certificate of inspection. The commercial invoice, bill of lading and insurance policy constitute the chief shipping documents in international trade. They are indispensable in almost every instance of export and import consignment. This unit mainly deals with commercial invoice, proforma invoice, bill of lading, packing list, weight memo, inspection certificate and insurance policy.

9.3.1   Commercial Invoice

Definition of Commercial Invoice

An invoice is a statement sent by the seller to the buyer giving particulars of the goods being purchased, and showing the sum of money due. Different parties require such statements for different purposes. There are various invoices, such as commercial invoices, banker's invoices, consular invoices, customer invoices and proforma invoices. Among these, the commercial invoice is the most common one and has to be provided for each and every consignment as one of the documents evidencing shipment. When speaking of an invoice covering a certain shipment, we usually refer to commercial invoice. It is a document which contains identifying information about merchandise sold for which payment is to be made. All invoices should show the name and address of the debtor, terms of payment, description of items, the price. In addition, the invoice should show the manner of delivery.

Contents of the Commercial Invoice

Commercial invoices vary in forms. However, no matter what forms they may take, the contents must be in full accordance with the contract. In general, a commercial invoice summarizes contract terms, and declares that shipments have been made on the basis of them. It contains, first of all, the names and addresses of the seller and the buyer; next, a full description of the goods dispatched, including the weights and numbers and marks of all the packages; thirdly, the price per unit and the total cost of the consignment. The commercial invoice will also state the port of shipment and the date, the terms of sale, such as CIF, and the terms of payment, such as by sight draft, perhaps under a letter of credit. Finally, it must be signed by an authorized employee of the seller, and may even quote import or export license numbers.

Sometimes the invoice price is broken down into such things as the cost of materials, the cost of processing and manufacture and the cost of packing and transport. The amount of detail on an invoice depends on the rules of the importing country. Some countries require a more detailed breakdown of the price.

Some foreign governments have special regulations for commercial invoices, such as requiring them to be translated into the local language or requiring the use of metric weights and other measurements. Several customs authorities and other regulatory agencies also insist on complete consistency between the different documents. Thus, the numbers and marks on the commercial and consular invoices, the insurance certificate, and the bill of lading must agree exactly.

Originals of commercial invoices must bear the signature of the seller, who is usually the shipper. The abbreviation "E. & O. E." standing for "Errors and Omissions Excepted" is usually printed at the foot of the invoice form. It means that the shipper is prepared to make correction in case errors and omissions are found.

Functions of Commercial Invoice

The invoice functions mainly as a record of the export transaction for buyers, sellers and customs authorities. Copies of the invoice are used by the exporters, their bank, the paying bank, the receiving agents at the port of discharge, the customs in the exporting country and the importers. The importer needs it to check up whether the goods consigned to him are in compliance with the terms and conditions of the respective contract. The banks need it together with the Bill of Lading and the Insurance Certificate to effect payment. The customs need it to calculate duties, if any. The exporters and importers need it to keep their accounts. In the absence of a draft, the commercial invoice takes its place for drawing money.

Illustrations of an Commercial Invoice

To understand and be able to write an invoice you should think about these points:

- Customer's name.
- (1) The office address; (2) The delivery address.
- The invoice number (for your records).
- The order number.
- The reference number and/or description of each item.
- The quantity.
- The price of each item.

- The total price of items and total of all items.
- The amount of discount allowed and the conditions.
- The method of freight, insurance and cost.
- The delivery address.
- The number of parcels, packages or crates.
- The markings on the parcels, packages or crates.
- Any other points.

To know the format of Commercial Invoice, please refer to the appendix of this text book.

9.3.2   Proforma Invoice

The Definition of Proforma Invoice

Proforma, in Latin, means "for the sake of form". Proforma invoice is a document such as an invoice, issued as a temporary statement, but ultimately to be replaced by a final statement which can only be issued at a later date. Outwardly, except the marked "proforma", it is like an ordinary commercial invoice containing the general particulars, for example, marks, number of goods, descriptions, quantities, quality, price, etc. However, in nature, it is a different form of invoice which treats "hypothetical" sales as though they had actually and contractually taken place. It is not a formal document but document without engagement, which is binding neither on the importer nor the exporter.

Functions of Proforma Invoice

Proforma invoices are required for various reasons. Primarily the importer requires them to comply with the regulations in force in his country. Moreover, the importer can require them in advance for information, or letter of credit purposes.

In many countries, especially in the developing countries of the Third World, foreign trade is under strict control. The governments of such countries usually enforce an Import License System or an Import Quota System. Importers must apply for the necessary Import License or Foreign Exchange and they may not import any goods without the approval of Import License or the allocation of Foreign Exchanges. Often their application has to be supported by an informal invoice, a proforma invoice, issued by the foreign exporter showing the name of commodity, specifications, unit price, etc.

The importer who asks for a proforma invoice is in fact making an enquiry, and the exporter who sends the proforma invoice is actually making an offer. If the exporter wants to make a firm offer, he must mention the term of validity in the covering letter when sending the proforma invoice.

9.3.3   Bill of Lading

Definition of Bill of Lading

The most important shipping document is the Bill of Lading (B/L or Blading). It is a document given by a shipping company, representing both a receipt for the goods shipped and a contract for shipment between the shipping company and the shipper. It is also a document of title to the goods, giving the holder or the assignee the right to possession of the goods.

Contents of Bill of Lading

A Bill of Lading is, firstly, a contract between the shipper and the shipping company; secondly, a receipt for the consignment; and thirdly, a document of title. A Bill of Lading doesn't only contain a full description of the consignment—numbers and weights and marks of packages—but a lot of other information as well. It quotes the name of the shipper and the carrying vessel, the ports of shipment and destination, the freight rate, the name of the consignee (unless the B/L is "to order", like a check), and the date of shipment, which is very important from a contractual point of view.

It may also contain a number of other clauses. Some bills of lading are marked "freight paid", when a shipper is selling CIF or CFR, others may allow transhipment, which means that the cargo may be transferred from one ship to another at some intermediate port. It is often important to a shipper that his bill of lading should be "clean" rather than "dirty", that is, that the shipping company should not have made any qualifications about the quantity or condition of the cargo actually shipped. This is because shipper's letter of credit may insist on clean bills, just as it may insist on "on board" as opposed to "alongside" bladings. Sometimes a mate's receipt is given to the shipper in advance of the B/L, which takes time to issue.

Functions of Bill of Lading

The Bill of Lading has three important functions. It is a receipt for goods signed by the shipping company and given to the shippers. It is also evidence of a contract of carriage between the shipping company and shippers. In addition, it is a document of title because the legal owner of the Bill of Lading is the owner of the goods.

For this reason the Bill of Lading can be used to transfer the goods from one owner to another. When the exporters complete it, they can write the buyer's name in the space, "consignee". This means the consignee is the legal owner of the goods, as named on Bill of Lading. Otherwise the exporters can write "to other" in the consignee space. Underneath "to order" they write, the name and address of the agent. Then the agent in the importing country can endorse the bill to the buyer. In this way the importers can transfer the consignment to their customers. This means that there has to be a separate Bill of Lading for each consignee and several consignments can not be consolidated on to one bill.

Use of Bill of Lading

The Bill of Lading is the central document of a sea export transaction. The form, provided by the shipping company, is filled in by the shippers as soon as they have all the details of the goods. Then it is sent to the ship where an officer of the shipping company checks that the goods are "in good order and condition" and sign the Bill when the goods are loaded over the ship's rails. The Bill must be in the hands of the shipping company or their agents by the time the consignment is ready to be loaded.

When a consignment is loaded, an officer or agent of the shipping company signs the Bill of Lading that the goods have been "received in apparent good order and condition". In other words, the consignment must be exactly as written on the Bill and not different. The cases should be undamaged and sacks, if any, should not be torn or stained. Drums of liquid should not be dented or leaking. The number and kind of packages should be the same as on the Bill.

If there is any difference between what it says on the Bill and the actual condition of the consignment, the shipping company has to write a clause on the Bill giving the damage or loss. In this case, it is no longer a clean Bill of Lading and the bank representing the importer may not accept it. So the exporters' bank may not be able to get payment for the goods. For this reason "foul" or "claused" Bill of Lading must be avoided at all costs and exporters must make sure their goods arrive at the docks in good order and condition.

Sometimes certain defects of the goods are unavoidable. For instance, timber often has "split ends". Chemicals cause discoloration on packing. In such cases, the exporters must get the agreement of the importers to certain clauses on the Bill of Lading. These clauses must be agreed before the export contract is agreed and the importers should tell their bank about the agreed clauses.

A Bill of Lading is usually made out in sets of three or four originals. The shipper may ask for several extra copies for his files. One copy of the Bill is kept for the ship. The other copies are sent to the exporters or directly to their bank. These negotiable Bills of Lading are used for payment. They pass to the buyers or their agents in the importing country.

Then the Bills and the other shipping documents are presented to shipping company when the ship arrives. The shipping company can then compare the negotiable Bills with their copy on the ship. In this way, the importers can show their legal right to the goods and obtain them from the ship.

In recent years, a considerable simplification of documentary practices has been achieved. Bills of Lading are frequently replaced by non-negotiable documents similar to those which are used for other modes of transport than carriage by sea. These document are called "sea waybills", "liner waybills", "freight receipts", or variants of such expressions.

Illustrations of a Bill of Lading

To understand and be able to draw up a Bill of Lading you should consider these points:

- How the goods are to be sent.
- Who will pay the freight charges.
- The name of shipper.
- The name of the consignee.
- The vessel and port of lading.
- The port of discharge.
- Type of packing, numbers and marks, gross weight and measurements.
- Any other points.

If a Bill of Lading only contains those particulars mentioned above, it is called a "short form B/

L" . However, if a Bill of Lading includes those particulars on the face of it, together with various clauses dealing with the rights and duties of the shipowner and shipper on its back, such a Bill is called "long form B/L" . The long form B/L is used much more frequently than short form B/L in international trade. To know the format of Bill of Lading, please refer to the appendix of this text book.

9.3.4 Packing List

Packing list is a document made out by a seller when a sale is effected in international trade. It shows numbers and kinds of packages being shipped, total of gross, legal and net weights of the packages, and marks and numbers on the packages. It is used to make up the deficiency of an invoice. It also enables the consignee to declare the goods at customs office, distinguish and check the goods when they arrive at the port of destination, thus, facilitates the clearance of goods through customs. What's more, packing lists can facilitate settling insurance claims in case of loss or damage. To know the format of Packing List, please refer to the appendix of this text book.

9.3.5 Weight Memo

Weight memo is a document made out by a seller when a sale is effected in foreign trade. It indicates the gross weight, net weight of each package. It is used to make up the deficiency of an invoice. It is also used to facilitate the customs formalities and the general check of the goods by the consignee on their arrival at the destination. Packing list and weight memo usually come out in a combined form.

9.3.6 Inspection Certificate

Inspection certificate or survey report is a document which shows the quality or quantity or other elements of the goods. It is issued by the manufacturer of the goods, chambers of commerce, surveyors, or government institutions. It mainly performs two functions: Firstly, as a document of quality or quantity, it can decide whether the quality or quantity of the goods shipped by the seller is in conformity with that stipulated in the contract. It is an important proof at the time of refusing payment, lodging or settling a claim. Secondly, it is one of the shipping documents used at the time of negotiating payment.

For the import commodities that are subject to inspection by the inspection authorities as stipulated in the foreign trade contract, upon their arrival, the receivers, and users or forwarding agents should apply to the inspection authorities at the arrival port/station in due time. A survey report or inspection certificate shall be issued to the applicant after inspection. These commodities shall be checked and released by the customs upon presentation of the seal of the inspection authorities affixed on the customs declaration.

For export commodities which are subject to inspection as stipulated in the contract or by law, the manufacturers or suppliers should apply for inspection before shipment. If they are proved up to the standard, the inspection authorities shall issue inspection certificates for them to clear the goods through customs.

If vessels and containers are used for carrying perishable goods, such as cereals, oils, foodstuffs, and frozen products for export, the carriers and container stuffing organizations should apply for inspection of the holds or tanks and containers to the inspection authorities at the port. They shall be permitted to carry the goods only after a certificate is issued after examination which proves that they conform with the technical condition for shipping. To know the format of inspection certificate, please refer to the appendix of this text book.

9. 3. 7　Insurance Policy or Certificate

An insurance policy or an insurance certificate is issued when goods are insured. An insurance policy (or a certificate) forms part of the chief shipping documents. A policy also functions as collateral security when an exporter gets an advance against his bank credit.

Insurance policy, issued by the insurer, is a legal document setting out the exact terms and conditions of an insurance transaction—the name of the insured, the name of the commodity insured, the amount insured, the name of the carrying vessel, the precise risks covered, the period of cover and any exceptions there may be. It also serves as a written contract of insurance between the insurer and the person taking out insurance. To know the format of insurance policy, please refer to the appendix of this text book.

Chapter 10

Cross-border E-commerce Practice

★LEARNING OBJECTIVES

- Understand the historical development of cross-border e-commerce.
- Be familiar with and grasp the concept of cross-border e-commerce.
- Understand the characteristics of cross-border e-commerce.

10. 1   Overview of Cross-border E-commerce

10. 1. 1   The concept of Cross-border E-commerce

Cross-border Electronic Commerce refers to as the Cross-border E-commerce and means a kind of international business transactions that trade subjects belonging to different customs areas make deals and conclude payment and settlement through e-commerce platform and deliver goods and complete deals via cross-border logistics.

Cross-border e-commerce can be divided into broad cross-border e-commerce and cross-border e-commerce in a narrow sense.

In a narrow sense, the cross-border e-commerce is substantially equivalent to the cross-border retail as a matter of fact. Cross-border retail refers to the transaction process that trade subjects belonging to different customs areas make deals, pay and settle through computer network and deliver goods to consumers via cross-border logistics by way of express mail, small packets and other luggage and mail. The popular way of cross-border electronic commerce in the international arena is Cross-border e-commerce which refers to cross-border retail actually. Generally, from the perspective of customs, the cross-border retail is equivalent to the online trading packet, both basically for consumers. In a strict sense, with the continuous development of cross-border e-commerce, consumers in cross-border retail will also include a portion of B-class business users with fragmented and small-amount transaction; however, it is quite difficult to distinguish such B-class businesses and C-class individual consumers and to identify the strict boundaries between the two. So, generally speaking, this part of sales targeting at B-class also attributes to cross-border retail.

Broadly speaking, the cross-border e-commerce is substantially identical to the foreign trade e-commerce which refers to an international business transaction that trade subjects belonging to different customs areas electronize the links of display, negotiations and transaction in traditional import and export trade via e-commerce means and deliver goods and conclude deals through cross-border logistics. In a broader sense, the cross-border e-commerce refers to applications of e-commerce in import and export trade, being the electronic, digital and networking version of traditional international business processes. It involves many activities in varied aspects, including electronic trading of goods, online data transfer, electronic fund transfer, electronic freight documents and others. In this sense, any aspect that involves e-commerce applications in international trade links can be included within this statistical category.

10. 1. 2 The Development of Cross-border E-commerce in China

In 1999, after Alibaba used Internet to connect Chinese suppliers and overseas buyers, China's foreign export and trade realized Internet-based. After that, it went through three stages, which realized cross-border e-commerce industrial transformation from information service, to online transaction and overall industrial chain service.

Cross-border e-commerce stage 1. 0 (1999 - 2003)

The main business mode of cross-border e-commerce 1. 0 era is foreign trade information service mode of online display and off-line transaction. The main function of third party platform in cross-border e-commerce stage 1. 0 is to provide network display platform for enterprise information and products, but not to involve any transaction links on the Internet. Currently, the profit mode is to ask enterprises which conduct information display for membership fees (such as annual service fee). During the development process of cross-border e-commerce stage 1. 0, auction promotion has been gradually derived. Counseling service and etc. have provided one package information flow value-added services.

In cross-border e-commerce stage 1. 0, Alibaba international station platform and Global Resource Website are typical representative platforms. Alibaba was founded in 1999. At first, Chinese suppliers of Alibaba are only the yellow pages on the Internet, which displays the product information of Chinese enterprises to global customers and positions in B2B block trade. Buyers have access to product information of sellers through Alibaba, and then both parties negotiate and make deals off-line, so most of transactions are finished off-line at that time. Global Resource Website was founded in 1971, formerly known as Asian Source, which is the earlier trade market information provider in Asia, and is listed on the NASDAQ stock exchange on April 28, 2000, and the share code is

GSOL.

During this period, many cross-border e-commerce platforms with supply and demand information transactions came up, such as Made in China, Korean EC21 net and Kellysearch. Although cross-border e-commerce stage 1.0 uses Internet to solve the global buyers' problems of Chinese trade information, it still cannot finish online transaction. As for the integration of foreign trade e-commerce industrial chain, only information flow integration link is finished.

Cross-border e-commerce stage 2.0 (2004 – 2012)

In 2004, with the online launch of DHgate, cross-border e-commerce stage 2.0 is coming. In this stage, cross-border e-commerce platform began to get rid of displaying information yellow page, realized electronization of off-line transaction, payment and logistics, and gradually realized online transaction platform.

Compared with the first stage, cross-border e-commerce 2.0 can better reflect the nature of e-commerce with the help of e-commerce platform, and go through the upstream and downstream of supply chain effectively with service and resource integration, including B2B (platform to small transaction of enterprises) platform mode, and B2C (platform to users) platform mode. In cross-border e-commerce stage 2.0, the platform mode of B2B is the main mode of cross-border e-commerce, through directly connecting medium-sized and small-sized enterprise merchants to realize the further shortening of industrial chain and enhancing sales profit space of commodity. In 2011, DHgate announced the enterprise made profits, and in 2012, and DHgate kept making profits.

In cross-border e-commerce stage 2.0, the third-party platform realizes the diversity of revenue, and backward charging mode. It will change "membership fees" to charging transaction commission, which is charging percent commission due to turnover effects. At the same time, using marketing promotion, payment service and logistic services to gain value-added profits.

Cross-border e-commerce stage 3.0 (2013 to present)

2013 became an important transformation year for cross-border e-commerce, the whole industrial chain of cross-border e-commerce had changes of business mode. With the transformation of cross-border e-commerce, the "big time" of cross-border e-commerce is coming.

First of all, cross-border e-commerce has five aspects of characteristics—the launch of large-scale factory, B type of buyers becoming scale, the enhancement of medium and large order proportion, the participation of large-scale service provider and the outburst of mobile users. At the same time, the service of cross-border e-commerce 3.0 service has upgraded comprehensively, the carrying capacity of platform is stronger. And the online of whole industrial chain service is an important characteristic of 3.0 era.

In cross-border e-commerce stage 3.0, user groups transform from grass-roots entrepreneurship to factory and foreign trade company, and have incredibly strong production design management ability. Using platform to sell products transform from network operators and second-hand sources to first-hand resources of good products.

For stage 3. 0, the main seller groups are in the tough transformation period from traditional foreign trade business to cross-border e-commerce business. Manufacturing production mode transforms from large-scale production line to flexible manufacturing, the requirements of agent operation and the supporting service of industrial chain are relatively high. On the other hand, main platform modes of stage 3. 0 transform from C2C and B2C to B2B and M2B. The medium and large transactions of wholesale buyers become main orders of platform.

10. 1. 3　The Characteristics of the Cross-border E-commerce

Compared with the traditional international trade, cross-border e-commerce presents five new features: global, intangible, anonymous, instant and paperless.

Global

Cross-border e-commerce depends on the e-commerce platform as the basis for international trade, which determines its characteristic of globalization. When compared with the traditional trade, cross-border e-commerce breaks the geographical restrictions and time limits: the sellers in one country can post information about products and services, and communicate, negotiate and make deals with buyers in another country through the Internet; what's more, buyers in one country can search for sellers through the Internet, and then make an inquiry, bargaining, payment and settlement, and ultimately purchase inexpensive products or services. The global feature of cross-border e-commerce brings information sharing to global buyers and sellers to the greatest extent; however, there are some risk of payment and settlement as well. Anyone who masters a certain basic knowledge of network can enter the network information and perform online transactions whenever and wherever he is. For example, a smaller Chinese foreign trade company can offer products and services to businesses or consumers in any country through e-commerce platform once they have access to Internet and met corresponding demands. Such a way largely facilitates international trade, but meanwhile, it brings some troubles to national tax.

Intangible

The traditional foreign trade is mainly for barter trade, but with the development of Internet, transactions on a number of digital products and services (such as e-books, movies and copyrights, etc.) becomes more and more. Digital transmission is conducted in a global network environment, being intangible; while cross-border e-commerce develops based on network, thus it inevitably possesses the invisible feature of the network. Take books deal as an example: traditional foreign trade makes the deal by treating a book (physical items for transaction) as the subject matter, while in cross-border e-commerce transactions, buyers in one country can simply purchase right to data of the book online so to get the appropriate information, being convenient. The intangible characteristic of e-commerce brings new challenges to a country's tax authority and legal department: its transaction records are in the form of data code, making relevant authorities difficult to define the trading activities, thus they cannot carry out effective monitoring and taxes collection.

Anonymous

Due to the globalization and intangibility of cross-border e-commerce, both parties of the transaction can use the Internet to trade anytime and anywhere. Consumers, who use e-commerce platform for trading, usually do not expose their true information, such as real name and the exact geographic location, out of the purpose to avoid transaction risks. But this does not affect their trading smoothly: the anonymity of the Internet provides consumers with such favorable conditions, allowing them to do so. The anonymity of cross-border e-commerce causes the extreme asymmetry between consumers' rights and obligations—consumers can enjoy maximum rights and interests in a virtual network environment, but they bear minimal responsibilities and obligations, and even some of them are trying to evade responsibilities, all of which make relevant authorities impossible to learn the true interest income and trading conditions of traders, thus they cannot calculate the tax payable and tax the taxpayer legitimately.

Instant

In traditional foreign trade, two trading subjects communicate mostly by mail and fax; the sending and receiving of information pose different levels of time difference, and there are may be some obstacles during the transfer process, making the information not able to transfer in a smooth and timely way and affecting the international trade to some extent. Unlike the traditional mode of foreign trade, cross-border e-commerce has real-time transmission of information, that is to say, regardless of how far away between the actual locations, sellers' sending message and buyers' reception of information are almost simultaneous, being no time difference, which is equivalent to face-to-face communication in traditional foreign trade to a certain degree. For transactions of some digital goods (such as software, movies, etc.), placing orders, payment, delivery and settlement can be done through the network instantly, bringing great convenience to both parties.

Real-time feature of cross-border e-commerce reduces the middleman link in traditional foreign trade and makes exporters directly face the end consumers, which improves the efficiency of trade, but also hides legal crisis. Such crisis is presented in the taxation area as follows: due to the immediacy of e-commerce activities, buyers and sellers can start, change and terminate trading activity at any time, which increases the randomness of trade while reduces the effectiveness of trade. This makes the tax authorities unable to ascertain the real situation of transactions by the parties and supervise ineffectively, bringing some difficulties to the collection of taxes in tax authorities.

Paperless

In traditional foreign trade, the whole process, from the inquiry, bargaining, negotiations, contracting and payment settlement, requires a series of written documents as the basis for the transaction. While in e-commerce, trading subjects mainly employ the paperless operation, which is the typical characteristic of cross-border e-commerce being different from the traditional trade. Sellers send information through network, and buyers receive information through the network, thus the entire electronic information transmission process is paperless. Paperless trade, on the one hand, makes information transmission get rid of limitation of paperwork, being more efficient; on the other hand, it also causes chaos in legal system. Since the majority of existing laws and regulations are based on "paper-trading" as a starting point, and do not apply to cross-border e-commerce "paperless" transactions.

The "paperless" transactions in cross-border e-commerce replace the written documents (such as a written contract, settlement documents, etc.) in traditional foreign trade for trading; in the context where no data is available, the tax authorities cannot be informed the real situation of taxpayers, which increases the difficulty for tax authorities to acquire taxpayer operating conditions, making a large part of tax revenues lost and not conducive to international tax policy. For example, stamp duty, as one of the traditional taxes levied universal, should be levied on the basis of written contract provided by trading parties. However, in the "paperless" e-commerce environment, there is no legal contracts and written confirmation in physical form, due to which the taxation of stamp duties by the nation is out of nowhere and lawless.

10.1.4   The Challenges Faced by Cross-border E-commerce

Currently, the development of cross-border e-commerce in China is mainly faced with several major challenges.

Obstacles in the aspects of privacy and cross-border data transmission

Electronic payment is a necessary condition for international e-commerce and an important link in the achievement of the transaction. If there is no third-party payment, there would not be what it is today in e-commerce industry. Cross-border e-commerce is also inseparable from the third-party payment. Therefore, the internationalization of third-party payment has become an important condition for the future occupation of the consumer market.

Currently, overseas buyers fraud is a big concern for China's small-and-middle-sized foreign trade businesses. As for the issue of transaction security, the survey shows that more than half of the respondents indicate their worries about fraud when trading with overseas customers and 27% of businesses concern the payment systems they currently use are not secure enough when carrying out cross-border transactions. In addition, 25% of businesses believe that within the next three months, the risk of overseas buyers' refusing to pay will increase.

For the issue of e-commerce application security in international trade, we should follow the progress of e-commerce technology and continue to improve laws and various regulatory measures, while all the industries and relevant organizations should formulate a variety of industry standards and certification systems to promote the resolution of security issues of e-commerce in international trade.

Urgent demand for unity of credit evaluation and identification

E-commerce, with a virtual feature, has not only the risk of traditional business activities, but also its own unique characteristics of being open, global, low cost, high efficiency. The behavior of trading subjects, and the behavior of market intermediaries have a great deal of uncertainty, thus the behavior of breaking one's words is more prominent in the field of electronic commerce.

There are many constraints in the establishment of e-commerce credit guarantee system, such as lack of credit awareness and credit ethics, unsound internal e-commerce credit management system, laggard credit intermediary services and the lack of legal protection and effective incentive mechanism. Cross-border trading requires a more comprehensive, cross-regional, cross-cultural and cross-institutional credit system to support more complex trading environment, putting forward higher demands for the parties' commercial credit.

National absence of formulating regulations on e-commerce and cross-border e-commerce

The impact of electronic commerce on the aspect of international trade law is mainly caused by the unsound laws applicable to international trade currently. The formulation of relevant legal system lags far behind the development of information industry in China. Concerning cross-border e-commerce services, China only has several relevant laws and regulations, including *Administration of Internet Information Services Procedures* and *Electronic Signature Law*; while there is no special norms and standards for the aspects of transactions, taxation and consumer protection involved in cross-border e-commerce. Consequently, a problem in need of urgent solution is to develop some corresponding e-commerce laws to resolve disputes on electronic commerce.

Given that the development of electronic commerce is hindered by technological and legal defects and lagged law, e-commerce legislation should both make up for the deficiency in technology and credit with definitive arrangements and create a relaxed and free environment for its development. To construct such a legal system needs new laws and requires a reasonable interpretation of existing laws and creating a supporting legal norms favorable for the development of electronic commerce.

On July 23, 2014, the General Administration of Customs released the Announcement No. 56 Announcement on the regulatory issue of inbound and outbound goods and articles in cross-border e-commerce trade; the file is a summary of cross-border e-commerce pilots carried out some time ago and the document makes clear the legal status of cross-border e-commerce and emphasizes the mentality of follow-up e-commerce port customs' systematic regulation on e-commerce, which means cross-border e-commerce will step to promotion stage from the pilot phase.

Lack of cross-border e-commerce professionals

Being different from domestic e-commerce, the payment and logistics of cross-border e-commerce are much more complicated, and small and medium foreign trade enterprises face many risks in the development of e-commerce. China's small and medium foreign trade enterprises are difficult to attract highly-skilled and capable senior e-commerce professionals who are relevantly short in the market, because these enterprises are small in scale, weak in strength and small in development space. The shortage of e-commerce talents poses as a serious impediment to the development of China's small and medium foreign trade e-commerce. This aspect needs to enhance urgently.

10. 1. 5   Cross-border E-commerce Talent Demand

Compared with the traditional international trade, foreign trade, electronic commerce, international business and other professional talents, cross-border commercial talents not only to master e-commerce professional knowledge, also need to have a higher level of foreign language, to online timely skillfully with foreign clients for communication and negotiation, can manage many language versions of the shop; to have a certain knowledge of international trade, can use a variety of international settlement and international logistics; Also understand culture, customs, laws and regulations in many countries and regions; can make use of the latest international network marketing tool, based on social media to communicate with customers, and analysis outside customers' demand etc.

Analysis from the investigation to the relevant enterprises, enterprises of cross-border business personnel professional ability requirements include: international trade skills, e-commerce skills, international logistics skills, cross-border online marketing skills, skill of communication and expression in foreign languages and professionalism capacity.

International trade skills

Business scope of cross-border e-commerce enterprises are worldwide. It belongs to the scope of international trade from the nature of business, which would inevitably require practitioners of cross-border e-commerce to have considerable foreign trade business abilities—being familiar with the business processes of import and export and foreign trade and relevant laws and regulations, skillfully filling and formulating all kinds of foreign trade documents, responding and handling customer orders and foreign trade disputes in a timely manner; actively carrying out online transactions, using EDI for customs clearance, inspection and tax rebates, and following up related business processes of international logistics, international insurance and international settlement.

E-commerce skills

Cross-border e-commerce is to make use of the e-commerce platform built by the Internet to conduct business activities. It belongs to the category of e-commerce from the perspective of media, which in turn requires practitioners have certain e-commerce capabilities—having a strong ability in e-commerce information retrieval, collection, production and publication; strong capability of marketing and online marketing the products and services; strong ability in the operation and maintenance of a small e-commerce system; ability of initial e-commerce project and management; formulating cross-border e-commerce implementation plans based on preliminary organizational situation; understanding the development situation and development trend of cross-border e-commerce.

International logistics skills

Cross-border e-commerce requires practitioners to master knowledge on cross-border procurement management and supply chain management, to be proficient in using modern logistics technologies for the management of international logistics business, to learn the rules of customs clearance in different countries so to choose a reasonable mode of transport for the import and export goods, to track and sign contract of carriage or agency contract of carriage and to arrange reasonable import and export goods insurance for goods purchased or sold by cross-border e-commerce.

Cross-border online marketing skills

The talent cultivation for cross-border e-commerce should meet the demands for global multi-level cross-border online marketing. It requires practitioners to be capable of international market research and forecast and to accurately conduct cross-border network planning, information collection and big data analysis, and ultimately explore the international market online.

Skill of communication and expression in foreign languages

During operation, employees working in cross-border e-commerce platform often encounter a large number of foreign materials, which requires them to have a higher foreign language reading and writing skills, and be able to cope with correspondences based on national business etiquette; to use English to correctly describe the product details and serve customers in foreign languages fluently on the online platform. Meanwhile, it requires people to collect and learn a lot about the real development environment in foreign countries—market, culture and economy and so on, so to express and communicate in a more native way.

Professionalism capacity

Cross-border e-commerce now faces customers from around the world and is a global open Internet-based business. Employees in cross-border e-commerce have to face the complex and changing international business environment, which will require that employees have the flexibility in negotiation skills, good communication adaptability, practical and serious working attitude and team spirit of positive cooperation, understand international business laws and regulations, and properly handle cross-border business disputes.

10. 2   The World's Major Cross-border E-commerce Platforms

Currently, the main cross-border e-commerce platforms around the world are AliExpress, Wish, DHgate, eBay and Amazon.

AliExpress

AliExpress is officially launched in April 2010. It is a foreign trade online transaction platform that Alibaba helps medium and small enterprises to connect with overseas terminals, to expand profit space and creates for integrating order, payment and logistics, which is called by sellers as "international version of Taobao". Currently, AliExpress is the largest B2C transaction platform in China, covering more than 220 countries and districts. The flow of overseas buyers per day is over 50 million.

From the launch in 2010 to 2014, the annual turnover growth of AliExpress remains from 300% to 500%, and the online product amount has reached hundred million level. The orders successfully cover more than 220 countries and districts around the world, and the platform sellers are 0. 2 million. The registered AliExpress accounts, which do not include those do not open the shops are close to 2 million. On November 11, 2014, it is the first time that AliExpress participated in global November 11, 6. 84 million transaction orders were created within 24 hours. The valid orders in that day cover 211 countries and districts. As for the ranking of flow, Brazil, Russia and Turkey rank the second in shopping website, the US tanks the fifth, Spain ranks the first, and Indonesia ranks the sixth. And the monthly flow is about 600M, the main distributions are mainly in these several countries, and the proportion of flow is about 50% of the total amount.

The home page of AliExpress as shown in Figure 10. 1.

Figure 10. 1   The home page of AliExpress

Wish

Wish was founded on December 1, 2011 in Silicon Valley of San Francisco in the US, and it is a commercial platform based on mobile App. At first, Wish platform is used to communicate and share, which does not involve product transaction. In March 2013, Wish online transaction platform is officially launched, and mobile APP is launched in June 2013. The operation revenue of 2013 is over 0. 1 billion dollars, which becomes a rising star in cross-border e-commerce mobile terminal platform.

Currently, Wish has 100 million registered users. And the daily active users are 1. 2 million, whose family incomes are about 65, 000 dollars. The main consumer group of Wish is 15 to 28 years old, and the proportions of men and women are 3 : 7. The proportions of Wish merchants in Asia and Europe are 81 : 19, which means Wish is dominated by Chinese sellers. Wish ranks the first in download amounts of the US. And its founder is named as one of the best founders, and APP is rated as the best APP for Android.

Compared with other e-commerce platforms like Amazon, eBay and AliExpress, Wish will not charge additionally through key words to recommend products for users. The system of Wish is calculated by data of buyers' behaviors to evaluate the hobby of buyers, interesting product information and choose corresponding product recommendation for buyers.

Mobile application page of Wish as shown in Figure 10.2.

Figure 10. 2   Mobile application page of Wish

DHgate

DHgate (*www. dhgate. com*) is the first B2B cross-border e-commerce platform in China, which was founded in 2004. It helps medium and small enterprises in China to enter into global market through e-commerce platform. Currently, it has more than 1. 2 million domestic suppliers with more than 30 million products in 224 countries and districts around the world and with the scale of 10 million online buyers. 0. 1 million buyers do online purchases per hour, and each order is created within each 3 seconds. DHgate mainly uses profit modes like commission, value-added service and advertising. The advantaged programs of DHgate are cell phone and electronic products.

With the mission of "promoting global trade and creating entrepreneurial dream", DHgate helps medium and small enterprises utilize e-commerce platform by a large scale and seek for new development space for Chinese foreign trade and expand new picture of foreign trade and transaction. The values of DHgate bringing to small and medium enterprise customers lie in two aspects: the first one is "to pay for success", which makes medium and small enterprises pay few commissions for trans-action and not pay for high annual fees; the second one is DHgate helps enterprise customers to look for overseas buyers and helps customers to solve the problems of overseas marketing. At the same time, DHgate has professional marketing teams of foreign staff from Facebook, YouTube and Google for promotion, uses the latest marketing method——community marketing for customers from medium and small enterprises to look for overseas buyers.

The home page of DHgate as shown in Figure 10. 3.

Figure 10. 3　The home page of DHgate

eBay

eBay is an online auction and shopping website for global citizens to buy and sell products. It was founded on September 4, 1995 by Pierre Omidyar using the name of Auctionweb in San Jose, California. The initial purpose of eBay is to help founder's girlfriend to find out the fans of Pez candy box in the US for communication. It was surprising that eBay became popular soon. At the beginning of establishment, eBay is positioned as an online auction and shopping website for global net citizens to buy and sell products.

There are two main selling methods for sellers to release products in eBay platform: "auction" and "fixed price". eBay charges sellers differently due to different selling methods, it is usually calculated as "publication fee" plus "transaction fee", which is: production release fees and trans-action commissions.

The home page of eBay as shown in Figure 10. 4.

Amazon

Amazon is the largest e-commerce company, which is founded in the US in 1995. Its initial po-sition is an online bookstore selling books and audio-visual products. In 1997, it transformed to the largest comprehensive network retailer. In 2015, among 20 Internet companies with the highest mar-ket values around the world, Amazon ranked the fourth.

Currently, Amazon mainly sells brand new, refurbished and used items, including books, di-

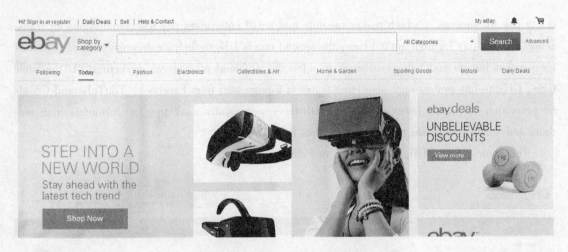

Figure 10. 4　The home page of eBay

gitals, home furniture, cookers, household appliances, beauty makeup, food, drinks, maternity stuff, toy, apparel and automotive supplies.

The home page of Amazon as shown in Figure 10. 5.

Figure 10. 5　The home page of Amazon

开展国际贸易前的准备

★学习目标

1. 理解设计国际贸易计划的重要意义
2. 掌握常用的国际贸易法律法规的基本内涵
3. 能制订国际贸易计划
4. 懂得从事国际贸易应具备的条件

国际贸易，通俗地讲又叫外贸，就是跟外国人做生意。开展国际贸易和任何其他创建自己的企业一样，都面临着同等概率的成功机会。因此，进行贸易前的计划准备工作尤为必要。

首先，我们要对产品和市场非常熟悉，找到一个市场前景好的产品，通过市场调查来分析机会和威胁，组建或者寻找一种合适的进出口组织形式，最终形成一个合适的国际贸易计划。

其次，要有充足的国际贸易方面的专业知识。虽然最近几年与国际贸易合同方面相关的法律变化不大，但是由于电子商务的广泛应用、交通方式以及通信手段的蓬勃发展，很多与国际贸易相关的流程、规则和单据也发生了很大的变化，尤其是与互联网紧密相关的一些单据文件需要及时更新。因此我们要知晓国际贸易方面的公约、惯例和各国的国内法律法规等。

最后，当我们做好上述准备，打算开展国际贸易时，我们需要经过一些程序来获得国际贸易从业资格，对于某些货物，还需要在进出口前取得相应的进出口配额或许可证。

第一节　设计并制订国际贸易计划

我能做国际贸易吗？这个问题实际上与"我能开创自己的事业吗？"一样，是很多人的梦想。在成千上万的创业者中，很多人尝试了，但失败了，也有很多人从某种程度上来说取得了成功，只有少数的人获得了极大的成功。因此，开始做任何一种生意，我们都需要做好设计和筹划方面的准备工作。

一、制定贸易计划的前期准备

本部分会探讨我们在打算进行进出口贸易前就应该考虑的一系列事项，这有利于减少将来可能发生的错误或困难。任何不做准备就马上进行贸易的公司迟早会遇到或大或小的问题，而计划可以帮助我们提前分析什么是能带来盈利的交易，并有助于国际贸易的顺利开展。有关制订商业计划的参考很多，我们可以从中国中小企业合作发展促进中心、相关的书籍、软件以及互联网来获得相关的建议。

通常贸易计划的内容包括贸易的形式、贸易的产品、潜在市场及其细分市场和竞争、促销计划、组织和人员、进出口物流安排和开业的安排（如果是新建企业）。要制订出这样一个完整的计划，首先要了解开展国际贸易前期准备工作的重要性，同时做好市场调研工作，还要考虑如何融资、如何找到供应商并签订协议、了解政府对进出口的管制情况、如何避免进出口欺诈等事项。

（一）开展贸易的基本条件

进出口贸易无疑是需要多方面的知识和技能的，尤其是对于那些想自己开公司做贸易的人士而言。采购、营销、销售、融资等知识必不可少，选修一门创业类课程或者自学一些如何创立和管理中小企业的书籍更好。

前期资金的储备同样重要。注册成立自己的企业、基本的办公设备、通信设施等都需要资金，同样，当采购商品用于出口或进口时，在卖出产品收款之前，通常需要先付款。另外，还需要安排运输、储存和其他服务时用的资金。即使是作为贸易代理商或者经纪人，虽然不需要取得货物的所有权，但也需要最基本的资金用于办公支出、通信支出等。

要对世界有充分的认识和了解。尤其是对世界地理知识的了解，它帮助我们了解不同国家的地理位置、哪些国家跟我们的关系良好、哪些国家的货币币值坚挺等，这些都可以通过阅读报纸（如国际商报），浏览一些网站，收看电视上的世界新闻和收听一些广播（如 BBC 世界新闻）等来获得。

从事国际贸易还需注意语言和文化差异带来的影响。由于思维方式、价值观念、风俗习惯和语言表达的差异，往往容易在跨文化交流中出现障碍。因此，要了解对方的文化、价值观和风俗习惯，在交流中尽量用简单、清楚、明确的用语，不要用易引起误会的多义词、双关语、俚语、成语，也不要用易引起对方反感的词句。在国际贸易活动中，遵守国际惯例和一定的礼节。当与相关国家的居民做生意时，尽量学一点儿那个国家的语言，如果确定要使用翻译或者口译，要确保找到能胜任的人员，切忌使用未经过专业人士校对过的以电脑或网络为基础的翻译。

耐心和坚持不懈是从事商业活动的一个优良品质。在开始阶段，通常可能几个月甚至半年都开不了单，这时候就要沉住气，一定要克服心理上的焦躁，静下心来一步一步把基础打牢，长久的积累一定会带来收获。当业务走上轨道后，也可能会出现亏损或者刚好保本的情形，这时候也需要时间来决定是否要进行改变或者是暂停一下来找到发展的方向。同样在跟顾客打交道的时候也需要耐心和毅力，比如每个月的定期联系和发展关系等。除此之外，注重细节也是从事国际贸易需要的良好品质。国际贸易比在国内的商业活动要复杂很多，比如把产品从一个国家卖到另一个国家，就涉及运输的问题，以船舶运输为例，我们首先要联系船运公司，船运公司将集装箱的安排交给货代，我们出货就要找货代，货物运出关境之前还需向政府报告，于是就有了报关，向海关申报；有些商品还需要报检等。这些环节的完成都需要细心。其他了解客户的需求，对产品本身的了解也需要注重细节。在关系支付的时候对细节的重视更为重要。还有就是善于识别买方并且有能力满足他们的需求。通常来说进出口贸易往往开始于发现一种好的产品然后寻找买方，在此过程中，卖出往往比买入更加困难，因为只有别人先把具体需求（如果有具体的价格更好）告诉给我们，我们才会找到产品并提供给他们。

（二）市场调研

市场调查对开展国际贸易非常重要，即使你已经具备国际贸易经验，也不可贸然进入一个没经过调查的市场。很多中小企业在开始创业时市场调查做得不够或根本就没做市场调研，这可能会给国际贸易的未来发展带来很大的隐患。而进行市场调研可以帮助我们分析收入的实现，帮助我们完成市场调研计划。贸易前的调研有助于分析怎么去找到市场，怎么去营销产品。

关于出口产品的市场调研大纲请参见英文部分的内容。熟悉和制作这样的营销调研大纲可以帮助我们尝试别的产品和市场，每一次的尝试都会使贸易的成功概率增加。而且一份可靠的市场调研报告能够让投资者和贷款人放心。

二、贸易计划的主要内容

（一）制作贸易计划的考虑事项

（1）考虑从事国际贸易的目标。为什么要开始你的事业？你对国际贸易感兴趣吗？你的目标是什么？如果目标是获取利润，那么你需要花更多的时间来好好计划可行性的财务报告。如果你的目标只是想在外贸行业中工作，获得实践经验，那么只是较低程度地参与国际贸易，工资低点儿也是可以接受的。

（2）考虑从事国际贸易的必备条件。即有没有相关的企业经营训练和经验、资金、产品知识、开发客户的能力、出差旅行经验或外语能力等。具备相关的经验比什么都没有要强得多。之后就要确定是打算做出口、进口还是两者都做。如果是刚开始自己的企业的话，最好是只选择一种来做。

（3）考虑贸易采取自营进出口还是贸易代理商的方式。两种方式各有优缺点。自营进出口可能会赚取更多的钱，但是需要投入更多的资金和承担更多的风险；而作为本国公司或者外国公司的代理商，对资金需求不多，只需签订代理合同，根据成交实绩获取佣金回报。不管采取哪种形式，自创企业还要考虑注册什么样的名字，需不需要以公司的形式存在，采用哪种形式的公司等。对这些问题的回答会关系负债和税收的承担。

（4）考虑目标市场。目标市场的确定要有实际意义，选择的市场是能明确辨认和联系的，与其将世界上所有人作为一种旅行箱包的顾客，不如把选定的一些国家的百货商店作为目标顾客的需求，这些商店将箱包卖给那些喜欢旅游的中等收入的顾客。所以不能把某个国家的每个人都作为你的目标顾客，而是要选择在什么地区、什么收入水平、什么生活方式的人会成为你的目标顾客。接下来是怎样满足目标顾客的需求，即用什么方式促销，用什么渠道销售给顾客。首先我们可以选用各种平台，如阿里巴巴、中国制造和环球资源等 B2B 平台；然后就是利用各种展会，如中国广交会和各种国外专业展会；或者绕过这些个中间环节，进行直销，这就是跨境电商，也是利用各种平台如亚马逊、ebay，各种国外本地网站，结合物流和媒体，目的是把产品卖出去。

（5）考虑交易的产品。如果你打算交易的产品现在已经有客户了，这是最好的情形。反之，就需要对将来要贸易的产品非常了解，知道这个产品具有哪些优势，有足够的供应数量，能够被运送到并通过海关进入目标国市场。不管选择什么样的产品，你也许会发现本国的公司都有进口、出口或者进出口都有。几乎每一种产品都可以进行国际贸易，要做的是找到合适的方式来进行。

（6）考虑产品的供应商和国家。与从中间商那里购买比较起来，不管是进口商还是出口商通常都愿意跟能自己生产产品的公司做生意，因为中间商发挥的作用都需要利润和成本来补偿，这些都提高了最终产品卖给消费者的价格。因此，如何在国内国外找到制造商也是我们需要掌握的知识。选择从哪个国家或地区购买的主要影响因素有质量信誉、成本结构、运输成本和海关税收等。

（7）考虑运输方面的事项。不同的产品可以有不同的运输方式。小的物件可以采用快递的方式，稍微大点儿的可以采用航空运输方式，更大的采用海运或者陆运的方式，但具体还要根据货物的价值和易受损性等因素来做选择。多式联运也很普遍，比如中国涪陵榨菜要出口到印尼的目的地可能就要经过卡车、轮船、火车甚至是牛车运到一些小的商店。在国际货物的运输中，会用到许多的标准术语如 FOB、CFR 和 CIF 等，理解它们的含义就很重要。

（8）考虑采用的国际支付方式。省钱又安全的支付对做任何生意都意义重大。在国际贸易中需要额外考虑的就是货币必须跨越国界，不同国家用的货币系统不一样，货币的汇率也会经常发生变动，因此采用适当安全的支付方式至关重要。除了传统的支付方式，新的支付方式也要关注和掌握。另外，任何一个国家都可能存在奸商，他们会采取种种欺骗手段来达到买东西不付款的目的，要加以注意。

（9）考虑本国和外国的政府对国际贸易的管制。国际贸易经营者要受到多重法律制度的制约，包括进出口国家的全国性的和地方性的法律法规，还有国家之间达成的双边和多边协议。

（10）考虑一定的经营场所。很多人是在自己的家里开始做生意的，这在互联网时代下变得更为普遍，一台电脑加上几样办公设备包括电话、传真机、打印机等就足够了，而且家庭办公还可以节省大笔的租金费用。但这只是适用那些小的刚开始的公司。

（11）考虑要选择的服务公司。刚建立企业时需要会计师和或律师的帮助，然后需要银行、网络提供商、电信运营商及保险经纪人。当开始有订单了，需要货运代理商和报关公司、快递公司以及其他服务商来提供服务。当生意做到一定程度，还需要有偿地从一些专业的组织或者行业协会得到帮助和指导。

（12）考虑你的资金投入和来源以及收入来源和赢利能力。不管是以什么方式开始贸易的，都需要作相关的资金投入，包括启动成本，如设备投入、时间成本和盈利前的营运成本。进行最低限度的估算之后就可以确定需要多少资金，从哪里获得这些资金。资金问题解决之后就要预计销售收入，然后预计所得和开支。

（二）贸易计划的制订

如前所述，在开始一项事业前制订有用可靠的商业计划是非常重要的。开展国际贸易前的准备工作是一个非常复杂的过程，它涉及很多有关自己公司和目标市场的调研和分析。进出口贸易计划请参见英文部分的范例。我们也可以从其他地方如互联网获得类似的大纲。

第二节　知晓国际贸易法律法规

国际贸易法律体系主要由三部分组成：国际贸易条约、国际贸易惯例和各国的商事法律法规。在从事国际贸易过程中，当事人都希望采用自己熟悉的本国的法律适用合同，但由于大家来自不同的国家或地区，势必产生应该用哪个国家法律方面的争议，于是在长期的贸易实践过程中，大家愿意在合同中采用普遍被认可的国际贸易方面的条约或者惯例，有时候也愿意在合同中规定适用某一国的法律。因此，国际贸易从业者应熟悉国际贸易方面的条约、惯例和各国有关国际贸易方面的商事法律。

一、国内相关的商事法律法规

现有的国际商事条约和惯例还远不能包含国际贸易各个领域中的一切问题，而且即使是现有的国际贸易方面的条约和惯例也尚未被所有国家和地区一致承认或采用。另外个人或企业在从事跨越国境的经贸活动时，也可能选择适用某国的国内法，因此，国内法在国际经贸活动中的影响很大。

（一）认识英美法系和大陆法系

目前，对国际经贸方面的法律制定影响较大的是英美法系和大陆法系国家的法律。在大陆法系国家，商法主要采用法典化的形式。其编排体例又分为两种：一种是民商分立的形式，即将民法和商法分别编撰为两部法律，在这类国家，一般是将买卖法编入民法典，在民法典中列专章或在债篇中予以规定；此外，还在商法典中对买卖的特别事项做出规定。采取这种方式的有法国、德国、日本等国家。另一种是民商合一的国家，即没有单独的商法典，而将买卖法在内的商法内容并入民法典，如意大利等国家。在英美法系国家，商法主要采取单行法规的形式。判例是英美法系国家商法的重要渊源。

（二）中国有关国际商事的国内法

中国在现阶段已经存在若干调整商事活动的法律、法规，其特点是将调整商事关系的法规分散于《中华人民共和国民法通则》与其他一些单行法规之中。这些法律包括非常广泛的领域，其中涉及国际经贸活动的主要有：1992 年通过的《中华人民共和国海商法》、2013 年修订的《中华人民共和国公司法》、2004 年修订的《中华人民共和国票据法》、2015 年修订的《中华人民共和国保险法》。

（三）中国对外贸易管制的法律体系

我国对外贸易管制制度主要由海关监管制度、关税制度、对外贸易经营者管理制度、进出口许可制度、出入境检验检疫制度、进出口货物收付汇管理制度及贸易救济制度等构成。为保障贸易管制各项制度的实施，我国已基本上建立并逐步健全了以1994年颁布并实施的《中华人民共和国对外贸易法》为核心的对外贸易法律体系，并依照这些法律、行政法规、部门规章和我国履行国际公约的有关规定，自主实行对外贸易管制。

我国现行的与对外贸易管制有关的法律主要有《对外贸易法》《海关法》《中华人民共和国进出口商品检验法》《中华人民共和国进出境动植物检疫法》《中华人民共和国固体废物污染环境防治法》《中华人民共和国卫生检疫法》《中华人民共和国野生动物保护法》《中华人民共和国药品管理法》《中华人民共和国文物保护法》《中华人民共和国食品卫生法》等。我国现行的与对外贸易管制有关的行政法规主要有《中华人民共和国进出口管理条例》《中华人民共和国技术进出口管理条例》《中华人民共和国进出口关税条例》《中华人民共和国知识产权海关保护条例》《中华人民共和国野生植物保护条例》《中华人民共和国外汇管理条例》《中华人民共和国反补贴条例》《中华人民共和国反倾销条例》《中华人民共和国保障措施条例》等。

我国现行的与对外贸易管制有关的部门规章很多，例如《货物进口许可证管理办法》《货物出口许可证管理办法》《货物自动进口许可管理办法》《出口收汇核销管理办法》《进口药品管理办法》《放射性药品管理办法》《两用物项和技术进出口许可证管理办法》等。

我国目前所缔结或者参加的各类国际条约、协定，虽然不属于我国国内法的范畴，但就其效力而言可视为我国的法律渊源之一。主要有：加入世界贸易组织所签订的有关双边或多边的各类贸易协定、《关于简化和协调海关制度的国际公约》（亦称《京都公约》）、《濒危野生动植物种国际公约》（亦称《蒙特利尔议定书》）、《关于消耗臭氧层物质的国际公约》、《精神药物国际公约》、《关于化学品国际贸易资料交流的国际公约》（亦称《伦敦准则》）、《关于在国际贸易中对某些危险化学品和农药采用事先知情同意程序的国际公约》（亦称《鹿特丹公约》）、《关于控制危险废物越境转移及其处置的国际公约》（亦称《巴塞尔公约》）、《建立世界知识产权组织公约》等。

二、国际贸易公约

（一）有关国际货物销售合同方面的公约

1. 《联合国国际货物销售合同公约》

《联合国国际货物销售合同公约》（CISG）（以下简称《1980年公约》）是关于国际货物销售的统一公约，截至2015年12月，核准和参加该公约的共有84个国家，这些国家的贸易份额占有世界贸易额的很大比重，使得该公约成为最成功的调整国际贸易关系的统一规范之一。越南是最新加入公约的国家，于2015年12月18日加入。

《公约》共分四部分。第一部分是《公约》的适用范围和总则，关于适用范围，《公约》详细规定了适用本公约和不适用本公约的有关事项；第二部分是合同的成立，这部分主要是对要约和承诺的规则作了详细的规定；第三部分是货物买卖，如果说第二部分是合同法部分，那么这部分实际上就是买卖法内容，其主要内容是买卖双方各项权利和义务、违约及其补救措施等规定；第四部分是最后条款，这部分的内容主要是一些程序性和技术性的规定，如《公约》的签字、加入、批准、生效、退出、允许保留的事项等条款。

《公约》的适用范围：根据《公约》规定，适用本公约的合同，其主体必须具备以下条件：①双方当事人的营业地必须处在不同的国家；②双方当事人的营业地所在国必须是缔约国［第一条第1款（a）项］，或者虽然不是缔约国，但如果根据国际私法规则导致适用某一缔约国的法律，也可以适用本公约［第一条第1款（b）项］。比如，一个由中国贸易商和泰国贸易商签订的合同中规定，仲裁地点在悉尼，适用澳大利亚法律。根据第一条第1款（b）项的规定，就会导致公约的适用，因此很多国家加入公约时会有所保留。

1986年12月11日中国交存核准书，在提交核准书时，提出了两项保留意见：①不同意扩大《公约》的适用范围，只同意《公约》适用于缔约国的当事人之间签订的合同；②不同意用书面以外的其他形式订立、修改和终止合同。2013年1月中国政府正式通知联合国秘书长，撤回对《联合国国际货物销售合同公约》所作"不受公约第十一条及与第十一条内容有关的规定的约束"的声明，该撤回已正式生效。

2. 海牙两公约：《1964年国际货物买卖统一法公约》和《1964年国际货物买卖合同成立统一法公约》

1964年7月的两个海牙公约是由国际统一私法协会（简称UNIDROIT，也称"罗马协会"）制定的，指《国际货物买卖统一法公约》（简称《海牙第一公约》）和《国际货物买卖合同成立统一法公约》（简称《第二海牙公约》）。国际统一私法协会于1930年组织了一个"国际货物买卖统一法公约起草委员会"，着手公约的草拟工作。经过30多年的努力，1964年4月25日在海牙召开的由28个国家参加的外交会议上通过了《国际货物买卖统一法公约》及其附件《国际货物买卖统一法》，同年7月1日开始签字，1972年8月18日起生效。1964年海牙会议同时通过的还有《国际货物买卖合同成立统一法公约》。但是直到现在，只有9个国家参加或批准这两个公约，包括比利时、冈比亚、德国、以色列、意大利、荷兰、圣马力诺、英国和卢森堡。

3. 两个实效期限公约：《1974年时效期限公约》和《1980年时效期限公约》

《1974年时效期限公约》全称为《联合国国际货物买卖时效期限公约》，简称《时效公约》，是规定与国际货物买卖合同有关的权利消灭期限的实体法公约。1974年6月14日在纽约联合国总部召开的外交会议上通过，公约确立了关于国际销售合同所引起法律诉讼必须开始的时限的统一规则。为使1974年《时效公约》与1980年《联合国国际货物销售合同公约》相配套，在1980年4月的联合国维也纳外交会议上缔结《联合国国际货物销售合同公约》的同时，还通过了《关于修正〈联合国国际货物买卖时效期限公约〉的议定书》。《时效公约》与1980年《修正〈时效公约〉议定书》于1988年8月1日生效，截止到2005年上半年，前者有25个参加国，后者有18个参加国。

《时效公约》仅适用于国际贸易，其适用范围内的合同可免于诉诸国际私法规则。在《时效公约》适用范围之外的国际合同，以及服从于有效选择的其他法律的合同，不受《时效公约》的影响。单纯的国内销售合同也不受《时效公约》的影响，而受国内法规范。中国尚未加入这两个时效公约。

（二）有关国际支付方面的公约

1. 《日内瓦统一汇票和本票公约》

《统一汇票和本票法公约》，又称为《1930 年关于统一汇票和本票的日内瓦公约》，是关于统一各国汇票和本票的国际公约。1930 年 6 月 7 日由国际联盟在日内瓦召集的第一次票据法统一会议上通过，1934 年 1 月 1 日生效。

日内瓦统一法体系只解决了法、德两大票据法体系的冲突，而英美票据法体系国家因《日内瓦公约》的规定与其票据的传统和实践相矛盾，拒绝参加。因此，在国际上形成了票据法的两大法系，即日内瓦统一法系和英美法系。

2. 《联合国国际汇票和国际本票公约》

《联合国国际汇票和国际本票公约》，简称《国际汇票本票公约》，1988 年 12 月 9 日在纽约联合国第 43 次大会上通过，并开放供签署。按该公约的有关规定，该公约须经至少 10 个国家批准或加入后，方能生效。到现在为止只有 5 个国家批准加入，所以该公约目前尚未生效。

3. 《国际保付代理公约》

《国际保付代理公约》，于 1988 年 5 月国际统一私法协会通过。国际保理是 20 世纪 60 年代发展起来的一种国际贸易结算方式。目前，国际上参加国际保理联合会的国家已有 130 多个。

（三）有关货物运输的国际公约

1. 有关海洋运输的国际公约

在统一各国有关提单的法规方面起着重要作用或有关国际货物运输的国际公约有三个：

（1）海牙规则（Hague Rules）。海牙规则的全称是《统一提单若干法律规定的国际公约》，1924 年 8 月 25 日由 26 个国家在布鲁塞尔签订，1931 年 6 月 2 日生效。公约草案 1921 年在海牙通过，因此定名为海牙规则。迄今为止已经有包括欧美许多国家在内的 99 个国家先后加入了这个公约。海牙规则使得海上货物运输中有关提单的法律得以统一，在促进海运事业发展，推动国际贸易发展方面发挥了积极作用，是最重要的和目前仍被普遍使用的国际公约，我国于 1981 年承认该公约。海牙规则的特点是较多地维护了承运人的利益，在风险分担上很不均衡，因而引起了作为主要货主国的第三世界国家的不满，纷纷要求修改海牙规则，建立航运新秩序。

（2）维斯比规则（Visby Rules）。在第三世界国家的强烈要求下，修改海牙规则的意见已被北欧国家和英国等航运发达国家所接受，但他们认为不能急于求成，以免引起混乱，主张折中各方意见，只对海牙规则中明显不合理或不明确的条款作局部的修订和补充，维斯比规则就是在此基础上产生的。所以维斯比规则也称为海牙—维斯比规则（Hague—Visby Rules），它的全称是《关于修订统一提单若干法律规定的国际公约的议定书》，或简称为"1968 年布鲁塞尔议定书"，1968 年 2 月 23 日在布鲁塞尔通过，于 1977 年 6 月生效。目前已有英国、法国、丹麦、挪威、新加坡、瑞典等 30 多个国家和地区参加了这一公约。

最后的修改是 1979 年的特别提款权协议。许多国家拒绝采用海牙—威斯比规则，宁愿保持 1924 海牙规则，另外的一些国家，接受海牙—维斯比规则后，但是随后也未能接受 1979 年修订的特别提款权协议。

（3）汉堡规则（Hamburg Rules）。汉堡规则是《1978 年联合国海上货物运输公约》，1976 年由联合国贸易法律委员会草拟，1978 年经联合国在汉堡主持召开有 71 个国家参加的全权代表会议上审议通过。汉堡规则可以说是在第三世界国家的反复斗争下，经过各国代表多次磋商，并在某些方面做出妥协后通过的。汉堡规则全面修改了海牙规则，其内容在较大程度上加重了承运人的责任，保护了货方的利益，代表了第三世界发展中国家的意愿，这个公约已于 1992 年生效。汉堡规则共有 34 个缔约国，基本上都是发展中国家，还有一些是非洲内陆国家。

还有一些国家，未加入上述任何一个公约，而在制定本国的海商法时参照和借鉴三个公约的内容。此外，还存在一些国家既未加入任何一个公约，也没有明确的相关国内法。运输规则的不统一给国际贸易带来诸多不便，影响了货物自由转让，增加了交易成本。这种现象引起了国际社会的高度重视，在国际海上货物运输领域构筑一个统一规则的呼声日益高涨。

（4）鹿特丹规则（Rotterdam Rules）。2008 年 12 月 11 日，在纽约举行的联合国大会上，《联合国全程或部分海上国际货物运输合同公约》正式得到通过，并且大会决定在 2009 年 9 月 23 日于荷兰鹿特丹举行签字仪式，开放供成员国签署，因而该公约又被命名为"鹿特丹规则"。从内容上看，鹿特丹规则是当前国际海上货物运输规则之集大成者，不仅涉及包括海运在内的多式联运、在船货两方的权利与义务之间寻求新的平衡点，而且还引入了如电子运输单据、批量合同、控制权等新的内容，此外公约还特别增设了管辖权和仲裁的内容。从公约条文数量上看，公约共有 96 条，实质性条文为 88 条，是海牙规则的 9 倍，汉堡规则的 3.5 倍。因此，该公约被称为一部"教科书"式的国际公约。

鹿特丹规则的目标是取代现有的海上货物运输领域的三大国际公约，即海牙规则、海牙—维斯比规则和汉堡规则，以统一国际海上货物运输法律制度。根据公约规定，鹿特丹规则于第 24 份核准书、批准书、接受书或加入书提交之日起满 1 年后生效。目前共有 24 个国家签署该公约，西班牙是第一个完成批准加入该公约的法律手续的国家，中国、日本、德国、英国、意大利、加拿大、澳大利亚等航运贸易大国尚未签字。

2. 有关陆路运输的国际公约

现有的铁路货运国际规则体系分别是建立在《国际铁路货物联运协定》（简称《国际货协》）和《国际铁路运输公约》这两个规则基础上的。它们是作为规范国际铁路货物运输的两套主要规则体系。《国际铁路运输公约》是由《国际货约》（CIM）和《国际铁路旅客和行李运输公约》（CIV）合并而成。

《国际货约》（CIM），全称《关于铁路货物运输的国际公约》，1961 年在伯尔尼签字，1975 年 1 月 1 日生效。其成员国包括了主要的欧洲国家，如法国、德国、比利时、意大利、瑞典、西班牙及东欧各国，此外，还有西亚的伊朗、伊拉克、叙利亚，西北非的阿尔及利亚、摩洛哥、突尼斯等 28 国。《国际货协》（CMIC）全称《国际铁路货物联合运输协定》，1951 年在华沙订立。我国于 1953 年加入。1974 年 7 月 1 日生效的修订本，其成员国主要是苏联、东欧加上我国、朝鲜、越南等共计 12 国。《国际货协》的东欧国家又是《国际货约》

的成员国，这样《国际货协》国家的进出口货物可以通过铁路转运到《国际货约》的成员国去，这为沟通国际铁路货物运输提供了更为有利的条件。我国是《国际货协》的成员国，凡经铁路运输的进出口货物均按《国际货协》的规定办理。

3. 有关航空运输的国际公约

1929 年《统一国际航空运输某些规则的公约》，又称《华沙公约》，是关于规范由航空设施有偿运送国际人员、行李及货物责任的一个国际公约，1929 年 9 月 12 日订于波兰华沙。1933 年 2 月 13 日生效，后经多次修改，我国于 1957 年 7 月通知加入，1958 年 10 月对我国生效。

华沙公约生效后又分别于 1955 年以《海牙议定书》的形式和 1960 年以《瓜达拉哈拉公约》的形式被修订。修订 1929 年 10 月 12 日在华沙签订的《统一国际航空运输某些规则的公约的议定书》，又称《海牙议定书》，1955 年 9 月 28 日在海牙订立，1963 年 8 月 1 日起生效，1975 年对我国生效。我国没有加入《瓜达拉哈拉公约》。《海牙议定书》是对华沙公约的修订，虽然正式的《海牙议定书》与《华沙公约》是一个单一的文件，《华沙公约》中原有的 152 个缔约方中有 137 个国家批准了《海牙议定书》。公约有法文、英文、西班牙文等三种正本。有分歧时，应以公约原起草文本，即法文为准。还规定本议定书应交存波兰人民共和国政府保管。

《统一国际航空运输某些规则的公约》又称《蒙特利尔公约》，是国际民航组织成员国在 1999 年的外交会议上通过的一个多边条约。它修正了华沙公约关于空难受害者赔偿制度的重要规定。该公约试图重新建立关于国际运输旅客、行李和货物的统一性和可预测性规则。同时保留了国际航空运输界几十年来一直依赖的核心规定（即华沙体系），新条约在若干关键领域做出了现代化的规定。到 2015 年年底，已经有 119 个国家加入此公约。国际民用航空组织 191 个成员国中的 118 个国家和欧盟已经批准加入了《蒙特利尔公约》，包括美国、欧盟各国、澳大利亚、加拿大、中国、印度、日本、韩国和墨西哥等。

4. 有关国际多式联运的公约

《联合国国际货物多式联运公约》是 1980 年 5 月 24 日在日内瓦举行的联合国国际联运会议第二次会议上，经与会的 84 个贸发会议成员国一致通过的。《联合国国际货物多式联运公约》全文共 40 条和一个附件。该公约在结构上分为总则、单据、联运人的赔偿责任、发货人的赔偿责任、索赔和诉讼、补充规定、海关事项和最后条款 8 个部分。中国已在《联合国国际货物多式联运公约》上签字。但是该公约尚未生效，因为依据该公约本身的规定，它将在第 30 个国家成为其成员国之日起的 12 个月后开始生效。

三、国际贸易惯例

国际贸易惯例，又称为"国际商务惯例"或"国际经贸惯例"，是指在国际贸易中，被买卖双方和其他从事相关国际贸易活动的人们所广泛认可和接受的习惯做法、规则和解释，并在与法律不抵触时作为判断争议的规范。比如贸易术语 CIF 和 FOB，就是典型的国际贸易惯例。因为国际贸易涉及的领域非常广泛，所以国际贸易惯例的内容和种类也比较多。下面是几种主要的国际贸易方面的惯例。

（一）有关国际贸易术语的国际惯例

为了规范买卖双方对国际贸易合同中自身义务的理解，各种术语以缩写的形式已经被使用，比如 ex–works，FOB，CIF，landed 等。虽然这些缩写的字母非常有用，但也可能引起贸易纠纷。国际商会（ICC）制定了"国际贸易术语解释通则，"最新修订的是 2010 年的通则。还有美国对外贸易定义修订本和华沙规则。尽管这些术语的缩写非常相似，但他们的意思与运作互有差异，所以在合同中使用术语缩写时明确它来自哪个惯例是极为重要的。主要的国际贸易术语的惯例包括《1932 年华沙—牛津规则》《1941 年美国对外贸易定义修订本》《国际贸易术语解释通则》等。

在中国进出口贸易业务中，采用国际商会的规定和解释的居多，如按 CIF 条件成交还可同时采用《华沙—牛津规则》的规定和解释。如从美国和加拿大按 FOB 条件进口时，在规定合同条款和履行合同时，还应考虑《1941 年美国对外贸易定义修订本》对 FOB 术语的特殊解释与运用。

（二）有关国际货款的收付方面的惯例

1. 国际商会的《跟单信用证统一惯例》

跟单信用证统一惯例（UCP）是一套关于信用证发行和使用的统一规则。国际商会为明确信用证有关当事人的权利、责任、付款的定义和术语，调和各有关当事人之间的矛盾，减少各国试图在信用证实践中推行本国的规则而引起的争议和纠纷，于 1930 年拟订一套《商业跟单信用证统一惯例》，并于 1933 年正式公布。并随着国际贸易变化进行了多次修订，称为《跟单信用证统一惯例》，国际商会定期对 UCP 规则进行修订，目前的版本是 UCP600，是由国际商会银行委员会于 2006 年 10 月 25 日的巴黎会议上通过的，于 2007 年 7 月正式开始实施。

UCP 被各国银行和贸易界所广泛采用，目前有超过 175 个国家的银行和商业人士在贸易融资中使用信用证。国际贸易中大约 11%～15% 的业务采用信用证方式结算，每年总额超过一万亿美元。UCP 已成为信用证业务的国际惯例。

2. 国际商会制定的《托收统一规则》1995 年修订本（国际商会第 522 号出版物）

国际商会为统一托收业务的做法，减少托收业务各有关当事人可能产生的矛盾和纠纷，曾于 1958 年草拟《商业单据托收统一规则》，又于 1978 年和 1995 年两次修订，最近的版本是《托收统一规则》国际商会第 522 号出版物（简称《URC522》）。《托收统一规则》自公布实施以来，被各国银行所采用，已成为托收业务的国际惯例。

（三）运输与保险方面的惯例

1. 英国伦敦保险协会制定的《伦敦保险协会货物保险条款》

在国际海运保险业务中，英国所制定的保险规章制度，特别是保险单和保险条款对世界各国影响很大。目前世界上大多数国家在海上保险业务中直接采用英国伦敦保险协会制定的"协会货物条款"（Institute Cargo Clause，I. C. C.）。

19 世纪，劳埃德和伦敦保险协会一起对海上保险条款进行了标准化，并一直保持至今。这些条款之所以被称为协会货物条款，是因为伦敦保险协会支付了保险条款的出版费用。"协会货物条款"最早制定于 1912 年，后来经过多次修订，最近一次的修订于 1981 年完成，

1983 年 4 月 1 日起生效。

2. 国际海事委员会制定的《约克—安特卫普规则》

《约克—安特卫普规则》最早于 1890 年制定，经过了多次修订，最近的一次修订是 2004 年 5 月至 6 月，国际海事委员会在加拿大温哥华对《1994 年约克—安特卫普规则》进行修改，修改后的规则称为《2004 年约克—安特卫普规则》，该规则于 2005 年 1 月 1 日生效。每次修改该规则都不废止旧规则，国际上常用的《约克—安特卫普规则》有 1974 年和 2004 年的规则，供有关方面选择使用。《约克—安特卫普规则》是国际社会为统一各国海损制度而努力的成果。虽然该规则不是国际公约，但是由于海运提单以及租船公司大多订有按照《约克—安特卫普规则》进行共同海损理算的条款，该规则在实践中已成功避免了各国共同海损制度差异造成的消极影响，已成为国际海运以及保险界广泛接受的国际惯例。

（四）国际仲裁方面的惯例

《联合国国际贸易法委员会仲裁规则》（以下简称《贸易法委员会仲裁规则》）是由联合国国际贸易法委员会于 1976 年制定的并于 2010 年修订，新版本于 2010 年 8 月 15 日生效。《贸易法委员会仲裁规则》可由合同的双方或一方选择作为其合同的一部分，或发生争议后，选择其作为解决纠纷的仲裁程序的行为准则。

在国际立法层面上商事仲裁的法律框架是由 1958 年的《关于承认和执行外国仲裁裁决的公约》（以下简称《纽约公约》）、1976 年的《联合国国际贸易法委员会仲裁规则》以及 1985 年的《联合国国际贸易法委员会国际商事仲裁示范法》（以下简称《示范法》）共同构成的。而其中 1976 年的《联合国国际贸易法委员会仲裁规则》自通过以来一直被公认为是最成功的契约性国际文件之一，30 余年来一直被广泛应用于临时仲裁、投资者与东道国之间的投资争议仲裁、国家间仲裁以及常设仲裁机构管理下的商事仲裁。

第三节　获得对外贸易经营者资格

一、开展国际贸易采取的形式

就像在第一节中谈到的国际贸易需要那些具备专业知识的人士，不同大小的公司从事国际贸易时需要的人才和在企业内部的安排也是有差异的，有时候同样的一个人既从事出口又从事进口方面的工作。在较小的公司，一个人可能全权负责所有贸易事务，而在进出口贸易额很大的公司，就需要很多国际贸易方面的专业人员。随着贸易规模增长，公司内部专业化分工会越来越细，每一个人所负责的事项也会越来越细。通常从事贸易采取的部门安排有以下几种：

（一）出口业务部

通常很多制造公司的出口业务部是从他们的销售部或营销部发展而来的。出口部的形成主要是基于开拓国外的客户。由于出口订单的履行在程序方面比较特殊，比如制造要求、信用审核、保险、装箱、运输和收款等，所以就需要配备特定的人员与货代公司、快递公司、银行、包装公司、班轮公司、航空公司、翻译、政府部门、国内运输公司及律师等进行联系，这个不同于公司原来与国内各种公司的联系。出口业务部所需要的人员及其承担的责任的大小取决于公司规模和出口额的大小。

（二）进口业务部

一般来说，一家制造企业的进口业务部往往从其采购部发展而来，其成员要承担购买制造过程所需的原材料和零部件。从事进口成品的贸易公司往往是一开始作为外国制造商的分销商，或购买那些在本国市场有销售潜力的外国产品发展而来的，比如最近增加较多的是这样一种情形：某些中国制造企业由于海外国家更为便宜的劳动力而将其以前在国内生产的部分转移出去，进口这些产品都需要与货代公司、本国报关经纪人、银行、海关、保险公司和其他服务公司等有业务上的联系。

（三）进出口业务部

在一些公司，进出口业务部将进口和出口方面的部分或全部功能合并在一起。规模小的公司只需要配备一两个人员就可以承担起进出口的业务工作。当公司成长壮大之后，就需要设置更多的进出口业务部来承担这些功能，那些进出口都需要与之打交道的业务或功能（比如银行、货代等）就会由专门的人员来负责，而其他人就专门从事出口或进口。

二、取得对外贸易经营资格

在中国要想开展国际贸易，需要先获得外贸经营权。国家规定，只有具备进出口经营权的企业，才能直接经营进出口业务。没有进出口权的企业则可以采用委托对外贸易经营者代为办理对外贸易业务。

对外贸易经营者，是指依法办理工商登记或者其他执业手续，依照《对外贸易法》和其他有关法律、行政法规、部门规章的规定从事对外贸易经营活动的法人、其他组织或者个人。对外贸易经营者管理制度是我国对外贸易管理制度之一。

目前，我国对对外贸易经营者的管理实行备案登记制。法人、其他组织或者个人在从事对外贸易经营前，必须按照国家的有关规定，依法定程序在商务部备案登记，取得对外贸易经营的资格，在国家允许的范围内从事对外贸易经营活动。对外贸易经营者未按规定办理备案登记的，海关不予办理进出口货物的通关验放手续，对外贸易经营者可以接受他人的委托，在经营范围内代为办理对外贸易业务。

开办外贸公司及申请进出口经营资格的步骤如下：

（一）公司成立

（1）获得经营场所，一般就是租好办公室。

（2）构思好公司的名字，到工商局进行名称核准（市级公司到市工商局，省级公司到省工商局），领表，填表，并按规定准备材料（主要是股东身份证复印件、授权书、简历等）。

（3）拿到名称核准通知书后，到银行开立临时账户，将注册资金存入该账户。

（4）找一间会计师事务所，准备出具验资报告。

（5）拿到验资报告后，到工商局办理营业执照。

（6）到技术监督局办理代码证及 IC 卡。

（7）到国税、地税办理税务证并申请一般纳税人资格。

（8）公司成立。

（二）获取进出口经营权

（1）网上申请进出口经营资格，提交后将按要求准备的材料交到区外经委，转市外经委，转省外经委批准，取得进出口经营资格。

（2）到工商局申请经营范围变更。

（3）到税务局进行经营范围变更。

（4）到当地海关备案。

（5）到外汇管理局办理进出口核销登记并开立美元账户。

（6）退税局办理退税登记，领取退税证。

（7）电子口岸登记，先到技术监督局审核盖章，再到工商局审核盖章，再到税务局审核盖章，再到海关领取操作系统、读卡器、IC 卡等。再拿 IC 卡到外经委、外汇管理局办理 IC 卡备案，到海关备案，到电信公司办理上网手续。（请注意 17999 上网卡是电子口岸专用的，有拨号和宽带两种。）

（8）到进出口检验检疫局办理登记，申请产地证、普惠制产地证注册。

（9）在网上申请核销单，再到外汇管理局买核销单。

（10）最后到海关买报关单。

上述手续办妥之后，就可以开展业务了。在办理手续期间需要准备的材料种类繁多，需要的公章很多，最好将各种证件原件、公章带在身边，并复印多份证件。

在这些环节中，如果申请为增值税一般纳税人，申请海关注册、进出口经营权、退税登记等都需要提供相应的文件和材料，具体需要准备什么资料请参阅各政府部门的网站。

三、获得国际贸易配额或许可文件

进出口许可制度作为一项非关税措施，是世界各国管理进出口贸易的一种常见手段。它是国家外贸相关部门通过发放许可证的方式管制商品进出口的一项行政措施，是国家对外贸易政策不可或缺的一个重要组成部分。货物、技术进出口许可管理制度是我国进出口许可管理制度的主体，其管理范围包括禁止进出口货物和技术、限制进出口货物和技术、自由进出口货物和技术。如果想判断某种货物是属于禁止、限制还是自由进出口类，就必须根据该进出口货物的种类（归类）来确定，即需要找到它的税号类别或商品名称协调制度编码（HS code）。有时候政府部门还可能会根据国际贸易形势的变化对货物的类别进行调整，所以对外贸易经营者需要不断关注中华人民共和国商务部、海关总署及其他相关网站发布的最新的政策和规定。

申请进出口配额或许可证的程序如下：

（一）申请者通过互联网申请配额或许可证

（1）申请者登录到中国商务部的网上申请系统（网址：http：//egov. mofcom. gov. cn），然后申请者选择进入系统中某一种具体配额或许可证。

（2）申请者根据系统提示和相关法律法规在网上填写申请表并提交。

（3）申请者在收到发证机构网上的通过通知后，向发证机构提交所需材料。

（二）申请者以书面形式申请配额或许可证

申请者可以直接将打印好的申请表和其他所需材料寄交给指定机构。例如，如果申请者想申请进口某些杀虫剂，则需要向农业部递交书面的申请；如果想申请某些美术品的出口，则需要向文化部提交书面的申请。

（1）指定机构审核申请书和跟申请书一起提交的符合相关法律和规定的文件材料，然后以公告方式或网上系统方式通知申请者结果。

（2）在收到许可通知结果后，申请者从指定机构获得配额证书或许可证。

本章小结

开展国际贸易之前需要进行一系列的准备工作，包括设定贸易目标、对潜在市场进行市场调研，制订贸易计划，还要知晓国际贸易的法律体系，熟悉取得对外贸易经营资格及申领进出口配额或许可证件的程序和做法。

国际贸易术语

1. 了解国际贸易术语的性质和作用
2. 了解国际贸易术语的国际惯例
3. 理解《Incoterms 2010》对各种贸易术语的解释
4. 掌握六种常用贸易术语的运用

国际贸易的买卖双方，一般来说很少一手交钱，一手交货。所以需要明确在哪里交货、交货的价格以及在货物交接过程中的风险、责任和费用等。上述问题往往通过使用特定的贸易术语来确定，如"FOB""CFR""CIF"等。不同的贸易术语对应不同的交货地点，由此所形成的买卖双方的权利和义务关系也就不同，相关的责任、费用和风险的界定也不一样。当买卖双方在合同中确定采用某种贸易术语时，就要求合同的其他条款都与其相适应，例如"FOB"术语，意味着卖方应负责租船订舱的责任，买方则需支付运费，并负责购买保险和支付保费等，故不同贸易术语的价格构成也不尽相同。总的来说，在国际贸易中，一般都以合同中规定的贸易术语来确定合同的性质，并以此来确定买卖双方各自承担的责任、费用、风险以及各自的权利和义务。

第一节 国际贸易术语及其惯例

一、国际贸易术语的性质

（1）贸易术语表示交货条件。采用某个专门的贸易术语，就能明确买卖双方在货物交接过程中各自承担的责任、费用与风险，即确定了交货条件。例如，按装运港船上交货条件FOB成交同按完税后交货条件DDP成交比较，由于交货条件不同，买卖双方各自承担的责任、费用和风险也有很大区别。

（2）贸易术语表示价格的构成。采用不同的贸易术语，货价中所包含的从属费用也不相同，因此报价也就不一样。如按 FOB 价成交与按 CIF 价成交的情况比较，前者不包括从装运港到目的港的运费和保险费，而后者则包括在内，由于两者价格构成因素不同，所以买卖双方确定成交价时，FOB 价比 CIF 价低。

二、国际贸易术语的作用

（1）简化交易手续，促进迅速成交。在国际贸易中，使用贸易术语对明确简化买卖双方洽商的内容，缩短交易洽商的进程与促进成交，节省业务费用和时间都有积极的作用。国际贸易涉及很多程序，所有这些程序需要进出口商来履行，例如，谁来负责办理保险？谁来负责租船订舱？谁来负责申请进口和/或出口许可证等。如果进出口商就每笔业务进行商务谈判时都要协商进出口每个步骤的费用和责任由谁来负担，这样就会浪费太多时间。而只要买卖双方商定按何种贸易术语进行成交，就可明确彼此在交易过程中应承担的责任、费用和风险，从而简化了交易手续，有利于买卖双方节省磋商时间，迅速达成交易。

（2）便于买卖双方核算成本和价格。由于贸易术语表示价格构成因素，采用不同的贸易术语，货价中所包含的从属费用也不相同，报价也就不一样，所以买卖双方在确定成交价格时，必然会考虑所采用的贸易术语中包含哪些从属费用，如运费、保险费、装卸费、关税等，这就有利于买卖双方进行价格比较和加强成本核算。

三、与国际贸易术语相关的国际贸易惯例

国际贸易惯例，是指在国际贸易实践中逐步形成的、具有普遍指导意义的一些习惯做法或解释。其范围包括：国际上的一些组织、团体就国际贸易的某一方面，如贸易术语、支付方式等问题所作的解释或规定；国际上一些主要港口、码头的传统惯例；不同行业的惯例。此外，各国司法机关或仲裁机构的典型案例或裁决，往往也视作国际贸易惯例的组成部分。有关贸易术语的国际贸易惯例主要有三个，它们是《1932 年华沙牛津规则》（Warsaw-Oxford Rules 1932）；《1941 年美国对外贸易定义修订本》（Revised American Foreign Trade Definitions 1941）；《2010 年国际贸易术语解释通则》（Incoterms 2010），现介绍如下：

1. 《1932 年华沙牛津规则》

本规则是国际法协会专为解释 CIF 合同而制定的。1928 年国际法协会在波兰华沙开会，制定了有关 CIF 买卖契约统一规则，称为《1928 年华沙规则》，后经 1932 年牛津会议，对华沙规则进行了修订，定名为《1932 年华沙牛津规则》，一直沿用至今，全文共 21 条。该规则主要说明 CIF 买卖合同的性质和特点，并具体规定了采用 CIF 贸易术语时，有关买卖双方责任的划分以及货物所有权转移的方式等问题。

当事人在订立合同时，只要明确说明采用《华沙牛津规则》，就表明双方同意使其合同为 CIF 合同，《华沙牛津规则》对其就有了约束力。但如本规则与合同发生矛盾，应以合同为准。由于《华沙牛津规则》只是对 CIF 的解释和规定，在实践中使用的人并不多。

2. 《1941 年美国对外贸易定义修订本》

1919 年美国的 9 个商业团体制定了《美国出口报价及其缩写条例》（The U. S. Export

Quotation and Abbreviations），1941 年又对它作了修订，并改称《1941 年美国对外贸易定义修订本》。该修订本在同年被美国商会、全国进口商协会和全国对外贸易协会采用。它对 ex-point of origin 、FOB、FAS、C&F、CIF、ex－dock 等 6 个术语作了解释。上述"定义"多被美国、加拿大以及其他一些美洲国家采用，不过由于其内容与一般解释相距较远，国际很少采用。

近年来美国的商业团体或贸易组织也曾表示放弃它们惯用的这一"定义"，将尽量采用国际商会制定的《国际贸易术语解释通则》。但应注意的是该规则仍在美洲国家中采用，因此在与这些国家进行国际贸易时，如双方同意采用此规则，则应在合同中作出明文规定，否则是没有约束力的。

3. 《2010 年国际贸易术语解释通则》

国际商会于 1936 年首次公布了一套解释贸易术语的国际规则，其后进行了七次修订共有 8 个版本。Incoterms 作为一种国际贸易惯例，在长期的国际贸易实践中，越来越普遍地得到承认和应用，成为当今国际贸易的双方当事人签约、履行及解决业务纠纷的依据。

连续修订 Incoterms 的主要原因是使其适应当代商业实践。《Incoterms》于 1953 年修订后为 8 个贸易术语，于 1967 年又补充了两个贸易术语，即"边境交货"（delvered at frontier，DAF）与"完税后交货"（delivered duty paid，DDP）；1976 年又补充了"启运地机场交货"（FOA）；1980 年又增加"货交承运人"（FRC）和"运费、保险费附至（目的地）"（CIP），到 1980 年第四次修订《Incoterms》，已有 14 个贸易术语。1990 年的修订是为了适应广泛使用的电子数据交换（EDI），于 1990 年 7 月 1 日正式生效，共有 13 个贸易术语。2000 年的修订则考虑了 20 世纪 90 年代出现的无关税区的广泛发展。

目前的最新修订版是《2010 国际贸易术语解释通则》，该通则考虑了免税贸易区的不断增加，电子沟通在商务中的不断增多，以及更加重视的货物运输中的安全和变化等问题。《Incoterms 2010》删去了《Incoterms 2000》D 组术语中的 DDU（delivered duty unpaid），DAF（delivered at frontier），DES（delivered ex ship），DEQ（delivered ex quay），只保留了 DDP（delivered duty paid），同时新增加了两个 D 组贸易术语，即 DAT（delivered at terminal）与 DAP（delivered at place），以取代被删去的术语。

毫无疑问，Incoterms 的每次修订都是改进其术语，为了使 Incoterms 更利于实务操作。值得注意的是，尽管新国际贸易术语解释通则已经生效，但是《2000 年通则》没有被废除，仍然可以使用。因此，在合同中应明确规定适用于哪个版本的国际贸易术语解释通则，这是非常重要和必不可少的。例如，"所选择的国际贸易术语的规则（含制定地点）适用《2010 年通则》"。

尽管国际贸易惯例在解决贸易纠纷时起到一定的作用，但应注意以下几个问题：

（1）国际贸易惯例并非法律，因此对买卖双方没有约束力，可采用也可不采用。

（2）如果买卖双方在合同中明确表示采用某种惯例时，则采用的惯例对买卖双方均有约束力。

（3）如果合同中明确采用某种惯例，但又在合同中规定与所采用的惯例相抵触的条款，只要这些条款与本国法律不矛盾，就将受到国家有关法律的承认和保护，即以合同条款为准。

（4）如果合同中既未对某一问题做出明确规定，也未订明采用某一惯例，当发生争议付诸诉讼或提交仲裁时，法庭和仲裁机构可引用惯例作为判决或裁决的依据。

在进出口业务中，我们应该多了解和掌握一些国际贸易惯例，对交易磋商、签订合同、履行合同和解决争议等是完全必要的。当然发生争议时，我们可以援引适当的惯例据理力争。

第二节 《2010 年国际贸易术语解释通则》

《2010 年国际贸易术语解释通则》更新并整合与"交货"相关的规则，将术语总数由原来的 13 条减至 11 条，并对所有规则做出更简洁、明确的陈述。同时，《国际贸易术语解释通则》2010 也是第一部使得所有解释对买方与卖方呈现中立的贸易解释版本。

一、2010 贸易术语的分类

新通则分类与《2000 年通则》有所不同。《2000 年通则》把 13 种价格术语分为 E、F、C、D 四组。第一组为"E"组（ex works），指卖方仅在自己的地点为买方备妥货物。第二组"F"组（FCA、FAS 和 FOB），指卖方需将货物交至买方指定的承运人；按 F 组术语签订的合同属于装运合同。第三组"C"组（CFR、CIF、CPT 和 CIP），指卖方须订立运输合同，但对货物灭失或损坏的风险以及装船和启运后发生意外所产生的额外费用，卖方不承担责任；按 C 组术语签订的合同也属于装运合同；C 组术语区别于其他组术语的最大特点是：风险划分点与费用划分点相分离。第四组"D"组（DAF、DES、DEQ、DDU 和 DDP），指卖方须承担把货物交至目的地国家所需的全部费用和风险；D 组术语是到达合同。

《2010 年国际贸易术语解释通则》将 11 个贸易术语按照所适用的运输方式划分为两大类：

（一）第一类：适用于任何运输方式的七个术语

这七个术语分别是：EXW、FCA、CPT、CIP、DAT、DAP 和 DDP。

EXW（ex works）：工厂交货

FCA（free carrier）：货交承运人

CPT（carriage paid to）：运费付至

CIP（carriage and insurance paid to）：运费/保险费付至

DAT（delivered at terminal）：目的地或目的港的集散站交货

DAP（delivered at place）：目的地交货

DDP（delivered duty paid）：完税后交货

这七个术语，不论选用何种运输方式，也不论是否使用一种或多种运输方式，均可适用。而且当船舶用于部分运输时，也可以使用这些术语。

（二）第二类：适用于海上和内河水上运输方式的四个贸易术语

这四个术语分别是 FAS、FOB、CFR、CIF。

FAS（free alongside ship）：装运港船边交货

FOB（free on board）：装运港船上交货

CFR（cost and freight）：成本加运费

CIF（cost，insurance and freight）：成本、保险费加运费

第二类术语，卖方将货物交至买方的地点都是港口，因此，被划分为"适于海运及内河水运的术语"。其中 FOB、CFR 和 CIF 三个术语中，删除了 Incoterms 2000 中以"船舷"作为交货点的表述，取而代之的是货物置于"船上"时构成交货。这样的修改更具有操作性。

如图 2-1 所示从卖方的角度，呈现了"2010 通则"中 11 种贸易术语的风险和费用转移点。

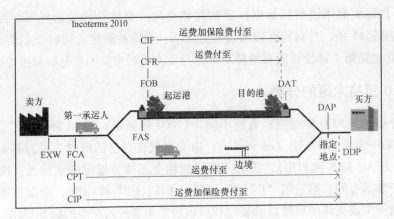

图 2-1　国际贸易术语示意图

二、2010 贸易术语的主要特点

国际贸易术语的数量从 13 个减至 11 个。该变化是通过使用两个适用于任何运输方式的新术语：即 DAT（运输终端交货）和 DAP（目的地交货）取代《国际贸易术语解释通则》2000 中的 DAF（边境交货）、DES（目的港船边交货）、DEQ（目的港码头交货）和 DDU（未完税交货）来实现的。

在这两个新术语中，交货都在指定目的地发生。使用 DAT 时，货物已从到达的运输工具中卸下，交由买方处置（与以前的 DEQ 术语相同）。使用 DAP 时，货物同样交由买方处置，但需做好卸货准备（与以前的 DAF、DES 和 DDU 术语相同）。

在新《通则》指导性解释中，要求货物的买方、卖方和运输承包商有义务为各方提供相关资讯，知悉涉及货物在运输过程中能否满足安检要求。此举将帮助船舶管理公司了解船舶运载的货物是否触及危险品条例，防止在未能提供相关安全文件下，船舶货柜中藏有违禁品。

《通则》亦因国际贸易市场的电子货运趋势，指明在货物买卖双方同意下，电子文件可取代纸质文件，具有同等效力。

值得注意的是：在新《通则》中，不再有"船舷"的概念。换言之，在原先的 FOB、CFR 和 CIF 术语解释中，"船舷"的概念被删除，取而代之的是"装上船"（placed on board）。之前关于卖方承担货物越过船舷为止的一切风险，在新术语环境下改变为"卖方承担货物装上船为止的一切风险，买方承担货物自装运港装上船后的一切风险"。

考虑到一些大的区域贸易集团内部贸易的特点，规定 Incoterms 2010 不仅适用于国际销售合同，也适用于国内销售合同，并进一步与《联合国国际货物销售合同公约》及《鹿特丹规则》衔接。

新的《国际贸易术语解释通则》必将进一步促进国际货物贸易的发展，并有助于解决国际货物贸易中的纠纷。

三、主要贸易术语的解释

（一）FOB（free on board）　（…named port of shipment）——装运港船上交货价（……指定装运港）

1. FOB 的含义

FOB 是 free on board 的缩写，采用这一贸易术语时，须在 FOB 后面注明装运港的名称，例如，装运港为广州，即 FOB 广州。该术语是指卖方在指定装运港将货物装上买方指定的船舶或通过取得已交付至船上货物的方式。买方自该时刻起，承担一切费用和货物灭失或损坏的风险，卖方办理货物出口清关。该术语仅适用于海运或内河运输。

2. FOB 下买卖双方的义务

（1）卖方义务。①在合同规定的装运港和规定的期限内，将货物装上买方指派的船只，并及时通知买方；②卖方要自负风险和费用，取得出口许可证或其他官方证件，并负责办理出口海关手续；③负担货物装上买方指定装运港"船上"为止的一切费用和货物灭失或损坏的风险；④卖方还要自费提供证明他已按规定完成交货义务的证件，包括商业发票、清洁的已装船单据以及合同规定的其他单据或具有同等效力的电子信息。

（2）买方义务。①必须自付费用订立自装运港运送货物的合同，负责租船定舱，支付运费，并将船期和船名及时通知卖方，以使卖方准备好货物装船；②负担货物在装到"船上"之后的一切费用和货物灭失或损坏的风险；③自负风险和费用，取得进口许可证或其他官方文件，办理进口报关手续，并办理货物进口及经由他国过境运输的一切海关手续；④接受卖方提供的有关货运单据，并按合同规定支付货款。

3. 采用 FOB 术语应注意的问题

（1）装船概念和风险转移点的问题。卖方必须在合同规定期限内，在指定的装运港内的装船点（如果有），"以将货物置于买方指定的船舶上的方式交货"或"以取得已在船上交付货物的方式交货"，这是卖方交货的义务。其后发生的风险和费用由买方负担。这里的风险是指货物灭失或损坏的风险，而费用是指正常运费以外的费用。

（2）船货衔接问题。按 FOB 条件成交，因为是买方派船，故一定要注意船货的衔接问题。一方面，买方需及时安排船只或者舱位；另一方面，卖方将货物送到指定的港口。为了做到船货衔接，买方应在其租船订舱后就船名、装载点和预计开船日期等信息给予卖方及时的通知（即"装船指示"）。

如果买方指定了船只，而未能及时将船名、装货泊位及装船日期通知卖方，或者买方指派的船只未能按时到达，或未能承载货物，或者未能在规定的期限内装货，买方要承担由此产生的一切风险和损失。前提是货物已清楚地分开或被固定为供应本合同之用。买方派船是卖方履行合同的前提条件，如果买方未能在规定时间内指派船只，卖方可以请求损害赔偿，致严重后果时可撤销合同。如果船只按时到港，因卖方货未备妥而不能及时装运，则卖方应承担由此而造成的空舱费或滞期费。

卖方在装船后应及时地给予买方充分的通知。因为在 FOB 条件下，货物的运输保险是由买方办理的，而货物装至船上之后的风险由卖方转移至买方，因此，如果卖方装运后不及时通知买方，可能使买方投保过迟，以致不能对货物取得保险。卖方发出装运通知便于买方及时安排收取货物以及货物的入库等事项。

采用 FOB 贸易术语，在买方要求并由买方承担风险和费用的情况下，卖方可给予一切协助，如替买方租船订舱，取得提单或其他运输单。

（3）关于装船费用的负担问题。按 FOB 定义，货物装至船上之前的一切费用都由卖方承担，装上船之后的费用由买方承担。但装船是一个连续的过程，这给实际费用的划分造成了困难。如果采用班轮运输，装船费用包括在班轮运费中，与装船有关的各项费用自然由负责办理运输事项的买方承担。而如果采用租船运输，按照航运惯例，通常在租船合同中规定船方不负担装船费用。在这种情况下，买卖双方应在合同中明确装船、理舱、平舱费用的负担问题。为了避免买卖双方在装船等费用的负担问题上发生争议，使装船费用的划分更加明确，我们往往在 FOB 之后加列各种附加条件，在贸易习惯做法中便产生了 FOB 的种种变型：

FOB 班轮条件（FOB liner terms），其含义是装船的有关费用按照班轮的做法办理。也就是卖方不负担这些费用，而由船方（实际上是买方）负担。

FOB 吊钩下交货（FOB under tackle），其含义是卖方仅负责把货物交到买方指派船只的吊钩所及之处，以后的装船费用概由买方负担。

FOB 包括理舱费（FOB stowed），这一条件是指卖方要负责把货物装入船舱并负担包括理舱费在内的装船费用。

FOB 包括平舱费（FOB trimmed），这是指卖方要负责把货物装入船舱，并为了保持船身的平稳，对装入船舱的散装货物进行填平补齐，上述各项费用均由卖方负担。

FOB 包括理舱和平舱（FOB stowed and trimmed 或 FOBST）：装船费、理舱和平舱都由卖方承担。

以上五种变型一般用于承租船运输方式，在使用班轮运输时并不需要在 FOB 后面画蛇添足，因为班轮运输本身就默示了运费是由租船方支付的。一般而言，FOB 术语的变形仅用于区分装船费用问题，并不改变 FOB 术语的风险转移点，即风险还是在货物装至船上时从卖方转移至买方。

（4）个别惯例对 FOB 的不同解释。《国际贸易术语解释通则》与《美国对外贸易定义修正本》对 FOB 的分类及有关办理出口手续付费有很大不同。所以在实际操作中需加注意。《1941 年美国对外贸易定义修正本》将 FOB 分为六种，只有第五种是装运港船上交货。对 FOB 的解释是运输工具上交货，其适用范围很广。因而在同美国商人和其他美洲国家商人签订 FOB 合同时，为了使其与《通则》FOB 规定相接近，应在 FOB 后面加"船舶"（vessel）字样，并标明装运港名称，如 FOB vessel New York。如果只订为 FOB New York，则指卖方只负责将货物运到纽约城内的任何地方，就算完成交货义务，不负责将货物运到纽约港口并交到船上，特别是在进口合同中更应注意。在费用的负担上，即使在 FOB vessel 下，出口报关的责任在买方而不在卖方，只有在买方提出请求并由买方负担费用的情况下，卖方才有义务协助买方取得由出口国签发的为货物出口所需的各种证件，并且出口税和其他税捐、费用也需由买方负担。所以我国在与美国、加拿大等国家洽谈进口贸易使用 FOB 方式成交时，除在 FOB 后注明 vessel 外，还应明确由对方（卖方）负责办理出口结关手续。在风险的划分上，是以船舱为界，即卖方负担货物装到船舱为止所发生的一切灭失与损坏。在费用的负担上，即使在 FOB vessel 下，出口报关的责任在买方而不在卖方，只有在买方提出请求并由买方负担费用的情况下，卖方才有义务协助买方取得由出口国签发的为货物出口所需的各种证件，并且出口税和其他税捐、费用也需由买方负担。在风险的划分上，是以船舱为界，即卖方负担货物装到船舱为止所发生的一切灭失与损坏。

（二）CIF（cost, insurance and freight）（…named port of destination）——成本加保险费、运费（……指定目的港）

1. CIF 的含义

CIF 是 cost insurance freight 的缩写。采用 CIF 这种术语，在其后面应注明目的港，例如目的港是青岛，则表示为 CIF 青岛。该术语是指卖方负责租船定舱，按期在装运港将合同规定的货物装上运往约定目的港的船上，办理保险手续，并负责支付运费和保险费。至于有关风险和责任的划分则同 FOB 条件一致，即卖方仅负责货物在装运港装上船以前发生的风险。该术语仅适用于海运或内河运输。

2. CIF 下双方的义务

（1）卖方义务。①负责租船定舱，在合同规定的装运港和期限内，将货物装上船并支付至目的港的运费，装船后通知买方；②负担货物装至船上以前的一切费用和风险；③负责办理保险并支付保险费；④负责办理出口手续；⑤负责提供有关货运单据如商业发票、保险单和货物运往约定目的港的通常运输单据，或具有同等效力的电子信息。

（2）买方义务。①负担货物装至船上以后的一切费用和风险；②接受卖方提供的有关货运单据，并按合同规定支付货款；③办理在目的港的收货和进口手续；④支付除正常运费、保险费之外的有关货物在运输途中所产生的额外费用。

3. 采用 CIF 术语应注意的问题

（1）CIF 属于装运合同。按 CIF 条件成交时，卖方在规定的装运港和规定的期限内将货物装上船，即完成了交货义务，因此按 CIF 术语订立的合同属于"装运合同"。但是由于在 CIF 术语后所注明的是目的港，所以在业务上有人称 CIF 为"到岸价"，CIF 合同常被误解为"到货合同"。必须明确的是，卖方在装运港完成交货义务，卖方承担的风险也是在装运港货物装至船上以前的风险。货物装至船上以后的风险由买方承担。按 CIF 条件成交时，虽然卖方负责装运和投保，但并不承担保证把货物安全送到目的港的义务。因此在任何情况下，在 CIF 合同下，都不应该规定抵达目的港的交货时间，而只就货物的装运做出规定。

（2）关于保险险别的规定。在 CIF 术语下，卖方负责办理从装运港到目的地的海运货物保险，支付保险费。由于货物装至船上以后的风险将由买方承担，而运输途中的保险又是由卖方办理的，实际上是卖方为买方的利益代办保险的。如果合同未作具体规定，卖方只负责投保最低险别。最低的保险金额是合同价加 10%，并按合同规定的货币提供。但世界各国对此并无明确统一的规定，故为避免日后纠纷，双方在合同中就应明确卖方应投保的险别和保险金额。

（3）象征性交货。CIF 是一种特殊类型的贸易术语，采用该术语时，交单的重要性远远高于交货，是一种最典型的"象征性交货"和凭单付款。象征性交货是指卖方只要按合同规定将货物装到运往目的港的船上，并向买方提交约定的、代表货物所有权的装运单证，就算是完成了交货义务，无须保证到货。

凭单付款是指买方一旦受领装运单据，就必须按合同规定的条件支付货款，即使在交单时，货物已经灭失或受损，也不得例外，只能在付款之后，凭单据向有关责任方提出索赔，所以 CIF 也被称作"单据买卖"。在 CIF 合同下，卖方是凭单交货，买方是凭单付款。只要卖方按期向买方提交了合同规定的全套合格单证，不管货物是否损坏或灭失，买方必须履行付款义务；反之，如果卖方提交的单证不符合合同规定的要求，即使货物完好无损地运达目的地，买方仍有权拒绝付款。

（4）关于租船订舱的问题。以 CIF 术语签订合同，卖方的基本义务之一是租船订舱，办理从装运港到目的港的运输事项。根据《通则》规定，卖方必须自行负担费用订立运输合同，将合同规定的货物，按惯常航线，用通常运输此类货物的海轮运至目的港。所以，如果卖方不能及时租船或订舱，导致不能按合同规定装船交货，即构成违约，将承担有关的法律责任。买方一般无权对班轮公司或其他船只等提出要求。但在实际业务中，如买方提出了合同中没有明确规定的某些要求，如船舶的国籍、船龄、船级等，在卖方能够办理且又不增加额外费用的情况下，卖方也可以考虑接受。但在合同中做出明确规定的，必须严格执行。

（5）关于卸货费用的负担问题。如 FOB 术语中所述，如果采用班轮运输，装卸费用包括在班轮运费中，此时运费由 CIF 合同的卖方支付，在目的港的卸货费用自然也由负责办理运输事项的卖方承担。而如果采用不定期租船运输，按照航运惯例，通常在租船合同中规定船方不负担装卸费用。以 CIF 方式成交，在装运港的装货费用应由卖方支付，至于卸货费用究竟由谁来负担的问题，仍然存在较大分歧。为了明确卸货费用到底由谁负担，在国际贸易实践中，产生了 CIF 的几种变型，主要有：

CIF 班轮条件（CIF liner terms），卸货费与班轮运输的收取方式一样，由负责订立运输合同并支付运费的一方，即卖方负担。

CIF 卸到岸上（CIF landed），卖方负担将货物卸到目的港岸上的费用。

CIF 吊钩交货（CIF ex tackle），卖方负责将货物从船舱吊起到吊钩所及之处（码头上或驳船上）的费用。如果船舶靠不上码头，那么应由买方自费租用驳船，卖方只负责将货物卸到驳船上。

CIF 舱底交货（CIF ex ships hold），买方负担将货物从舱底吊卸到码头的费用。

（三）CFR（cost and freight）（…named port of destination）——成本加运费（……指定目的港）

1. CFR 的含义

CFR 是 cost and freight 的缩写，其中该术语的成本（cost）是指 FOB 价格。采用该术语时，在其后面应注明目的港，例如目的港是鹿特丹，则表示为 CFR 鹿特丹。该术语是指卖方负责租船定舱，按期在装运港将合同规定的货物装到运往约定目的港的船上，负责货物在装运港装上船前发生的一切费用和风险，支付运费。该术语仅适用于海运或内河运输。该术语中的成本是指 FOB 价格。

2. CFR 下买卖双方的义务

（1）卖方义务。①负责租船定舱和支付运费，在合同规定的装运港和期限内，将货物装上船并及时在装船后通知买方；②负担货物装至船上以前的一切费用和风险；③负责办理货物出口清关手续；④负责提供货运单据等有关单证或具有同等效力的电子信息。

（2）买方义务。①承担货物装至船上以后的一切费用和风险；②办理在目的港收货和进口清关的手续；③接受卖方提供的有关货运单据，并按合同规定支付货款。

3. 采用 CFR 术语应注意的问题

（1）关于卸货费用。上述的关于 CIF 术语为解决目的港卸货费用的负担而产生的变形，同样适用于 CFR 术语。如 CFR 班轮条件（CFR liner terms）、CFR 吊钩交货（CFR ex tackle）、CFR 卸到岸上（CFR landed）、CFR 舱底交货（CFR ex ship's hold）。

（2）装船通知问题。CFR 由于是由买方办理保险，故卖方一旦将货物装上船，必须立即以 E-mail、传真等快速通信方式向买方发出装船通知，以便买方办理投保。如果由于卖方疏忽致使买方未能投保，则卖方必须承担运输途中的风险。因此 CFR 术语下的装船通知尤为重要。

（3）关于 CFR 术语下的欺诈问题。在进口业务中，如采用 CFR 术语应特别慎重，因为在此术语下更易发生欺诈。当按 CFR 术语成交时，由外商安排装运，中方购买保险，有可能出现外商与船方勾结出具假提单、租用不适航的船舶、伪造品质证明书与产地证明书等情况，从而使我方蒙受损失。实践中应注意选择资信好的客户成交，并对船舶提出相关要求。

FOB、CIF 和 CFR 这三个贸易术语为装运港交货常用的三个传统的贸易术语。其相同之处在于买卖双方的风险划分都是以装运港"船上"为界的；它们的区别在于买卖双方在所承担的手续和费用方面的责任有所不同。

（四）FCA（free carrier）（...named place of delivery）——货交承运人（……指定地点）

1. FCA 的含义

FCA 是 free carrier 的缩写。在使用该术语时，要在 FCA 后加上指定地点，例如指定地点是长沙，则表示为 FCA 长沙。该术语是指卖方只要在合同规定的时间、地点，将货物交给买方指定的承运人监管，并办理出口清关手续，就算完成交货义务。FCA 术语的适用范围最广，它适用于各种运输方式，其中包括多式联运，但卖方只需负责将货物交给第一承运人即可。这种贸易术语在国际贸易业务中占有十分重要的地位，发挥的作用越来越大。

2. FCA 下买卖双方的义务

（1）卖方义务。①负责在合同规定的交货期内，在指定地点，将符合合同的货物交至买方指定的承运人；②负责办理货物出口手续，取得出口许可证或其他核准书；③承担货物在货物交给承运人以前的一切费用和风险；④负责提供商业发票和证明货物已交至承运人的通常单据或具有同等效力的电子信息。

（2）买方义务。①负责订立运输合同，支付运费，并将承运人的名称、要求交货的时间和地点及时通知卖方；②承担货物在交给承运人后的一切费用和风险；③自负风险和费用取得进口许可证或其他核准书，并办理货物进口以及必要时经由另一国过境运输的一切海关手续；④负责按合同规定支付价款，收取卖方按合同规定交付的货物，接受与合同相符的单据。

3. 采用 FCA 术语应注意的问题

（1）关于交货地点的问题。由于 FCA 适用于各种运输方式，它的交货地点需按不同的运输方式和不同的指定交货地点而定。

在 FCA 条件下卖方的交货在下列情况才算完成：

①若指定的地点是卖方所在地，则当货物被装上买方指定承运人或代表买方的其他人提供的运输工具时；②若指定的地点不是卖方所在地而是其他任何地点，则当货物在卖方的运输工具上，尚未卸货而交给买方指定的承运人或者其他人，或由买方指定的承运人或者其他人支配时。

可见，FCA 术语下装货和卸货的义务是：如交货地在卖方所在地时，卖方负责装货；交货地在卖方所在地之外时，卖方不负责卸货。

（2）运输合同问题。FCA 术语下，买方必须自负费用订立从指定地点发运货物的运输合同，并将有关承运人的名称、要求交货的时间和地点及时全部通知卖方。若卖方能协助取得更好的效果时可由卖方协助订立运输合同，但有关费用和风险由买方负担。

（3）风险的转移问题。卖方要承担货物被交由承运人监管为止的一切费用和货物灭失或损坏的风险。鉴于采用 FCA 术语时，货物大都作了集合化或成组化处理，如装入集装箱或装上托盘。因此，卖方应考虑将货物集合化所需的费用，也计算在价格内。

（五）CPT（carriage paid to）（...named place of destination）——运费付至（……指定目的地）

1. CPT 的含义

CPT 是 carriage paid to 的缩写。该术语是指卖方将货物在双方约定地点（如果双方已经约定了地点）交给卖方指定的承运人或其他人，并须支付从装运地至约定目的地的运费，而买方承担交货之后的一切风险和费用。CPT 术语适用于包括多式联运在内的任何运输方式。

2. CPT 下买卖双方的义务

（1）卖方义务。①办理出口结关手续，订立运输合同并支付运费，将货物运至指定目的地约定的地点，并向买方及时发出货物已交付的通知；②承担货物交给承运人以前的一切费用和货物灭失与损坏的一切风险；③向买方提供约定的单证或具有同等效力的电子信息。

（2）买方义务。①从卖方交付货物时起，承担货物灭失和损坏的一切风险；②支付除通常运费之外的有关货物在运输途中所产生的各项费用及卸货费；③在目的地从承运人那里受领货物，并按合同规定受领单据和支付货款。

3. 采用 CPT 术语应注意的问题

（1）风险划分和费用划分。按照 CPT 术语成交，虽然卖方要负担从装运地到约定目的地的运输责任和正常运费，但正常运费之外的其他费用，由买方负担。CPT 和 CFR 有许多相似之处，如卖方承担的风险并没有延伸至目的地，自卖方货物交付给承运人时起，货物发生灭失或损坏的一切风险，即转移给买方，可见货物在运输途中的风险，一概由买方承担。

（2）关于装运通知。CPT 术语下，卖方将货物交给承运人后，应及时向买方发出货已交付的通知，以便买方及时办理货物运输保险及在指定目的地从承运人那里受领货物。如果交货地点未约定或习惯上未确定，卖方可在给定目的地选择最适合其要求的地点。装运通知的重要性与上述的 FOB、CFR 术语条件下一样。在使用 CPT、CIP、CFR 或 CIF 术语时，当卖方将货物交付给承运人时，而不是当货物到达目的地时，即完成交货。

（六）CIP（carriage and insurance paid to）（…named place of destination）——运费、保险费付至（……指定目的地）

1. CIP 的含义

CIP 是 carriage and insurance paid to 的缩写。该术语是指卖方将货物在双方约定地点（如双方已经约定了地点）交给其指定的承运人或其他人，并须签订运输合同，支付从装运地至约定目的地的运费，还须订立保险合同以防买方货物在运输途中发生灭失或损坏风险。也就是说卖方除应订立运输合同和支付通常运费外，还应负责订立保险合同并支付保险费。在此需提请买方注意，按 CIP 条件成交，如果买方没有特殊要求并负担费用的话，卖方只需投保最低险别的保险，而买方承担卖方交货之后的一切风险和额外费用。CIP 术语的适用范围同 CPT 术语完全一样，它适用于各种运输方式，包括多式联运。

2. CIP 术语下双方的义务

（1）卖方义务。①办理出口结关手续，自费订立运输合同和保险合同，按期将货物交给承运人，并向买方及时发出货物已交付的通知；②承担货物交付承运人以前的一切费用和货物灭失与损坏的一切风险；③向买方提交约定的单证或具有同等效力的电子信息。

（2）买方义务。①从卖方交付货物时起，承担货物灭失和损坏的一切风险；②支付除通常运费之外的有关货物在运输途中所产生的各项费用和卸货费用；③在目的地从承运人那里受领货物，并按合同规定受领单据和支付货款。

3. 采用 CIP 术语应注意的问题

按 CIP 条件成交时，卖方除负有与 CPT 术语相同的义务外，还须办理运输途中应由买方承担的风险的货运保险，因此同 CIF 一样，卖方也是为买方的利益代办保险，卖方之所以自费办理保险是因为货物的售价中包括保险费。在一般情况下，卖方只按约定的险别投保。如未约定险别，卖方也按惯例投保最低限度的险别，保险金额一般在合同基础上加成 10%。如有可能，卖方应按合同货币投保。按 CIP 条件成交，是否加保战争、罢工、暴乱及民变险，由买方决定，卖方并无加保此险的义务。若买方要求加保，卖方应予以办理。不过，加保此类险的费用，如事先未约定计入售价中，应由买方另行负担。

四、装运港船上术语（FOB、CFR、CIF）与货交承运人术语（FCA、CPT、CIP）的比较

（一）共同点

（1）都是象征性交货，相应的买卖合同为装运合同。

（2）都由出口方负责出口报关，进口方负责进口报关。

（3）买卖双方所承担的运输、保险责任相对应。即 FOB 和 FCA 一样，由买方办理运输；CFR 和 CPT 一样，由卖方办理运输；而 CIF 和 CIP 一样，由卖方承担办理运输和保险的责任。由此产生的相应的注意事项也是相同的。

（二）不同点

（1）适合的运输方式有区别。FOB、CFR、CIF 适合于水运（包含海运和内河运输），其承运人一般是船公司。FCA、CPT、CIP 不仅适合于水运，而且也适合于陆运、空运等各种运输方式的单式运输，以及两种或两种以上不同运输方式相结合的多式运输，承运人一般是船公司、铁路局、航空公司和多式联运经营人。

（2）交货地点和风险划分界限有区别。FOB、CFR、CIF 方式中，卖方交货的地点和风险划分的界限是装运港船上。而 FCA、CPT、CIP 则是自货交承运人处置时起。

（3）装货费用和卸货费用的承担责任有区别。FOB、CFR、CIF 方式中，如果采用班轮运输，负责运输的一方就负责装卸，即负担装卸费用。而如果是租船运输，装卸费用的承担就通过贸易术语的变形来明确。在 FCA、CPT、CIP 方式下，一般是由承运人负责装卸，因而不存在贸易术语变形的问题。

（4）涉及的运输单据有区别。FOB、CFR、CIF 一般提交已装船清洁提单，而且应具有物权凭证的性质，而 FCA、CPT、CIP 则视运输方式的不同，提交相应的单据。且航空运单和铁路运单等是不具有物权凭证的性质的。

（5）投保险别不同。在使用 FOB、CFR 和 CIF 术语时投保的都是海洋货运险；而使用 FCA、CPT 和 CIP 术语时则要根据具体采用的运输方式投保相应的海运险、陆运险和空运险等。

五、其他贸易术语的解释

（一）EXW（ex works）（…named place of delivery）——工厂交货（……指定地点）

该术语是指卖方在其所在地或指定地点（如工场、工厂或仓库）将货物交给买方处置时，即履行了交货义务。按此贸易术语成交，卖方既不承担将货物装上买方备妥的运输工具，也不负责办理货物出口结关手续，除另有约定外，买方应承担自卖方的所在地提取货物之后的一切费用和风险。因此，EXW 是卖方承担责任最小、买方责任最大的贸易术语。

在 EXW 术语下，交易双方按工厂交货条件成交，货物出口结关手续不是卖方负责办理，而由买方办理。在此情况下，买方须了解出口国家的政府当局是否接受一个不住在该国的当事人或其代表在该国办理出口结关手续，以免蒙受不必要的损失。如果在买方不能直接或间接地办理货物出口手续的情况下，就不应使用这一术语，而应选用 FCA 术语。该术语适用于任何运输方式，特别是陆地接壤国家之间应用得比较普遍。

（二）FAS（free alongside ship）（…named port of shipment）——装运港船边交货（……指定装运港）

该术语是指卖方把货物送到指定的装运港，买方所指派的船只的旁边（例如码头上或驳船上），即完成了交货。按装运港船边交货条件成交，买卖双方费用和风险的划分，以船边为界，货物交至船边前的一切费用和风险（其中可能包括驳运费用和驳运过程中发生的货物风险损失）一概由卖方负担；当货物有效地交到船边后，费用风险即由卖方转移给买方。FAS 术语要求卖方办理出口结关手续。它只适用于海运或内河运输方式。

（三）DAT（delivered at terminal）——终点站交货（……指定目的港或目的地）

该术语是"2010 通则"新增的两个术语之一。"终点站交货"是指卖方在指定的目的港或目的地的指定的终点站卸货后将货物交给买方处置即完成交货。"终点站"包括任何地方，无论约定或者不约定，包括码头、仓库、集装箱堆场或公路、铁路或空运货站。卖方应承担将货物运至指定的目的地和卸货所产生的一切风险和费用。本术语适用于任何运输方式或多式联运。

（四）DAP（delivered at place）——目的地交货（……指定目的地）

DAP 是《国际贸易术语解释通则 2010》新添加的术语，取代了"2000 通则"的 DAF（边境交货）、DES（目的港船上交货）和 DDU（未完税交货）三个术语。目的地交货是指卖方在指定的目的地交货，只需做好卸货准备无须卸货即完成交货。该术语与 CPT 术语很相似，都要求卖方负责将货物运至指定目的地的费用，但他们之间的显著区别是，DAP 还须卖方承担将货物运至指定目的地的一切风险。DAP 术语下，交货时卸货的费用和风险均由买方负责。本术语适用于任何运输方式及多式联运方式。

（五）DDP（delivered duty paid）（...named place destination）——完税后交货（……指定目的地）

"完税后交货"是指卖方在指定的目的地，将仍处于抵达运输工具上，但已完成进口清关，且可供卸载的货物交由卖方处置时，完成交货。卖方承担将货物运至指定目的地的一切风险和费用，并有义务办理出口清关手续与进口清关手续，对进出口活动负责，以及办理一切海关手续。DDP 术语下卖方承担最大责任。

如果卖方不能直接或间接地取得进口许可，不建议当事人使用 DDP 术语。如果当事方希望买方承担进口的所有风险和费用，应使用 DAP 术语。除非买卖合同中另行明确约定，任何增值税或其他应付的进口税款由卖方承担。例如：如果当事人希望买方承担货物进口应支付的某项费用（如增值税）的义务，则应明确规定："delivered duty paid，VAT. unpaid（... named place destination）"——完税后交货，增值税未付（……指定目的地）。

本术语适用于任何运输方式及多式联运方式。

以上贸易术语中买卖双方各自应承担的风险、责任和费用见表 2-1。

表 2-1　2010 国际贸易术语解释通则风险、责任和费用对照表

贸易术语		交货地点	风险转移界限	出口清关责任、费用由谁负担	装运合同由谁负责签订	保险合同由谁负责签订	进口清关责任、费用由谁负担	适用的运输方式
EXW	工厂交货	商品产地、所在地	货交买方处置时起	买方	买方	买方	买方	任何方式
FCA	货交承运人	出口国内地、港口	货交承运人处置时起	卖方	买方	买方	买方	任何方式
FAS	装运港船边交货	装运港口	货交船边后	卖方	买方	买方	买方	水上运输
FOB	装运港船上交货	装运港口	货物装至装运港船上	卖方	买方	买方	买方	水上运输

续表

贸易术语		交货地点	风险转移界限	出口清关责任、费用由谁负担	装运合同由谁负责签订	保险合同由谁负责签订	进口清关责任、费用由谁负担	适用的运输方式
CFR	成本加运费	装运港口	货物装至装运港船上	卖方	卖方	买方	买方	水上运输
CIF	成本加保险费加运费	装运港口	货物装至装运港船上	卖方	卖方	卖方	买方	水上运输
CPT	运费付至	出口国内地、港口	货交承运人处置时起	卖方	卖方	买方	买方	任何方式
CIP	运费保险费付至	出口国内地、港口	货交承运人处置时起	卖方	卖方	卖方	买方	任何方式
DAT	目的地交货	进口国内地、港口	指定目的地货交买方为界	卖方	卖方	卖方	买方	任何方式
DAP	目的地交货	进口国内地、港口	指定目的地货交买方为界	卖方	卖方	卖方	买方	任何方式
DDP	完税后交货	进口国内地、港口	指定目的地货交买方为界	卖方	卖方	卖方	卖方	任何方式

第三节　合同中通用的贸易术语

一、合同中国际贸易术语举例

（1）每打 100 美元，CFR 纽约价。

（2）每公吨 150 美元，CIF 伦敦价，包括我方 2% 的佣金。

（3）每箱 25 元人民币，CFR 新加坡价，减 1% 的折扣。

（4）每打 4.5 美元，FOB 上海净价。

（5）除非另有规定，价格采用 FOB 工厂价，运费付至威斯丁豪斯选定的美国港口。

（6）在工厂交货合同条件下，货物装运后由买方负责投保。在 CIF 合同条件下，卖方应负责按货物价值的 110% 投保综合险。

（7）在信用证规定的交货期内，（向银行）提交规定的装船单据后，信用证允许支付 CIF 价格的 100%。

（8）每吨 1 230 美元，CIF 上海价，含 3% 的佣金，佣金以卖方收付全部货款为条件。

（9）合同签订后，不管合同是依据 CIF，还是 FOB，运费的变化与出口税的增加变化均由买方负责。

二、选用贸易术语需考虑的因素

国际贸易术语解释通则的主要作用在于为贸易双方权利和义务的划分提供准则。贸易术语的选择取决于买卖双方对于风险、费用和责任的可接受程度。双方都不应该接受其能力范围所不能及的术语。因而正确选择贸易术语非常关键，需要考虑的因素主要包括：进出口的可能性、定价策略、运输方式、市场竞争力、风险控制力等。

　　值得注意的是，虽然"2010通则"对贸易双方的权利和义务都作了比较详细的规定，但是它不能取代销售合同中规定的任何条款。熟识贸易术语的规定有利于买卖双方的贸易磋商，而其他通过长期实践获得的经验也是必不可少的。

本章小结

　　贸易术语是在长期的国际贸易实践中产生和发展起来的专门术语，有关贸易术语的国际惯例主要有三种，其中运用较广的是国际商会不断修订的《国际贸易术语解释通则》。该解释通则最新修订版是《2010年国际贸易术语解释通则》。

　　《2010年国际贸易术语解释通则》与《2000年国际贸易术语解释通则》的最主要的差别在于：新《通则》增加了DAT和DAP两个贸易术语，删除了原来的DAF、DES、DEQ和DDU四个贸易术语，使国际贸易术语由13个减至11个；在新《通则》中，不再有"船舷"的概念，换言之，在原先的FOB、CFR和CIF术语解释中，"船舷"的概念被删除，取而代之的是"装上船"（placed on board）。之前关于卖方承担货物越过船舷为止的一切风险，在新术语环境下改变为"卖方承担货物装上船为止的一切风险，买方承担货物自装运港装上船后的一切风险"。

　　同《2000年国际贸易术语解释通则》一样，新《通则》对每个贸易术语都做出了更加简洁明确的陈述。实际业务中运用较多的主要是FOB、CFR、CIF术语以及FCA、CPT、CIP术语等。在国际贸易中，采用哪一个贸易术语，应在合同中说明。

出口商品价格核算

1. 了解影响进出口商品价格的因素
2. 知道佣金和折扣的计算方法和目的
3. 掌握出口商品的价格核算
4. 学会不同交易条件价格的换算
5. 掌握签订合同价格条款的技巧

在国际贸易中，如何确定进出口商品价格和规定合同中的价格条款，是交易双方最为关心的一个重要问题。上一章我们介绍了"2010 通则"中的 11 种国际贸易术语，合同中的价格条款与贸易术语有着密切的联系，采用不同的贸易术语，商品的出口价格也不尽相同。本章主要介绍如何对出口商品的价格进行核算，并定好进出口合同中的价格条款。

第一节　出口商品价格构成

一、出口商品的价格构成

在国际贸易中，合同中的价格条款包括商品的单价和总价。总价是指一笔交易货款的总金额。单价是由计价货币、单位价格金额、计价数量单位和贸易术语构成的。例如，价格条款可以写成"每件 100 美元 CIF 纽约价"，也可以写成"FOB 广州价每打 25 欧元"，贸易术语可以放在价格的前面或后面，由出口商决定，但需注意，贸易术语后面应注明正确的地点。

二、出口商品作价方法

在国际货物买卖中，可采取下列几种作价方法：

（1）固定价格。买卖双方按约定价格交接货物和收付货款，任何一方无权要求对约定价格进行变更。

（2）暂不固定价格。在价格条款中明确规定定价的时间和定价方法，如"在装船前60天，参照国际市场价格水平，协商议定正式价格"。或只规定作价时间，如"由双方在2016年3月1日商定价格"。

（3）价格部分固定，部分不固定。交易双方只约定近期交货部分的价格，远期交货部分的价格则待以后商定。

（4）滑动价格。在规定价格的同时，还规定价格调整条款。如"如果卖方对其他客户的成交价高于或低于合同价5%，那么对本合同的数量，双方协商调整价格"。

三、出口商品的定价策略

为了把握好商品的定价，这里主要介绍四种定价策略：

（1）成本加成定价法。就是出口商计算出成本，再加上预期的加成，以弥补总产或单产中的未分摊成本。这种方法的主要优点就是简单方便。

（2）边际成本定价法。就是使得出口一单位商品所增加的生产成本低于国内销售的平均成本，这种方法的关键在于确定损益平衡点，即出口商以某一特定价格销售而使得收入等于支出的最低销售数量。销售量超过损益平衡点越多，盈利就越多，低于该点的销售量，就会使卖方亏损。该方法适用于产能过剩，致力于通过调整出口数量来降低出口商品价格，以提高市场竞争力的企业。

（3）买方主导定价法。是根据买方的购买力和目标市场的需求潜力进行定价，该方法要求出口商对目标市场有充分的了解。

（4）竞争导向定价法。如果市场竞争激烈，出口商定价时必须采取以竞争对手或市场平均价格为基准的策略，但该方法的盈利率比较低。

一般而言，无论使用哪种定价方法，出口价格都必须包括成本和预期利润。否则，出口就没有意义了。

四、价格核算的影响因素

出口价格的定价与国内价格定价有所不同。如果对出口商品的成本核算不够全面和精准，则有可能导致原本有利可图的交易变得无利可图，甚至出现亏损，因而，国际贸易的价格核算显得尤为重要。本小节主要介绍出口商品价格核算中需要考虑的主要因素，出口商品的价格主要是由生产成本、费用和预期利润三大部分组成。

（一）生产成本

计算产品的实际成本是报价的关键要素之一。狭义的生产成本包括材料成本、人工成本、固定成本和包装成本等。此外，还需考虑出口的行政成本。如果出口商不是生产商，那么他可以简单地把这些成本归纳为"出厂价格"或"采购成本"。当然其他的综合成本也需考虑在内。

值得注意的是，我国为了鼓励出口，对出口产品实行增值税退税制度。现行的退税办法是，对国内流通环节统一征收的增值税，当商品出口后，由出口企业按当时国家规定的退税率获取一定的退税额。出口退税实际上是国家补贴出口商品、降低出口企业的出口成本、提高出口商品竞争力的一种做法。因此，如果一个商品有出口退税的话，成本中所包含的出口退税额必须减掉，才算该商品的实际成本。出口退税额的计算方法是：先确定出口商品的实际价格，然后用商品的实际价格乘以退税率。即：

出口商品的实际价格 = 出口商品进价（含增值税）/（1 + 增值税率）

出口退税额 = 出口商品的实际价格 × 退税率

出口商品实际成本 = 出口商品进价（含增值税） - 出口退税额

（二）费用

从工厂车间到进口国的目的地，国际贸易通常可能会产生四类费用。

1. 销售费用

主要指国际营销和销售活动产生的相关费用。为了促进产品在国际市场的销售，出口商经常要参加国际展览，印刷产品目录，或建立公司网站。为了进入某些市场，出口商可能愿意为当地的中介机构支付合作和援助费用。此外，营销过程中的佣金和折扣，也常常作为一种销售费用支出。这些费用是一笔不小的开支，不应该被忽视。

2. 交货费用

为了实现货物的交付，出口商要将货物从出口国所在地运送至进口国，其中涉及本国和海外的仓储费、运费、保险费、关税和其他税费、海关手续费和必要的单证费，如申请出口或进口许可证。当货物抵达目的港后，还会产生卸货费、终点费用，包括出口商必须向港务局缴纳的管理费、码头费和停泊费。复杂的交付成本是出口价格与国内销售价格差异较大的原因之一。

3. 融资费用

出口价格与国内销售价格不同的另一个原因是融资费用不同。出口业务通常涉及长时间的生产和海外运输。一笔交易通常需要三至六个月完成，有时甚至更长。为了维持企业正常运转，出口商需要借助各种渠道取得融资，如出口信贷，这可能会产生一些融资费用，如银行利息。不同的支付方式也会产生不同的银行费用，其中信用证支付的银行费用相对较高。因此，为了最大限度地减少这些费用，出口商必须合理地选择融资方式和付款方式。

4. 附加费用

这里还应考虑到那些意想不到的附加开支，如国际电报或电话、传真费用，附加的仓储费，甚至还要算上给外商的礼物。

报价单就是在以上各种费用基础上合计而成的。绝大部分中国的贸易公司都使用这些项目。出口价格是个多元的变量，是否能以上述成本核算出的价格成功出口，最后还得取决于外国消费者。精确的计算并不意味着出口一定成功，还要取决于你的定价策略和措施。另外，还有一些必须考虑的因素，例如：商品的质量与档次，成交的数量，运输的距离，交货地点，季节性需求变化等。

（三）预期利润

出口价格的另一个重要组成部分是预期利润。出口商想获得多少利润，直接影响产品的出口价格。这与国外市场的营销目标有直接关系，例如，一些公司可能会试图进入一个新的市场；有些可能寻求长期的市场增长；有的只是解决产能过剩或过时产品的出口问题；还有许多企业，把国外市场视为一个次要的市场，因此对商品的市场份额和销售量有较低的预期。所有这些不同的目标都会影响企业的定价决策。

对利润的核算方法有两种：一是取绝对数，如：每吨利润要求 10 美元；二是比例法，用某基数的百分比作为经营的利润，如：利润要求是出厂价格的 10%。

第二节　出口商品价格核算

一、主要贸易术语的价格核算

我们以成本加成定价法为例，以表格的形式罗列出了"通则 2010"中 9 个典型术语的价格构成（见表 3-1），只考虑采用水路运输的情况。表中的价格构成只包括主要的成本项目，买卖双方需视实际交易情况进行增减。

表 3-1　成本核算表

项目	分项合计	合计
采购价		
+出口包装费（取决于所选取的运输方式）		
+利润		
+总管理费用支出		
−折扣/出口退税/佣金		
=工厂交货价（EXW）		
+工厂至装运地（火车站或汽车站）的费用		
+出口清关费用		
=货交承运人价（FCA）		
+将货物运至装运港港口的费用		
+国内运输保险费		
=船边交货价（FAS）		
+仓储费，港口操作费（包括向港务局缴纳的管理费、码头费和停泊费等），装船费		
=船上交货价（FOB）		
+将货物运至目的港的海运费		
成本加运费价（CFR）		
+海洋运输货物保险费		
成本、保险加运费价（CIF）		
+全程运输的其他费用（运费、保费等）		
+运输终端的卸货费等		
运输终端交货价（DAT）		
+将货物运至进口国指定地点的运费		

项目	分项合计	合计
目的地交货价（DAP）		
+进口清关费用		
+进口关税以及其他进口所需的税费		
完税后交货（DDP）		

对价格构成有了一定的认识之后，以下重点介绍三种最常用术语的价格核算：FOB，CFR 和 CIF。

（一）FOB 价格核算

FOB 价格 = 实际成本 + 费用 + 利润

1. 成本

对于贸易公司来说，成本指出口货物的采购价格；对于自产自销企业来说，成本是指商品的出厂价格。如果一种商品有出口退税的话，成本中所包含的出口退税款必须减掉，才算该商品的实际成本。如果一种商品没有出口退税，那购货价或出厂价就是实际成本（具体公式参见前文）。

我国的增值税率通常为 17% 或 13%，每种出口商品是否有退税以及退税率是多少，可根据商品的协调制度编码从海关税则中查到。

2. 费用

出口商品价格中的费用主要是指商品流通费用。通常包括包装费、仓储费、国内运费、认证费、港杂费、检验检疫费、税费、利息、业务费、银行费用等。

关于费用的计算，一般有两种：一种是按照各项费用加起来，然后分摊到每一单位商品中去，即求和分摊法；另一种是按某一个基数，用百分比来乘以这个数，一般是按货价的 3% ~ 8%，这种叫作比例法。

以下介绍几种费用的计算公式：

①应纳关税：

$$应纳关税 = 出口货物完税价格 \times 出口货物关税税率$$

$$应纳关税 = \frac{FOB}{1 + 出口关税率} \times 出口货物关税税率$$

②银行垫款利息：

$$垫款利息 = 采购价格（采购成本） \times 贷款年利率 \times \frac{具体垫款天数}{一年的天数}$$

例：某商品的采购成本为人民币 100 元，银行贷款年利率为 9%，预计垫款周期为 2 个月，则单位产品的垫款利息为 $100 \times 9\% \times \frac{2}{12} = 1.5$（元）。

③银行费用。通常是货款结算产生的费用，如汇款手续费、托收手续费、信用证使用手续费等。通常按两种方式收取：按次收取（应平均分摊到单位商品上）和按委托金额的一定百分比收取（按报价）。

3. 利润

如前文所述，它的核算方法主要采用绝对数和比例法。

通常情况下，报价应该使用出口商所在国的货币。如果需要以外币报价，那么就应该按当前汇率转换成外币，公式为：

$$FOB（外币）＝FOB（本币）/汇率$$

（二）CFR 价格核算

根据《国际贸易术语解释通则 2010》规定，FOB 和 CFR 之间的价格差异，仅在于 CFR 要包括从装运港至目的港的海运费。因此，如果 FOB 已知，则：

$$CFR＝FOB＋海运费$$

海运费可以从船公司或货运货代理处获得。海运费的大小取决于所采用的运输方式和计价方法。

（三）CIF 价格核算

同样，已知 FOB 和 CFR 价格，可以计算出 CIF 价格：

$$CIF＝FOB＋海运费＋保险费$$
$$CIF＝CFR＋保险费$$

CIF 价格核算的关键是计算保险费的问题。一般而言，保险公司按照合同金额计算保险费，保险加成通常为 10%，保险费的计算公式（详见第七章）是：

$$保险费＝CIF×（1＋10%）$$
$$CIF＝CFR＋CIF×（1＋10%）×保险费率$$
$$CIF＝CFR／（1－110%×保险费率）$$

保险费率可以从保险公司处获得，已知 CFR 价，可以通过上述公式直接得到 CIF 价。

二、价格换算

买卖双方在洽谈交易时，经常会根据对方要求改变原报价的价格术语，这就涉及价格换算问题。我们将价格换算归纳为以下几种：

（1）FOB 价换算为 CFR 或 CIF 价：

$$CFR＝FOB＋F$$
$$CIF＝（FOB＋F）／［1－（1＋投保加成率）×保险费率］$$

（2）CFR 价换算为 FOB 或 CIF 价：

$$FOB＝CFR－F$$
$$CIF＝CFR／［1－（1＋投保加成率）×保险费率］$$

（3）CIF 价换算为 FOB 或 CFR 价：

$$FOB＝CIF×［1－（1＋投保加成率）×保险费率］－F$$
$$CFR＝CIF×［1－（1＋投保加成率）×保险费率］$$

三、佣金

合同中的价格如果是只由成本和利润加总所得，称之为"净价"。但贸易商为了促进贸易，常会对合同的价格进行一些调整，这其中就包括佣金和折扣。

佣金（Commission）是中间商为买卖双方提供贸易机会而收取的报酬。在货物买卖中，往往表现为出口商付给销售代理人的佣金，进口商付给购买代理人的佣金。

（一）表示方式

佣金的表示方式是在其价格术语后面用百分比表示，如"每码 200 美元 CFR 香港，包括 2.5% 佣金"，即"USD 200 per yard CFR Hong Kong including 2.5% commission"。也可以在价格术语后加注英文字母"C"，如"每打 2 000 美元 CIF 新加坡，包括 5% 佣金"，即"USD 2 000 per dozen CIF C5 Singapore"或者"USD 2 000 per dozen CIF C5% Singapore"。价格中所包含的佣金也可以用绝对数表示，例如"每公吨付佣金 25 美元"，即"USD 25 commission per m/t"。

凡是价格中含有佣金的称为"含佣价"。佣金在合同中有明确规定的，称为"明佣"；佣金没有在合同中明确规定出来的，称为"暗佣"。

（二）佣金计算

按照国际贸易的习惯做法，佣金可以按实际成交数量一定的百分比进行计算。计算佣金可以有以下几种方法：

（1）不管买卖双方以何种价格成交，均按 FOB 或 FCA 价计算佣金。这意味着对于 CIF 或 CIP 等其他贸易术语合同，应该先行扣除海运费和保险费，再计算佣金。理由是：运费、保险费是卖方固定支付的，而不是卖方销售收入，因此不应支付佣金。

例如：CIF 价格为 1 000 美元，运费为 100 美元，保险费为 10 美元，佣金率 2.5%。则佣金 =（1 000 - 100 - 10）× 2.5% = 22.25（美元）

这种方法在实际业务中很少采用。因为对卖方和中间商均无好处。对卖方来说，在计算佣金时虽然事先扣除了运费和保险费，可以少支付一点佣金，但由于佣金直接从原来的成交价格中扣除，减少了卖方的外汇收入。对中间商来说，按 FOB 价格计算佣金，中间商所得的佣金较少，挫伤了中间商的积极性。

（2）按发票价格或成交价格计算佣金。例如，合同的成交价格为 CIFC2.5% 伦敦价，1 000 美元，则：佣金 = 合同金额 × 佣金率 = 1 000 × 2.5% = 25.00（美元）。这种方法是最为常用的一种，它是按含佣价来计算佣金的，然后从含佣价中扣除佣金，即可得出净价。

磋商时，有时需根据净价来计算含佣价，则可以采用以下公式：

$$含佣价 = 净价 /（1 - 佣金率）$$

$$佣金 = 含佣价 × 佣金率$$

$$净价 = 含佣价 - 佣金$$

净价是指在进出口合同中订立的不包括佣金和折扣的价格，如"每公吨 1 000 美元 FOB 净价广州"。

【例 3-1】 已知某出口商品每计算单位 CIF 净价为 1 000 美元，佣金率为 5%，要求按 CIF 价格计算佣金。

CIF 含佣价 = CIF 净价 /（1 - 佣金率）= 1 000 /（1 - 5%）= 1 052.64（美元）

佣金 = 含佣价 × 佣金率　即：1 052.64 × 5% = 52.64（美元），或

佣金 = 含佣价 - 净价　即：1 052.64 - 1 000 = 52.64（美元）

（三）支付方法

佣金一般在出口方收到全部货款后再另行支付给中间商，但为了防止误解，对佣金在全部货款收妥后才予以支付的做法，出口企业与中间商应予以明确说明，并达成书面协议。否则，中间商可能在买卖双方交易达成后，即要求支付佣金。这样，以后合同能否得到履行，货款能否按时支付，就缺乏中间商的保证。

四、折扣

折扣（Discount）是指卖方按照原价给予买方一定的价格减让，或称价格优惠。折扣的高低可根据具体成交条件及买卖双方关系而定。

1. 表示方法

折扣一般用百分比表示，如"每打 200 美元 CIF 纽约减 1.5% 折扣"或"USD 200 per dozen CIF New York less 1.5% discount"。此外，也可以用绝对数表示，如"每打折扣 3 美元"，即"USD 3 discount per dozen"。

2. 折扣计算

折扣的计算较为简单，不存在按 FOB 价值还是按 CIF 价值计算的问题。一般按合同金额乘以约定的折扣百分率为应减去的折扣金额，即

$$折扣金额 = 合同金额 \times 折扣百分率$$
$$实际价格 = 合同价格 - 折扣 = 合同价格 \times （1 - 折扣百分率）$$

此外，折扣也可以按商品数量计算折扣金额。例如，每件商品折扣 5 美元，共 500 件商品，则折扣金额 $= 5 \times 500 = 2\,500$（美元）

3. 支付方法

折扣的支付方法与佣金不同，它由买方预先主动从货款中扣除。

【例 3-2】　广东星光进出口公司从红日冶炼总厂购进一批铅锭用于出口。该批货物购货价为人民币 15 000 元/吨，增值税为 17%，退税率为 13%，出口的各项费用是购货价的 6%，利润要求是购货价的 11%。客户要求星光公司报 CIF London 价。广州到伦敦海运费是每吨 115 美元。按照发票金额的 110% 投保一切险，保险费率为 0.7%。请问该公司应该报价多少？如果客户要改报 CIF C5%，应该报多少？假设汇率是：1 美元 = 6.42 元人民币。

$$FOB = [15\,000 - 15\,000 \div （1 + 17\%）\times 13\% + 15\,000 \times 6\% + 15\,000 \times 11\%] \div 6.42$$
$$= （15\,000 - 1\,666.67 + 900 + 1\,650）\div 6.42$$
$$= 2\,474.04 \text{ 美元}$$

$$CFR = FOB + F = 2\,474.04 + 115 = 2\,589.04 \text{ 美元}$$
$$CIF = CFR \div （1 - 1.1 \times 0.7\%）= 2\,589.04 \div 0.992\,3 = 2\,609.13 \text{ 美元}$$
$$CIFC5\% = CIF \div （1 - 5\%）= 2\,609.13 \div （1 - 5\%）= 2\,746.45 \text{ 美元}$$

五、出口效益衡量指标

1. 出口商品盈亏率

出口商品盈亏率用于计算出口商品的盈亏程度，用出口商品盈亏额与出口总成本的比率表示。计算公式如下：

出口商品盈亏率 = （出口销售人民币净收入 – 出口总成本）/出口总成本×100%

出口商品盈亏额 = 出口销售人民币净收入（人民币） – 出口总成本（人民币）

其中，出口总成本是指出口商品购进价（含增值税）加上定额费用，减去出口退税收入。出口销售人民币净收入是指出口商品的 FOB 价，按当天的外汇牌价（银行外汇买入价）折算人民币的数额。盈亏率大于 0 为盈利，反之为亏损。需要注意的是，计算出口销售人民币净收入和出口总成本时，需将海外运费和保险费剔除在外，因为出口商并不能从这些业务中获益。

2. 出口商品换汇成本

出口商品换汇成本也是用来反映出口商品盈亏的一项重要指标，它是以某商品的出口总成本与出口所得的外汇净收入之比，得出用多少人民币换取 1 美元，即该商品出口净收入 1 美元需要的人民币总成本。其计算公式如下：

出口商品换汇成本 = 出口总成本（人民币）/ 出口销售外汇净收入（美元）

出口商品换汇成本高于银行外汇牌价，则出口亏损；反之，则出口盈利。

所以，对于我国出口公司来讲，出口换汇成本越低越好。

换汇成本与盈亏率有着内在联系，即出口商品盈亏率为负时，数值越大，换汇成本越高；反之亏损率越小或有盈利，则换汇成本就越低。

盈亏率 = （折算率 – 换汇成本）÷ 换汇成本×100%

第三节　合同中的价格条款

国际贸易合同的价格条款是合同的主要交易条件之一，是确定买方支付货款数额的依据。条款内容应该完整、明确、具体、准确，一般包括商品的单价和总值两项基本内容，总值所使用的货币应与单价所使用的货币一致。

一、合同中的价格条款举例

（1）单价：FOB 新加坡每公吨 12.00 美元，包括平舱费和理舱费，总值 1 200 000.00 美元。

（2）单价：每套 3 000 日元 CFRD2% 东京，总值：5 880 000 日元

二、规定价格条款的注意事项

（1）根据拟采用的运输方式和销售意图，选择适当的贸易术语。

（2）争取选用有利的计价货币，必要时要加订保值条款。

（3）灵活运用各种不同的作价方法，力求避免承担价格变动的风险。

（4）参照国际贸易的习惯做法，注意佣金和折扣的合理运用。

（5）如对交货品质、交货数量订有机动幅度而又同意机动部分的价格另订的，必须明确规定另订价格的具体办法。

（6）单价中所涉及的计量单位、计价货币、装运港、目的港等都要写清楚、正确，以免影响合同履行。

本章小结 \\\\

　　货物的价格是买卖双方磋商的中心议题和矛盾的焦点，也是买卖合同中的主要交易条件。在长期国际贸易的实践中，为简化谈判程序，人们对不同的交货条件和成交价格的构成因素，采用不同的贸易术语来表示。影响货物国际价格的具体因素主要有：商品质量和档次的高低、运输距离的远近、成交量的大小、风险成本的高低、市场需求及季节性需求的变化、贸易术语的不同等。在国际货物买卖合同价格条款中，必须对计价货币做出明确的规定。为避免外汇风险，在出口业务中，一般尽可能争取多使用"硬币"，在进口业务中，则应争取多使用"软币"。另外还要注意作价方法的选择，国际贸易的作价方法主要有固定价格和非固定价格。在国际贸易中还涉及佣金和折扣问题，佣金和折扣运用得好，可达到促销的效果。

合同中的商品条款

1. 了解商品的品名、品质、数量和包装条款在国际贸易合同中的重要意义
2. 掌握如何缮制合同中商品的品名、品质、数量和包装条款
3. 掌握在制定上述条款时如何规避风险

在国际贸易中，商品的品名、品质和数量是买卖双方进行交易时首先要考虑和洽商的问题，而国际贸易中的绝大多数商品又都需要一定的包装。因而品名、品质、数量和包装条款都属于合同要件，必须在合同中具体订明，订立的原则是明确、具体、规范，从而有利于合同的顺利履行，尽量避免争议的发生。

第一节　商品的名称

根据贸易惯例，商品的名称是合同的重要组成部分，是买卖双方交接货物的依据，卖方如果交付不符合约定品名的货物，买方有权提出损害赔偿要求，甚至拒收货物或撤销合同。因此，列明合同标的物的具体名称具有重要意义。

一、商品品名（name of commodity）的含义

商品的品名就是商品的名称，是指能使某种商品区别于其他商品的一种称呼或概念。商品的品名应该能够高度概括地体现出该商品的自然属性、用途及主要性能特征。一般以其用途、使用原料、主要成分、制作工艺、外观造型和人名地名等来命名。

二、合同中的品名条款

国际货物买卖合同中的品名条款一般由买卖双方协商确定。品名条款取决于成交商品的

品种和特点，通常是在"商品名称"或"品名"的条款下列明成交商品的名称。有时为了简短，也可不单列商品名称这一条，只在合同开头部分列明买卖双方同意买卖某种商品的文字。为了明确，有时会把具体品种、等级型号概括性地描述进去，如"东北一级大豆"。有些合同甚至把规格型号也包括进去，这时就是品名和品质条款的综合了，如"海尔 32 英寸液晶电视"。

三、订立品名条款的注意事项

（1）商品的名称必须明确具体。商品的品名条款应该订得明确具体，切实反映交易标的物的特性。为了明确，有时还需增加商品的品名、型号、产地和等级。如四川榨菜、洛川红富士苹果、一级红枣等。

（2）商品的名称应该实事求是。商品的品名必须是卖方能够供应给买方的，凡是做不到或不必要的描述词句，都不要列入合同。

（3）尽可能使用国际通用名称。有些商品的名称，在各地叫法不一，例如，土豆在我国不同地区大概有七八种叫法。为了避免误会，应尽可能使用国际上通用的称呼。尽量采用《商品名称及编码协调制度》中的名称。

（4）尽量选用有利于降低关税的名称。应根据有关国家的海关税则和进出口限制的规定，恰当地选择有利于降低关税和方便进出口的名称。例如，命名商品为"苹果汁"可能会比"苹果酒"节省一部分关税开支。

（5）了解进出口商品的归类及编码。世界海关组织制定了《商品名称及编码协调制度》（Harmonized Commodity Description & Coding System）简称（协调制度）（HS），我国于 1992 年起采用《协调制度》。该制度将国际贸易涉及的各种商品分为 21 类 97 章，并采用六位数编码。前两位代表章，第三、第四位为商品在该章中的位置。这四位称为品目。第五第六位称为子目。前六位码各国均一致。第七位码以后各国可依本身需要而定，我国在六位数编码的基础上，使用的是十位数编码。

第二节　商品的品质条款

商品品质不仅关系商品的使用价值和商品价格，而且还影响着商品的销路和信誉。因此，合同中的品质条件是构成商品说明的重要组成部分，也是买卖双方交接货物的依据。各国商法均一致认同将品质条款作为合同的要件之一。

一、商品品质（specification）的含义

品质就是商品的内在素质和外观形态的进一步描述。内在品质包括商品的物理性能、机械性能、化学成分和生物特性等自然属性；而外观形态包括商品的外形、色泽、款式和透明度等。

二、表示品质的基本方法

表示品质的方法有很多，主要包括以实物表示和凭说明约定两大类。

（一）以实物表示商品品质

以实物表示商品品质时，又可分为看货买卖和凭样品买卖两种。在拍卖、展卖或寄售等国际贸易方式中，由于交易的特殊性，一般采用实际看货买卖的方式。但在一般情形下，由于买卖双方分处不同的国家和地区，看货买卖存在相当大的难度，因此，大多采用凭样品成交的方式。

在凭样品成交的方式下，根据样品提供者的不同又可分为凭卖方样品成交、凭买方样品成交和凭对等样品成交三种。

（1）凭卖方样品成交（quality as per seller's sample）。以卖方提供的样品作为交货的品质依据。在此情形下，一般须在合同中注明"Quality as per seller's sample"的字样。

（2）凭买方样品成交（quality as per buyer's sample）。以买方提供的样品作为交货的品质依据。在此情形下，一般须在合同中注明"Quality as per buyer's sample"的字样。

（3）凭对等样品成交（quality as per counter sample）。在贸易磋商过程中，有时是买方先行提供了样品，但如果按买方样品成交，对于卖方来说，容易因为原材料或加工过程与买方存在的差异，而造成所交货物的品质与买方要求不符，最终导致买方退货或索赔。因此，明智的卖方往往会依据买方样品制作一个"回样"，也称"对等样品"交买方确认，然后将确认后的样品作为日后交货的依据。

凭对等样品成交的实质是将"凭买方样品成交"转化为"凭卖方样品成交"。

在以样品作为主要品质依据的贸易磋商过程中，应该注意以下问题：

①以推销或介绍商品为目的而寄出的样品，最好注明"for reference only"的字样，以避免与日后成交时的标准样品相混淆。

②如果在磋商过程中，买方先行提供了样品，那么，卖方最好采用"凭对等样品成交"方式，这样将大大降低履约时出现交货品质不符的概率。

③由于凭样买卖在履约过程中极易产生争议，因此必须注意留存复样（duplicate sample），以备交货或处理品质纠纷时核对用之。

（二）凭说明约定商品的品质

凭说明约定商品的品质是指借助文字、图表和图纸等来说明商品的品质，具体又可分为四种类型：

（1）凭规格、等级或标准买卖（sale by specification，grade or standard）。

（2）凭牌名或商标买卖（sale by brand name or trade mark）。

（3）凭产地名称买卖（sale by name of origin）。

（4）凭说明书或图样（sale by descriptions and illustrations）。

（三）品名、品质条款示例

（1）以品名和规格表示品质。

Name of Commodity：White Rice

Specifications：　　Long – shaped

　　　　　　　　　Broken Grains：20% max.

　　　　　　　　　Moisture：0.20% max.

　　　　　　　Admixture（Impurities）：0.15% max.

（2）以规格结合产地来表示品质。

Name of Commodity：White Rice

Specifications：Long – shaped

　　　　　　　Broken Grains：20% max.

　　　　　　　Moisture：0.20% max.

　　　　　　　Admixture（Impurities）：0.15% max.

　　　　　　　Country of Origin：China

（3）结合样品表示品质。

Description：Cloth Doll Quality as per seller's sample No…dated…

　　　　　　　Quality to be about equal to the samples.

（4）允许存在品质偏差时的品质条款。

Commodity and Specifications：China Grey Duck Down with 90% down content,1% more or less allowed.

（5）增减价条款。

Bonus and Penalty：

According to the standards hereinbefore in the provision of Quality, each 1% over or below its max, price to be increased or decreased by 1% for the unit price.

三、规定品名与品质条款时的注意事项

　　规定品名与品质条款时，总的原则是尽可能地明确、具体，不要过于笼统。同时，注意避免不必要的描述词句，以免给履约带来麻烦。在具体业务中，可参考本国的海关进出口商品分类目录，也可参考国际上较为通行的《商品名称与编码协调制度》。对于新开发产品的名称确立，双方应事先通过商榷取得共识，尽量避免争议的发生。

　　对于卖方而言，尤其要注意：

　　（1）如果成交货物容易存在品质偏差时，最好在合同的品质条款中订明允许的品质机动幅度，以免品质稍有偏差就造成违约。此外，对于约定幅度内的品质差异，还可配合制定相应的增减价条款，以更好地体现按质论价。

　　（2）如果拟按买方提供的商标或品牌交货时，还应注意避免侵犯第三方的工业产权或其他知识产权以至引起纠纷。

　　（3）正确选择适当的品质表示法，尽量避免采用两种以上的方法规定品质，因为这意味着对卖方有更多的约束。比如，既采用文字说明，又注明凭样品，将意味着卖方实际提交的货物品质既要与样品一致，同时又要符合文字说明，只要其中任何一项不符合合同约定都将造成违约，买方都将有权索赔，甚至拒收货物。

　　（4）在凭样买卖中，由于样品与实际交付的货物难免存在品质上的细微差异，因此，最好在品质条款中注明"Quality to be about equal to the samples."以给日后交货预留一定的品质差异空间。

　　对于进口商而言，为了充分保障自己的利益，对于品质条款的订立应该尽量做到详明和

周全，必要时还应对货物的牌名，产地，制造商等做出规定。此外，对于大型设备的进口交易，要特别注意在合同中配合订立售后的质量保证条款，以确保品质要求的实现。

第三节　商品的数量条款

商品的数量是国际贸易合同中主要的交易条件之一，它由数量和计量单位两部分组成。

根据《联合国国际货物销售合同公约》的规定，按合同约定的数量交付货物是卖方的一项基本义务。如果卖方交付货物的数量大于合同规定的数量，买方既可以拒收多交的部分，也可以收取多交部分中的一部分或全部，但必须按合同价格支付相应的货款。如果卖方交付货物的数量少于合同规定的数量，卖方应在规定的交货期届满前补交，但不得由此使买方遭受不合理的不便或承担不合理的开支，并且买方仍可以保留公约所规定的要求损害赔偿的权利。

由于交易双方在合同中约定的商品数量是履约和交接货物的依据，因此，正确掌握订立合同中的数量条款，具有十分重要的意义。

一、计量单位和计量方法的采用

（一）计量单位

在国际贸易中，对于计量单位的确立通常采用以下几种方式：

（1）按重量（Weight）计算。按重量计算的常用单位有长吨（Long Ton）、短吨（Short Ton）、公吨（Metric Ton）、千克（Kilogram）、克（Gram）及盎司（Ounce）等。

（2）按数量（Number）计算。按数量计算的常用单位有件（Piece）、双（Pair）、打（Dozen）、套（Set）、卷（Roll）、令（Ream）、箩（Gross）、包（bale）、袋（bag）、桶（drum）、箱（box）等。

（3）按长度（Length）计算。按长度计算的常用单位有米（Meter）、码（Yard）、英尺（Foot）等。

（4）按面积（Area）计算。按面积计算的常用单位有平方米（Square Meter）、平方尺（Square Foot）、平方码（Square Yard）等。

（5）按体积（Volume）计算。按体积计算的常用单位有立方米（Cubic Meter）、立方尺（Cubic Foot）、立方码（Cubic Yard）等。

（6）按容积（Capacity）计算。按容积计算的常用单位有公升（Liter）、加仑（Gallon）以及蒲式耳（Bushel）等。其中，公升、加仑一般用作酒类、油类的计量单位，而蒲式耳在美国则用于各种谷物的计量。

需要注意的是，由于各国度量衡制度存在差异，有可能造成同一计量单位所表示的数量不一致。比如，同样是表示"吨"，若采用公制则表示为"公吨"，代表1 000千克；若采用英制则表示为"长吨"，相当于1 016千克；若采用美制则表示为"短吨"，仅相当于907千克。为了消除由于各国度量衡制度的差异给国际贸易带来的障碍，国际标准计量组织在各国广为通用的公制的基础上推出了国际单位制（SI），已为越来越多的国家所采用。根据《中华人民共和国计量法》的规定，我国采用的是国际单位制。

（二）重量计算方法的采用

在国际贸易中，当价款依据商品重量确定时，可能采用的重量计算方式通常包括以下几种：

（1）以毛重（Gross Weight）作为计价重量。毛重等于商品本身的重量加包装的重量。一般适用于低价商品。

（2）以净重（Net Weight）作为计价重量。净重是指除去包装物后的商品实际重量。以净重作为计价重量是国际贸易中使用最多的一种重量计算方法。但对于一些低值商品有时也采用"以毛作净"（Gross for Net）的办法进行计价，实际上就是按毛重计价。

（3）以公量（Conditioned Weight）作为计价重量。以公量作为计价重量的方法适用于吸湿性较强的商品，如棉花、生丝、羊毛等。这些商品由于吸湿性较强导致重量很不稳定，极易随着周围湿度的变化而变化。为准确计算这些商品的重量，便利国际贸易的开展，国际上通行的办法是以公量作为计价重量。具体计算公式为：

公量＝商品的干净重（商品烘去水分以后的重量）×（1＋国际公定回潮率）

（4）以理论重量（Theoretical Weight）作为计价重量。通常适用于那些按固定规格生产和销售的商品。由于这些商品按固定规格生产，因而每件的重量基本相同，依据总件数便可以推算出总重量。但由于推算出的总重量毕竟会和实际重量存在细微差别，所以，一般也仅作计重时的参考。

（5）以法定重量（Legal Weight）作为计价重量。通常在某些国家海关征收从量税时使用。其计算方法为商品净重加上直接接触商品的包装物的重量。

二、订立数量条款时的注意事项

（1）一般应在合同中明确规定成交商品的数量，不宜使用"大概""约"（about, approximately, or so, etc.）等意义模糊的字眼，以避免履约过程中的争议。但对于一些大宗的或不易准确计量的商品，应该规定机动幅度，亦即溢短装条款（More or less clause）。溢短装的数量一般由负责装运的卖方确定；在买方指派运输工具的情形下，也可由买方根据装载能力确定溢短装数量。例如，

Description：Chinese Walnuts

Unit Price：USD 300 per M/T CFR Hong Kong, Gross for Net.

Quantity：1,000 M/T, 3% more or less at seller's option

Total Amount：USD 300,000 +／－3% at Delivered Weight

（2）对于溢短装部分的计价，也应在合同中做出相应的规定。一般是按合同价，但为防止当事人利用机动幅度获取额外的利益，也可另行规定计价方法。比如，可规定溢短装部分按装运时的国际市场价格计算。

（3）如果计价数量是按照重量计算，则应在合同中注明是按毛重还是按净重。假如没有注明，依据《公约》的规定，将按净重计算。

（4）对于按信用证方式结算的交易涉及的相关规定。当数量条款中使用了"大概""约"（about, approximately, etc.）等含义模糊的字眼时，根据 UCP600 的相关规定，应解释为有关数量、金额允许有10%的增减幅度；对于不能以包装单位或个数准确计量的商品，除非信用证规定数量不得有增减，否则，准许有5%的溢短装。

第四节　商品的包装条款

商品包装是商品实现自身使用价值和附加值的必要手段之一。适当的商品包装，不仅便于运输、装卸、搬运、储运、保管、盘点等，而且可以提高产品的市场竞争力，扩大销路，提高售价。

商品的包装分为内包装（packaging or inner packing）与外包装（outer packing or packing）两大类。内包装亦称销售包装，外包装亦称运输包装。

商品内包装的主要功能是保护商品，促进销售和便于使用；商品外包装的主要功能则是为了方便运输和避免在运输途中的货损货差。相比之下，二者最大的共同点就是对商品的保护作用，而最大的差异则在于前者更突显其营销上的功效，而后者更偏重于运输上的便利性。

一、销售包装

由于销售包装的作用更多地体现在营销功能，因而在设计上除了考虑对商品的必要保护外，更要注意的是在款式、色彩、图案、用料、文字说明等方面迎合进口国目标市场的消费偏好，能有利地促进销售。比如，在色彩方面，法国人忌讳墨绿色，伊拉克人忌讳橄榄绿色，而意大利、奥地利、爱尔兰、埃及、巴基斯坦等国的消费者却喜爱绿色；在图案方面，东南亚国家的人民普遍喜爱大象，而英国人则忌讳用大象和山羊作图案等。由于进口商对母国市场相当熟悉，因此，出口商在这方面应多多征询进口商的意见。除此之外，在销售包装的设计上还要注意符合相关的法律法规，具体体现在：

（1）注意符合进口国对销售包装的相关管理规定。许多发达国家都对销售包装有严格的管理规定，出口商在设计内包装时应该注意遵守，如销售包装上应该标注的内容，应该使用的语种，应该使用的计量单位以及销售包装是否符合环保要求等。

（2）使用买方指定的牌名或商标。使用买方指定的牌名或商标，应特别注意避免侵犯第三方的工业产权或其他知识产权。

（3）注意印制条形码。我国已于1991年正式加入了国际物品编码协会IANA，出口企业应对拟出口商品申请设定条形码，或应买方要求在出口商品上印刷条形码。这样既有利于出口商品顺利进入进口国的各类零售市场，同时，也有助于提高出口商品在进口国公众心目中的形象。

（4）注意施加原产地标识。大多数发达国家的海关都要求在商品的内外包装上施加醒目的原产地标识，如不符合进口国海关的规定，进口商品将被拒绝放行。

二、运输包装

（一）运输包装设计的基本要求

对于出口商品运输包装的设计，主要本着便于运输、保护货物以及符合进口国的相关管理规定的原则来确定。重点应体现以下几个方面的要求：

（1）依据不同商品的特点来设计运输包装。比如易碎、怕湿、易渗漏的商品，其外包装就必须具备相应的防震、防潮、防渗漏的性能。

（2）依据所采用的运输方式的不同来设计运输包装。比如拟采用铁路运输的商品，其外包装必须具备防震性能；拟采用空运的商品，其单件包装不宜过大等。

（3）必须符合进口国的相关法律规定。比如有的国家因为担心稻草、藤条之类的包装物会夹带有害的病虫害入境，因此禁止采用这类材料作为包装的填充料；也有的国家对单件包装的数量和包装标志有特殊的规定等，这些都必须在设计运输包装时加以考虑。

（4）便利运输和流通环节的各种操作。在商品的运输和流通环节中，运输包装要经历装卸、搬运、存储、查验等诸多环节，为便于这些环节的业务操作，在外包装的设计时就必须考虑适合装卸、搬运和储存，方便清点和查验。

（5）在达到上述要求的情况下，尽可能地节省包装费用。

（二）运输包装的分类

由于运输方式、包装造型和包装材料的差异使得运输包装具有多样性，但一般而言主要分为单件运输包装和集合运输包装两大类：

（1）单件运输包装。单件运输包装是指货物在运输过程中作为一个计件单位的包装。常见的单件运输包装有纸箱（carton）、木箱（wooden case）、花格箱（crate）、夹板箱（plywood case）、金属箱（metal case）、木桶（wooden cask）、铁桶（iron drum）、琵琶桶（barrel）、塑料桶（plastic cask）、布袋（sack）、麻袋（gunny bag）、纤维袋（fibre glass bag）以及包（bale）、捆（bundle）、瓶（bottle）等。

（2）集合运输包装。集合运输包装是指由若干单件运输包装所组合成的一件大包装。如托盘（pallet）、集装箱（container）、集装包和集装袋（flexible container）等。

（三）运输包装的标志

运输包装上的标志，依据其用途的不同可分为指示性标志、警告性标志和运输性标志三类：

（1）指示性标志（indicative mark）。指示性标志是指根据商品特性，对容易破碎、损坏或变质的商品用简单、醒目的图形或文字做出标志，提醒有关人员在装卸、搬运、储存、清点和查验等操作过程中给予充分注意。

以下是国际标准化组织（ISO）核准的几个统一的指示性标志用语：

①Sling here 此处用链

②Fragile, Handle with care 易碎，小心搬运

③Use no hooks 禁止用钩

④This way up 此端朝上

⑤Keep away from heat 防热

⑥Keep dry 保持干燥

⑦Center of Gravity 重心

（2）警告性标志（Warning Mark）。警告性标志又称危险品标志，是对易燃易爆品、有毒物品、腐蚀性物品以及放射性物品等进行的标识。

我国已经颁布了危险品《包装储运指示标志》和《危险货物包装标志》，联合国海事协

商组织也颁布了《国际海运危险品标志》。我们在出口危险品时，既应在运输包装上刷写我国规定的危险品包装标志，又要刷写国际海运危险品标志，以免货到国外口岸后，不准靠岸卸货，造成不必要的损失。

（3）运输标志（shipping mark）。运输标志习惯上称为"唛头"，通常包括以下四个内容：

①收货人的名称。采用图形和字母的形式，通常在图形的内外刷上作为发货人或收货人代号的字母。标准化之后的唛头不再采用图形。

②参考号。也可以刷上发票号、订单号或 L/C 号码。

③目的地名称。通常刷上合同规定的目的地或目的港的名称，如果中间需要中转，也可列明中转港。

例如：

Rotterdam

Via

Hong – Kong

④件号。总件数及单件包装的数字标号。

唛头示例：

ABC

LC1 234

ROTTERDAM

VIA

HONG KONG

C/NO. 1 - -100

除了指示性标志、警告性标志和运输标志外，有时在外包装上还需根据进口国、出口国的规定或受货人的要求刷写附属标志。如制造国、重量、体积、进口许可证等。由于附属标志不属于运输标志，所以应与运输标志分开刷写。如，我国规定中国出口商品应注明"中华人民共和国制造"或"中国制造"的字样，但也有例外情况。

附属标志示例：

GROSS WEIGHT　　56 KG

NET WEIGHT　　52 KG

MEASUREMENT 42 CM × 48 CM × 17 CM

MADE IN THE PEOPLE'S REPUBLIC OF CHINA

三、中性包装（neutral packing）

中性包装是指在商品的内外包装上既不标明产地，又不标明出口厂商牌名的包装，又可分为定牌中性包装和无牌中性包装两种。

定牌中性包装是指在包装上标注买方指定的商标或品牌，但不注明生产国别。无牌中性包装是指在包装上既不标明商品的商标或品牌，也不注明生产国别。

采用中性包装一般都是为了打破进口国所实行的关税壁垒或非关税壁垒等限制进口的歧视性措施，目前已成为国际贸易中的一种习惯做法。我国在出口业务中也可酌情采纳这种做法，但务必注意不得违反有关商标使用权和专利使用权的法律和国际惯例，其图案或内容不得违反公共社会道德。对于受配额限制的商品和享受普惠制待遇的商品不能接受中性包装的要求。

四、包装条款示例

（一）制订包装条款

合同中的包装条款主要是对包装方式、包装材料、包装费用和运输标志等做出规定。一般有两种订法：

（1）笼统的订法。

①seaworthy packing（适合海运的包装）；

②customary packing（习惯包装）；

③seller's usual packing（卖方惯用包装）。

采用笼统订法时，一般都基于买卖双方对包装方式在事前已经过充分的沟通取得共识或由于长期的业务往来已达成共识。否则，仍以在合同中具体规定为好，以免引起不必要的争议。

（2）具体的订法。

①In wooden cases containing 20 pcs. of 30 yards each.

木箱装，每箱20匹，每匹30码。

②In press – packed bales of 200 lbs. net each.

机压包，每包净重200磅。

③In iron drums of 30 kgs net each.

铁桶装，每桶净重30公斤。

④In gunny bags of 50 kgs net each.

麻袋装，每袋净重50公斤。

⑤In cartons, then on pallet

先用纸箱装，然后装托盘。

⑥In international standard tea boxes, 20 boxes on a pallet, 10 pallets in one container.

国际标准纸箱装，20纸箱装一托盘，10托盘装一集装箱。

⑦To be packed in new wooden cases, suitable for long distance ocean transportation and well protected against dampness, moisture, shock, rust and rough handling. The sellers shall be liable for any damage to the goods on account of improper packing and for any rust damage attributable to inadequate or improper protective measures taken by the sellers, and in such case or cases any and all losses and/or expenses incurred (suffered) in consequence there of shall be borne by the sellers.

包装于新木箱内，适合长途海运，能有效防湿、防潮、防震、防锈和耐受粗暴搬运。凡由于卖方包装不善而给买方带来的一切损失和费用，或由于保护措施不当、不周导致货物生锈而给买方造成的一切损失和费用均由卖方承担。

（二）制定包装条款时的注意事项

（1）对于包装材料、包装方式等的规定最好采用具体订明的方法，尽量避免笼统订法，以免引起货损货差时带来不必要的争议。

（2）包装费用一般包括在货价内，无须另行订明。但如果买方要求采用特殊包装，则采用特殊包装的费用一般应由买方负责，并应在合同中具体订明相关的费用和费用支付方法。

（3）对于包装费用较高或包装技术性要求较强的商品，通常要在单价条款的后面注明"包括包装费用"（packing charges included）以免在最后结算时发生争议。

本章小结

国际贸易合同中的品名、品质、数量和包装条款都属于合同要件，必须本着明确、具体、规范、有利于合同的顺利履行和尽量避免发生争议的原则在合同中具体订明。

品名就是商品的名称，品质则是商品的内在素质和外观形态的进一步描述。一般情形下，品名与品质条款一起构成合同中的商品说明（Description）。

表示品质的方法主要包括以实物表示和凭说明约定两大类。以实物表示商品品质时，又可分为看货买卖和凭样品买卖两种。在凭样品成交的方式下，根据样品提供者的不同又可分为凭卖方样品成交、凭买方样品成交和凭对等样品成交三种。凭说明约定商品的品质又可分为凭规格、等级或标准买卖，凭牌名或商标买卖，凭产地名称买卖，凭说明书或图样买卖四种类型。

合同中的数量条款由数量和计量单位两部分组成，一般应在合同中明确规定，但对于一些大宗或不易准确计量的商品，则应该规定机动幅度，亦即溢短装条款，为交货数量的差异预留一定的空间。

商品的包装分为内包装与外包装两大类。内包装的主要功能是保护商品，促进销售和便于使用；商品外包装的主要功能则是为了方便运输和避免在运输途中的货损货差。

运输包装上的标志分为指示性标志、警告性标志和运输标志三类。运输标志习惯上称为"唛头"，通常由图形和字母、目的港名称以及件号等组成。

国际货物运输及合同中的装运条款

1. 了解海洋运输在国际货物运输中的作用
2. 学习班轮运输的运作过程
3. 了解其他重要的运输模式，如航空运输，集装箱运输
4. 理解海运提单在海洋运输中的作用
5. 学习常见的装运条款

国际货物运输是指货物在国家与国家、国家与地区之间的运输。在国际贸易中货物运输是国际商品流通过程的一个重要环节，具体表现为商品从发货方经营所在地到达收货方仓库的实际操作阶段，是完成商品进出口、实现安全收汇的关键。

第一节　海洋运输

一、海洋运输的特点

海洋货物运输是最常见、最普通的一种国际货物的运输方式，其是指使用船舶通过海上航道在不同国家或地区的港口之间运送货物的一种运输方式。据统计，全球三分之二以上的进出口货物皆通过海洋运输完成国际流通，而在我国进出口货物总量的90%以上都是利用海洋运输。海洋运输主要具有以下突出优势：

（1）运输量大。船舶的运载量远远超过火车、汽车、飞机。例如，油轮的载重量可达到60万吨，货船总载重量也可高达50万吨。有些轮船也可装载4 000多个标准集装箱。

（2）运费低。由于运载量大，具有较好的规模经济效应，所以海运的单位商品运费低。一般认为，海运的单位商品运费成本相当于铁路运输的1/20，相当于航空运输的1/30。

（3）适用性好。可适用于各种形状、性质和重量的货物运输，而且不受轨道和道路的规划限制。地球面积的70%被水域覆盖，本身就是四通八达的天然航道，轮船不像火车和汽车那样受到轨道和道路的限制，而且无须大量前期投资用于修建，也无须后期的维修。

当然海洋运输也有不足之处，例如，海洋运输容易受到季风、大浪、浓雾以及冰山等自然灾害的影响，因此航期的准确性低，货物风险大，航行速度慢。

二、船舶的主要类型

根据所运输的货物，船舶主要分为以下几种：

1. 集装箱船

集装箱船是以载运集装箱为主的运输船舶，装载20英尺标准箱（TEU）或40英尺标准箱（FEU）的集装箱。有些集装箱船可以通过宽敞的船首装载已载有集装箱的驳船，因此集装箱船必须依靠大型吊车装卸，故只适用于有集装箱设备的码头。

2. 散货船

散货船主要用于运输大宗货物，如粮、硫黄、铁矿、煤等。常分为粮谷船、煤船和矿砂船。

3. 普通货船

普通货船俗称为杂货船，一般而言这种船长度500英尺左右，一般有4~5个货舱，但有些机舱前面有4~5个货舱，后面还有2个货舱。货舱口装有装卸货的吊货杆。装载的往往是成捆、成包、成箱的各种货物，如水泥、糖、染料、化肥等。有时有些杂货船甚至会装有冷柜来保存容易变质的货物。

4. 油轮

油轮主要承运散装原油或燃油。油轮的运输重量占全世界贸易总重量的一半，其载重量可高达50万吨。

5. 冷藏船

冷藏船适用于运输冷冻商品，比如水果、生肉或鱼类等。

6. 滚装滚卸船

滚装滚卸船适用于运输带滚轮的商品，如汽车、公交车、大货车和挖掘机等。装载货物的拖车或其他带滚轮的工具可自行上船，无须重复装载。

7. 多功能船

可用于运输以上所述货物，功能齐全、用途广泛、适用于某些固定航线、常自带装卸设备而无需港口起货设备。

8. 载驳货船

载驳货船又称子母船，是先将货物装入驳船（子船），再把驳船放入母船体内的一种船舶。子船（驳船）于水位浅的内河承载货物，然后拖至海洋航线，载驳船利用自身设备装载子船至母船。载驳船适用于水位不足的内河运输。

三、海洋运输的经营模式

当前海洋运输的经营模式分为班轮运输和租船运输两种。

（一）班轮运输

班轮运输是指船舶按照规定的时间，一定的地理航线，以及规定的港口顺序，定期在各港口之间来回航行。不论货物量多寡，季节更变，定期定航线地运营，和火车、公交车或班车等其他公共运输工具一样。

1. 班轮运输的特点

班轮运输具有固定航线、固定的航海时间、固定的运输费用、沿途固定的挂靠港口的特点。这四个"固定"有利于发货方掌握船期，核算运输费用，促进出口成交。一般而言，班轮运输的承运人负责货物的装卸，托运人或收货人无须再另外支付此笔费用，也不涉及滞期费和速遣费。

2. 班轮运费

班轮运费根据船运公司所公开颁布的班轮运费本（freight rate tariff）进行计算。不同的班轮公司颁布的班轮运费本在内容和形式上也许稍有区别，但一般都包括了运输说明及有关规定、货物分级表、航线费率表、附加费率表、冷藏货及活牲畜费率表等。按费率形式不同，班轮运费本分为商品费率本（commodity rate freight tariff）和等级运价本（class rate freight tariff）。前者列出每项货物的基本费率，后者将承运的货物分为若干等级（一般分为20个等级），每一个等级的货物有一个基本费率，称为"等级费率表"。属于第一级的商品运费率最低，第二十级的商品，运费率最高。在实际业务中，大都采用等级费率本。

（1）等级运价表计算班轮运输的步骤：

①按照船务承运人颁布的等级运价本，根据货物名称在运价本上查找到运费计算的等级（class 1 至 class 20）以及其计算标准（W，M 或者 Ad. val）。

②在运价本上找到货物的运输航线、装运港和目的港，根据货物的实际情况计算出所需的基本运费（basic rate）。

③在运费本的附加费部分，根据货物的运输航线找到应收取的附加费种类、数额或计费标准（百分比或集装箱数量）和使用货币。

④总基本运费与总附加费相加之和就是总运费。

由上述可见，可以推断出班轮运费由两部分组成，即基本运费（basic rate）和附加费（surcharge）。

（2）基本运费的计算标准：

①重量吨（weight ton）：指货物 1 立方米的重量，以 W 表示，1 公吨（TNE）（1 TNE = 1 000 kgs）为一个运费吨。以重量计算的货物 1 立方米的重量往往超过 1 公吨。

②尺码吨（measurement ton）：又译为测量吨，指货物一公吨的体积，用 M 表示，一立方米（简称 1 MTQ 或 1 CBM 或 1 CUM）为一个运费吨。以体积计算的货物 1 立方米的重量往往不足 1 公吨。重量吨和测量吨皆称为运费吨（freight ton）。

③选择法：指基本运费率的计算在重量吨和测量吨之间选择运费较高者，以 W/M 表示。

④从价运费：按货物 FOB 价格的百分比进行计算，以 Ad val. 表示。

⑤按件法：指运费按货物的件数收取，比如活牲畜按每头（per head），集装箱按每箱（per ctr），起运费按每份提单（per B/L）。

⑥议价法：运费由船方与托运方临时讨论决定。

（3）附加费。由于船舶、货物、港口以及其他方面的种种原因，使得船方在运输货物时增加费用开支或蒙受经济损失，船方为补偿这些开支或损失规定另外收取的费用，就叫附加费（Surcharge 或 Additional）。附加费种类繁多，而且随着一些情况的改变，会取消或制定新的附加费。

常见的附加费有燃油附加费（BAF）、货币贬值附加费（CAF）、绕航附加费、苏伊士运河附加费（Suez Canal Surcharge）、转船附加费（Transshipment Surcharge）、直航附加费（Direct Additional）、港口附加费（Port Surcharge）、港口拥挤附加费（Port Congestion Surcharge）、超重附加费（Heavy – Lift Additional）、超长附加费（Long Length Additional）、洗舱费（Cleaning Charge）、选择卸货港附加费（Optional Fees or Optional Additional）、变更卸货港附加费（Alteration Charge）、码头装运作业费（THC）。

班轮运费的计算公式为：班轮运费 = 总货运量 × 基本运费率 × （1 + 附加费率）

【例5-1】 从我国广东黄埔港运往国外某港口的一批货物，运费计收标准为 W/M，共300箱，每箱毛重25千克，每箱箱长50厘米、宽30厘米、高20厘米，基本运费率为每运费吨60美元。特殊燃油附加费为5%，港口拥挤费为10%。试计算这300箱货物的运费为多少？

解： W = 25 千克 = 0.025 运费吨

M = 50 厘米 × 30 厘米 × 20 厘米 = 30 000 立方厘米 = 0.03 运费吨

由于 M > W，所以应采用 M 计算运费。

运费 = 总货运量 × 基本运费率 × （1 + 附加费率）

　　　= 300 × 0.03 × 60 × （1 + 5% + 10%）

　　　= 621 （美元）

3. 货运代理人

许多进出口商的货物量不足以使用租船运输，他们往往选择使用班轮运输货物，因为班轮运输经济、服务周到、简单方便。但是国际贸易货运业务本身相当繁杂，许多进出口商不能或不愿提供具有这种专业性职员的职位，于是他们往往求助于货运代理公司（shipping agent 或 forwarder，有时是无船承运人）帮忙代理其国际货物的运输业务。货运代理人在国际货运市场上，处于货主与承运人之间，他们熟悉国际货运的运输市场、运输成本以及相关的进出口政策。货运代理人本人并不参与运输，却是个物流专家，他为发货人安排把货物从生产地运往某一市场或某个目的地，与某个或多个承运人签订运输合同并完成货物运输。运输模式往往多样，涉及船只、飞机、卡车和火车。例如，货运代理人安排卡车把货物从工厂运至发货港，在目的地安排卡车把货物从目的港运至客户场所。其具体业务包括订舱、缮制有关证件、报关、报验、保险、拆装箱和其他国际货物运输业务。

（二）租船运输

租船运输（charter shipping）又称不定期船运输，是指租船人向船东租赁船舶用于货物运输。与班轮运输相反，租船运输没有固定的船期表、航线及挂靠港口，而是根据船东与租船人双方签订的租船合同（charter party），按照贸易需求安排船期、航线和港口，适用于大宗货物运输。有关航线和港口、运输货物的种类以及航行的时间等，都按照承租人的要求，由船舶所有人确认，租船人与出租人之间的权利义务以双方签订的租船合同确定。租船的方式主要有以下几种：

（1）定程租船。定程租船是指船东按双方事先议定的运价及条件向租船人提供船舶全部或部分舱位，在指定的港口之间进行一个或多个航次运输指定的货物。程租的情况下，船舶的营运调度工作，仍由船舶所有人负责，包括燃料费、港口费、船员工资等在内的一切费用，都由船舶所有人支付，租船人只需按合同规定将货物装上船后便可在卸货港等待提货。程租可分为单趟程租、往返程租或连续程租。可见程租在一趟或数趟航次完成之时，便是程租结束之时。在这种情况下，缩短航次时间直接关系到船方利益。程租合同一般包括以下内容：

①发货人即托运人，也就是承租人；承运人即船方也就是出租人；船舶名称及其旗号。

②货物类型，包装状况以及数量。

③所涉港口。

④受载期。受载期是指所租船舶到达指定装货港或地点并已做好装货准备，随时接受货物装船的期限。如果船只在规定的受载日之前到达，租船人无须立即装载，但如果船舶迟于受载日到达，即船未能在受载期限抵达指定装货港或地点，合同可以取消。

⑤运费的支付方式、时间及地点。运费可按运费吨计算或一次性付清，可以装运前支付，也可以货到目的地之后支付。

⑥装卸费的责任方。装卸费用由租船人和船东协商并确定在程租合同中。做法通常有五种：船方负责装卸费用（Gross Terms/ Liner Terms）、船方管装不管卸（F.O.）、船方管卸不管装（F.I.）、船方不管装卸（F.I.O.）、船方不管装卸、理舱和平舱（F.I.O.S.T）。

⑦受载期、滞期费和速遣费。受载期是货物装卸的一段期限，如10~15天，以适应海上船舶航行和货运活动实际情况与要求。主要有自然日、工作日及晴天工作日三种表述方式。滞期费是船方为了约束租船人在规定的时间内（即受载期）完成装卸，以避免实际装卸时间超过合同规定的装卸时间而使得船方遭受船期损失。滞期费实际上是租船人应支付给船东补偿金；但如果情况相反，实际装载时间少于合同规定的装载时间，使得船舶节省了船舶的在港费用并获得了船期利益，则租船人可以从船东方获得一定的奖金，即速遣费。通常，速遣费是滞期费的一半。

（2）定期期租。定期期租又称定期租船，是指船舶所有人将船舶出租给他人使用一定期限。在承租期内承租人可以利用船舶的运载能力安排货物的运输、选择所需港口和决定船舶的去向。同时，不论是否经营，也须按天或月向出租方支付租赁费和船舶所需费用如燃料费、港口费等经营费用。而在此期间，出租人则负责船舶的维修、保险、配备船员和船员工资等固定费用。期租合同一般包括以下内容：

①发货人即托运人，也就是承租人，承运人即船方也就是出租人。

②船舶的描述，如其载重、船舶名称及旗号。

③使用限制，如不得装运的货物，不能航行的区域。

④租用的期限。

⑤船只的交付。

⑥租船费用。租金的高低主要取决于船舶的装载能力、租期长度以及租船的时间。

（3）光船租船。光船租船是期租的一种特殊租船方式，也是按一定期限租船，但船方不提供船员，仅将一条船出租给租船人。租船人自行配备船员，负责船舶的经营管理和航行的各项事宜，承担船舶的全部固定及变动的费用。实际上租船人在租赁期间对船舶有着支配权和占有权。这种方式不适用于一般货主与船东之间。

第二节　其他运输方式

一、航空运输

随着国际贸易的迅速发展，航空运输作为国际货物运输的一种方式越来越被广泛地应用，在国际货物运输中占的比重越来越大。全球航线 1 000 多条，飞机场 30 000 多个，其中 1 000 多个拥有国际航线。据国际航空运输协会（IATA）统计，2014 年国际航空完成了超过 5 亿运费吨的货物，为航空运输创造了约 625 亿美元的价值，2004 至 2014 年，国际航空运输量的增长速度平均约为 3.9%。

（一）航空货运的特点

（1）运送速度快。采用航空运输，可保证易腐商品的新鲜成活率，也适用于季节性商品和其他应急物品。

（2）破损率和失窃率低，安全性好。航空运输与其他运输模式相比，安全技术高，因此货损少，可以简化商品包装，还可降低保险率。

（3）加快资金周转，减少库存费用。由于运输速度快，商品周转期短，存货量可相对降低，加快了资金回收，大大节省了储存费和利息费用。

（4）可开辟运距较远的新市场。例如，如果没有发达的航空运输，荷兰的郁金香不可能在它的存活期内到达中国城市。

（5）托运手续统一又简单，服务优质。国际航空运输协会拥有 262 家航空会员，占全世界航空运输量的 83%。这些会员都必须在遵循统一的相关条例和规则的同时不断地提高服务质量以获得更多的运输量。

（二）航空公司和航空货运公司

（1）航空公司就是承运人，拥有飞机，实际办理乘客和货物运输业务的责任人，如我国三大货运航空公司，即位于北京的中国国际货运航空有限公司，位于上海的中国货运航空有限公司和位于广州的南方航空公司货运部。

（2）航空公司的主要业务是保障飞行安全，难以直接接待众多的一般客户。所以在现实操作上有专门的航空货运代理公司为航空公司处理航运前和航运后的繁杂服务项目，具体为揽货、组织货源、出具运单、收取运费、进口疏港、报关报验、送货中转等工作。航空货运代理公司作为货主和航空公司之间的桥梁和纽带，即可以是货主的代理，代替货主向航空公司办理托运或提取货物；也可以是航空公司的代理，代替航空公司接受货物，出具航空公司的总运单和自己的分运单。我国的中国外运空运发展股份有限公司就是一家与许多航空公司和物流公司合作的货运代理。

（三）航空货运方式

空运的运输方式主要有班机运输、包机运输、集中托运等。

（1）班机运输（Scheduled Airline）。班机运输指具有固定开航时间、航线和停靠航站的飞机。班机通常为客货混合型飞机，货舱容量较小，运价较高，但由于航期固定，有利于客户安排鲜活商品或急需商品的运送。据估计50%的空运货物通过客运班机运输。

（2）包机运输（Air Charter）。包机运输是指航空公司按照约定的条件和费率，将整架飞机租给一个或若干个包机人（包机人指发货人或航空货运代理公司），从一个或几个航空站装运货物至指定目的地。包机运输适用于需要专门的个人航线的客户，货物往往讲求时间或急需，或货物需要特别运输。

（3）集中托运（Consolidation）。集中托运可以采用班机或包机运输方式，是指航空货运代理公司将若干批单独发运的货物集中成一批向航空公司办理托运，填写一份总运单送至同一目的地，然后由其委托当地的代理人负责分发货代提单给各个实际收货人。这种托运方式，可降低运费，是航空货运代理的主要业务之一。

（四）空运货物的运价

航空货物运价一般按托运货物的实际重量、体积重量或等级重量计算收取。

1. 实际重量（Actual Gross Weight）

实际重量是指包括货物包装物在内的货物的实际重量。实际毛重是按每1千克毛重作为计费单位，重货（High Density Cargo）按实际重量作为计费重量（即每千克体积不足6 000立方厘米或366立方英寸，或每不足166立方英寸），如金、机器、金属部件等。

2. 体积重量

将货物的体积按一定的比例折合成的重量就是体积的重量。轻泡货（Low Density Cargo）是按体积重量作为计费重量（即每1千克体积超过6 000立方厘米或366立方英寸，或每超过166立方英寸），如膨化小食品、帽子皮鞋等。无论货物形状如何，丈量每件货物的最长、最宽、最高的部分，三者相乘算出体积。将体积折算成千克（或磅），通常按每6 000立方厘米折1千克为标准（注：国际航空货物运输组织是按7 000立方厘米折1千克），体积重量的计算公式为：体积重量（kg）=货物体积（cm^3）÷6 000（cm^3/kg）。

3. 等级重量

等级重量即较高重量分界点的重量。如果托运人所托运的货物接近于较高重量的分界点，用较高重量分界的较低运价计算出来的运费低于按适用的运价计算出来的运费，可按较低运价收费。那么，其中的较高重量的分界点重量就是等级重量。这实际上是航空公司的优惠运价。

二、铁路运输

铁路运输是国际贸易中陆地运输的一种重要方式，也是我国初期的对外贸易中的一种重要的运输方式。它在速度上仅次于航空运输，在运输量上仅次于海洋运输，安全、可靠、风险低，不易受气候影响，时间准点性高，而且价格相对低廉。我国对外贸易中，铁路运输主要有国内铁路运输和国际铁路联运两种方式，以下介绍国际铁路联运。

（一）国际铁路联运

国际铁路联运是指把两个国家以上的铁路连接起来，完成一批货物从出口国向进口国转移的运输。它使用一份统一的国际联运单据，由铁路部门负责办理铁路的出入境，在由一国铁路向另一国铁路移交货物时无须发货人、收货人的参与。

国际铁路联运中，按托运货物的数量、体积等，可分为整车运输（full car load, FCL）和零担运输（less than car load, LCL）。整车运输是指按一张运单办理的一批货物，需要单独车辆运送的，作为整车货物。中国铁路规定：凡一批货物的重量、体积、形状或性质需要一辆或一辆以上的 30 吨货车运输的，均应按整车办理托运。并规定以下 7 类货物必须按整车托运：①需要冷藏、保温或加温运输的货物；②某些危险货物；③易于污染其他货物的污秽品；④蜜蜂；⑤不易计件的货物；⑥某些未装容器的活动物；⑦单件货物重量超过 2 吨、体积超过 3 立方米或长度超过 9 米的货物。零担运输是指一张运单办理的一批货物，其重量小于 5 000 千克，其体积又不需要单独一辆货车运送的货物，即为零担货物。

（二）国际铁路联运运费

我国的国际铁路联运货物运费的计算，其主要依据是 1951 年苏联与其他国家签订的《国际铁路货物联运协定》（简称国际货协）和中国铁道部颁布的《铁路货物运价规则》（简称价规）。运费由三部分组成：第一部分是进口国家所在的铁路费用；第二部分为过境国铁路运费；第三部分为出口国家铁路运费。

（1）发送国和到达国铁路的运费，均按铁路所在国家的国内规章办理。

（2）过境国铁路的运费，均按承运当日统一货价规定计算，由发货人或收货人支付。如由参加国际货协铁路的国家向未参加国际货协铁路的国家之间运送货物，则有关未参加货协国家铁路的运费可按其所参加的另一种联运协定计算。我国出口的联运货物，交货共同条件一般均规定在卖方车辆上交货，因此我方仅负责至出口国境站一段的运送费用。但联运进口货物，则要负担过境运送费用和我国铁路段的费用。

（3）过境运费的计算流程：①根据运单上载明的运输路线，在过境里程表中，查出各通过国的过境里；②根据货物品名，在货物品名分等表中查出其可适用的运价等级和计费重量标准；③在慢运货物运费计算表中，根据货物运价等级和总的过境里程查出适用的运费率。其计算公式为：

$$基本运费额 = 货物运费率 × 计费重量$$
$$运费总额 = 基本运费额 × 加成率$$

加成率指运费总额应按托运类别在基本运费额基础上所增加的百分比。快运货物运费按慢运运费加 100%，零担货物加 50% 后再加 100%。随旅客列车挂运整车费，另加 200%。

（4）国内段运费的计算流程：①根据货物运价里程表确定发到站间的运价里程。一般应按最短路径确定，并需将国境站至国境线的里程计算在内；②根据运单上所列货物品名，查找货物运价分号表，确定适用的运价号；③根据运价里程与运价号，在货物运价表中查出适用的运价率；④计费重量与运价率相乘，即得出该批货物国内运费，其计算公式为：运费 = 运价率 × 计费重量。

（三）中国的国际铁路联运概况

（1）泛亚铁路网：2006 年 11 月 10 日，17 个亚洲国家在我国签订了建设泛亚铁路网的协议。该铁路网包括四大路线，分别是北回线、南回线、东南亚线和南北线。北回线经由德国、波兰、白俄罗斯、俄罗斯、哈萨克斯坦、蒙古、中国及韩国连接欧洲及太平洋地区。已建成的西伯利亚铁路是该线的主要组成部分，全长9 250千米，连接莫斯科至符拉迪沃斯托克。南回线是从欧洲直到东南亚，经由马来西亚和新加坡连接土耳其、伊朗、巴基斯坦、印度、孟加拉国、缅甸、泰国及中国云南省。东南亚线主要连接中国云南昆明至新加坡。南北线是指北连北欧南接波斯湾，始于芬兰赫尔辛基越过俄罗斯直至里海。这个项目借印古代贸易的"丝绸之路"而被称为"铁丝绸之路"。

（2）新亚欧大陆桥实际上是亚欧大陆桥的南段分支。亚欧大陆桥贯穿东亚与欧洲，从中国连云港到欧洲荷兰鹿特丹，穿越哈萨克斯坦、俄罗斯、白俄罗斯和波兰，长达11 870千米。2013年起生产商如惠普开始使用快车专线从中国内陆城市经由哈萨克斯坦、俄罗斯、白俄罗斯和波兰到达欧洲。由于白哈俄三国海关结盟减少了检验手续，大大降低了偷窃和货物延迟的风险。中国中西部城市如重庆或成都的货物通过铁路运输到达欧洲只需3周，航海需要5周，运输费用只增加了25%；航空1周，运费成本是铁路的7倍，而且释放的二氧化碳量是铁路运输的30倍。

（3）香港与内地连接的铁路终点是香港九龙。由于香港特别行政区系自由港，许多内地的货物经由香港进出。利用东部铁路，分别开辟了从北京、上海和广东（肇庆）到香港的三条直达路线。经由深圳从广州到香港的快线已批准建设而且广州段部分已经开始。这条快线的路程将会由原来的2小时缩短为1小时。香港特别行政区的铁路运输是由境内段铁路运输和港段铁路运输两部分构成的。具体做法是：从发货地运至深圳北站后，由中国对外贸易运输总公司深圳分公司接货，然后办理港段铁路托运手续，并向海关申报。报关后，由深圳外运分送至在中国香港特区的代理人——香港中国旅行社收货后，转运至九龙目的站，交给香港收货人。

三、集装箱运输

（一）集装箱运输的含义与特点

据统计60%以上的海洋货物都是通过集装箱来装运。集装箱是一种柜状大型密封性容器，也称为货柜、货箱、标箱或标柜。集装箱运输是指以集装箱为载体将货物集合组装成集装单元，以便运用大型装卸机械和大型载运车辆进行机械化装卸和堆放，适合长途运输，货物从某一运输模式转换到另外一种不同的运输模式时无须再次拆装，是一种新型、高效率和高效益的运输方式。其主要特点如下：

（1）简化包装，大量节约包装费用。为避免货物在运输途中受到损坏，必须有坚固的包装，而集装箱具有坚固、密封的特点，两侧和上顶是瓦楞式耐候钢，其本身就是一种极好的包装。

（2）减少货损货差，提高货运质量。集装箱是一种坚固密封的箱体，其本身就是一个坚固的包装。货物装箱并铅封后，途中无须拆箱倒载，一票到底，即使经过长途运输或多次换装，也不易损坏箱内货物。集装箱运输可减少货物被盗、潮湿、亏损等引起的货损和货差的概率，进而减少保险费。

（3）降低装卸和经营成本。没有集装箱之前，一船货物需要20～22个码头装卸工人，但有了集装箱之后港口无需大量的装卸工人。由于集装箱码头的装卸不容易受恶劣气候的影响，装卸时间缩短了，并提高了航行率，对承运人而言降低了承运人的船舶运输成本。

（二）集装箱的类型

集装箱种类繁多，常用的有干货集装箱、散货集装箱、液体货集装箱、冷藏箱集装箱以及一些特种专用集装箱，如汽车集装箱、牧畜集装箱、兽皮集装箱等。

干货集装箱也称为普通集装箱，主要用于运输一般杂货，适合各种不需要调节温度的货物。这种集装箱四周密封，占全世界集装箱的80%。

散货集装箱的顶部可开启或不固定，以适用于顶部装载。这种集装箱适用于运输或贮存散装液体或粉粒状货物，如化学用品、食品、溶剂或者药物。

冷藏集装箱是一种内部附有温度调节器，或通过船舶的中央制冷系统获得冷气。它适用于需要保温或冷藏的货物，如水果、蔬菜、肉类或奶制品等。

标柜或标箱用于测量船舱承载量，或港口的吞吐量。一个标柜（TEU）长20英尺约6米，宽8英尺约2.44米，高8.6英尺约2.6米，20英尺集装箱的载货重量约为17 500千克，有效容积约为25立方米。同高同宽的集装箱长度是标柜的2倍，即12.192米，称为40英尺标柜，与此同宽同长但高度为9.6英尺即2.896米的集装箱称为高柜。40英尺集装箱的载货重量约为24 500千克，有效容积约为55立方米；40英尺高柜集装箱的载货重量约为26 000千克，有效容积约为67立方米。

框架集装箱只有底板结构，由钢架、软质木材底板和两尾端构成。具有很高的承载力，两端可固定也可拆卸，但很牢固，可安装保证货物安全的设备，适用于超重、不规则货物，如机器、钢材和木材。

（三）集装箱运输术语

1. 整箱货物运输与拼箱运输

（1）整箱货物运输（FCL）是指货物的装卸及风险由某一发货人负责，只有一个收货人。实际业务上是指整箱货物只发送给一个收货人。整箱货物运输相比分批发货同量货物而言运费要便宜。FCL本意是指尽量利用集装箱容量把货物装满，但实际上FCL不仅仅是指将集装箱装满，更多的是指发货人把未满的集装箱作为一批集装箱货，这样就可以简化许多物流手续，比与其他人适用同一集装箱要安全。

（2）拼箱运输（LCL）是指货物未满一标准箱。LCL本来是用于火车运输，是指为了方便把不同发货人的货物或去不同目的地的货物拼凑在一个车厢里。实际操作上，LCL不仅仅指货物不足一个标准箱，也指不满一标箱的货物需与其他发货人的货物组成一满箱发给不同的收货人。

2. 集装箱存放地

分为集装箱堆场和集装箱货运站，前者为货物上船之前或目的地下船之后用于收集、分发、堆放、运转整箱装货的场所；后者是处理拼箱货集散场地，办理拼箱货的交接和配箱积载。

3. 集装箱交接方式

集装箱运输中，整箱货和拼箱货在船货双方之间的交接方式有以下几种：

（1）门到门（Door to Door）：由托运人负责装载的集装箱，在其货仓或工厂仓库交承运人验收后，由承运人负责全程运输，直到收货人的货仓或工厂仓库交箱为止。这种全程连线运输称为"门到门"运输。

（2）门到场（Door to CY）：由发货人货仓或工厂仓库至目的地或卸箱港的集装箱装卸区堆场。

（3）门到站（Door to CFS）：由发货人货仓或工厂仓库至目的地或卸箱港的集装箱货运站。

（4）场到门（CY to Door）：由起运地或装箱港的集装箱装卸区堆场至收货人的货仓或工厂仓库。

（5）场到场（CY to CY）：由起运地或装箱港的集装箱装卸区堆场至目的地或卸箱港的集装箱装卸区堆场。

（6）场到站（CY to CFS）：由起运地或装箱港的集装箱装卸区堆场至目的地或卸箱港的集装箱货运站。

（7）站到门（CFS to Door）：由起运地或装箱港的集装箱货运站至收货人的货仓或工厂仓库。

（8）站到场（CFS to CY）：由起运地或装箱港的集装箱货运站至目的地或卸箱港的集装箱装卸区堆场。

（9）站到站（CFS to CFS）：由起运地或装箱港的集装箱货运站至目的地或卸箱港的集装箱货运站。

4. 集装箱运输费用

集装箱运输费用除了海运费用，有时还包括内陆或装运港市内运输费、拼箱服务费、堆场服务费、集装箱租赁费及相应设备使用费等。集装箱货物海运费用根据货量的大小，有两种计算方法：

（1）整箱货。以每个集装箱为计费单位，收取包箱费率（Box Rate）。当运载货物体积吨达到集装箱内部容积的85%，或重量吨达到集装箱最大载重量的95%，就可以申请用整箱装载。装箱后直接送往集装箱堆场（Container Yard，CY）

（2）拼箱货。当货物体积吨或重量吨不足一个集装箱时，必须将货物送船公司指定的集装箱货运站（Container Freight Station，CFS）与他人的货物合并成整箱装运。拼箱费用由船方按货品的体积吨或重量吨二者收费较高者计算。运价表上常用 M/W 或 R/T 表示。

【例5-2】 1 000 辆儿童脚踏车要出口到法国，目的港是马赛港。纸箱包装，每箱装6辆，每箱体积为 0.057 6 CBM，毛重为 21 KGS。求该批货物的集装箱海运费。

第一步：计算应选集装箱的种类：

总箱数 = 1 000 ÷ 6 = 166.6，取整 167 箱

总体积 = 167 × 0.057 6 = 9.6 CBM

总毛重 = 1 000 ÷ 6 × 21 = 3 500 KGS = 3.5 TNE

根据集装箱规格比较可知，由于达不到一个 20 尺集装箱的限重或容积，宜采用拼箱方式装运。

第二步：经查询，到马赛港的拼箱运费为：

基本运费：每体积吨（MTQ）USD 151.00，每重量吨（TNE）USD 216.00

第三步：根据上述资料，计算海运费，公式如下（假设美元与人民币的汇率为6.3）

按体积计算运费 $= 9.6 \times 151 = 1\,449.6$（美元）

按重量计算运费 $= 3.5 \times 216 = 756$（美元）

两者比较，体积运费较大，船公司按较大者收取，则运费为1 449.6美元。

总运费 $= 1\,449.6 \times 6.3 = 9\,132.48$（元）

四、国际多式联运

国际多式联运（International multimodal transport）简称多式联运，是在集装箱运输的基础上快速发展起来的，是指发货人与某承运人只签订一份运输合同，以至少两种不同的运输方式完成货物的整个运输流程，比如铁路运输加海洋运输，或公路运输加海洋运输。多式联运经营人一般都不拥有以上所有的工具，而只是把某一部分的运输转包给另外一些承运人，但仍然对货物的整个运输过程负责；多式联运因为适用水路、公路、铁路和航空多种运输方式，实现了门到门的运输服务。多式联运经营人可以是有船承运人，他在接受货物后，不但要负责海上的运输，还须安排汽车、火车与飞机的运输。有船承运人往往委托给其他相应的承运人来完成内陆运输任务，对交接过程中可能产生的装卸和包装储藏业务，也委托给有关行业办理。但是，这个经营人必须对货主负整个运输过程的责任。多式联运经营人更多的是货运代理人，由于经营人不拥有船舶，所以称为无船承运人。这个货运代理人业务的扩展，承担了更多的承运义务。经营人在接受货物后，也是将运输任务委托给各种方式运输承运人进行，负责安排海上运输和内陆地区的汽车、火车甚至是飞机的运输，同时经营人本人仍对货物的整个运输航程负责。

五、陆桥运输

集装箱的兴起促进了多式联运的发展，但后者的繁荣同时也受益于发达的陆桥运输。陆桥运输是指利用横贯大陆的铁路（公路）运输系统作为中间桥梁，把大陆两端的海洋连接起来的集装箱连贯运输方式。陆桥运输一般有三种模式：第一种是大陆桥，两边是海运，中间是陆运，陆地把海洋连接起来，形成海—陆—海联运，而大陆起到了"桥"的作用。比如货物从中国运往德国先在中国港口上船，到达美国加州洛杉矶港口，然后通过铁路把货物陆运至美国纽约港口，最后经海路运到德国，模式即为海—陆—海。第二种是小陆桥运输，从运输组织方式上看与大陆桥运输并无大的区别，只是其运送的货物的目的地为沿海港口，即陆—海。比如，货物从武汉运往洛杉矶，先经由铁路把货物运至上海上船，直接抵达洛杉矶港口。第三种是微型陆桥运输交货地点在内陆地区，即海—陆。比如货物于中国港口上船，抵达加州洛杉矶下船，然后经由铁路把货物运至目的地科罗拉多州的丹佛市。

第三节　运输单据

为了保证进出口的安全交接，在整个运输过程中需要缮制各种单据。这些单据各有其特定的用途，彼此之间又有相互依存的关系，它们把船方、港口和货主各方联系在一起，分清各自的权利和义务。

一、海洋运输单据

（一）海洋运输的主要单据

（1）托运单（booking note，B/N 或 shipping note，S/N）：是托运人根据贸易合同和信用证条款内容填写的，向承运人或货运代理人办理货物托运的单证。托运单的内容应该包括货物名称、重量、体积、装运港口、目的港口、发货日期等。

（2）装货单（shipping order，S/O）：是承运人根据托运单的内容，并结合船舶的航线、挂靠港、船期和船舱等条件，接受了托运申请而签发给托运人的单据。装货单既可以用作装船依据，又是货主凭已向海关办理出口货物申报手续的主要单据之一。所以，海关验货后在装货单加盖放行印章的装货单又称为"关单"。

（3）集装箱装箱单（container load plan，CLP 或 unit packing list，UPL）：集装箱装箱单是详细记载每一个集装箱内所装货物名称、数量、尺码、重量、标志和箱内货物记载情况的单证。如果在集装箱货运站装箱，则集装箱装箱单由装箱的货运站缮制；如果由发货人装箱，则集装箱装箱单由发货人或其代理人的装箱货运站缮制。集装箱装箱单是集装箱运输的辅助货物舱单，其用途很广，可用作报关材料、货物交接单、集装箱装卸作业依据和索赔依据等。

（4）装货清单（loading list，L/L）：是承运人根据装货单留底联，将全船待装货物按卸货港（货物的目的港口）和货物性质归类，依航次、挂靠港顺序排列编制的装货单的汇总单。其内容包括装货单号码、货名、件数及包装、重量、估计的体积及特种货物对运输的要求或注意事项的说明等，故为积载计划提供依据，同时也是理货等业务的单据。

（5）大副收据（mate's receipt，M/R）：又称收货单，是船上大副根据理货人员在理货单上的签注日期、件数以及舱位等，并与装货单进行核对无误而签发的单据。这是船方收到货物的收据以及货物已上船的凭证。

（6）海洋提单（ocean bill of lading，B/L）：是海洋运输中的货物承运人（或其代理人）在收到货物后签发给托运人的凭证。

（7）提货单（delivery order，D/O）：提货单是收货人凭正本提单或副本提单随同有效的担保向承运人或其代理人换取的、可向港口装卸部门提取货物的凭证，用来进口报关。

（二）海运提单

1. 海运提单的性质和作用

海洋提单是整个货物运输最重要的单据，是指海洋运输中的货物承运人（或其代理人）签发给托运人的凭证。提单对货物的运输有详细的说明，其有三个功能：

（1）货物收据：证明承运人或其代理人已收到托运人的货物。

（2）货物所有权凭证：收货人在目的港提货必须向承运人或其代理人提交正本提单。正因为提单具有物权凭证效用，所以可用来向银行结算货款甚至转让或抵押。

（3）运输合同的凭证：海洋运输中的货物承运人有责任把货物安全送至目的地。如果双方就运输出现的问题有争议，海运提单上的合同条款是解决争端的主要依据。同时提单还是处理索赔与理赔以及或进行议付的重要单据。

2. 海洋提单的种类

按不同角度海运提单重点分为以下几类：

（1）根据货物是否已经装船，可分为已装船提单和备运提单。前者是指承运人已将货物装上指定船舶后签发出来的提单；后者是指承运人在收到货物存放在码头仓库等待装运期间所签发的提单，其往往不被议付银行接受。

（2）根据提单是否对货物外表或状况有不良批注，分为清洁提单和不清洁提单。前者是指货物在装船时表面状况良好，没有不良批注；后者是指承运人在提单上对货物表明状况或包装有不良批注，如"packages in damaged condition"或"short shipment"。这种提单往往遭到议付银行的拒付。

（3）根据提单收货人抬头不同，常分为记名提单和指示提单。前者是指在"收货人"一栏内填写收货人名称和地址，只能由收货人提货，不得转让给他人，不具有流通性；后者是指在"收货人"一栏内填写"to order"或"to the order of …"字样。这种提单可背书转让，具有流通性。

（4）按不同的运输方式分类，可分为直达提单、转船提单、联运提单和多式联运提单。直达提单是指船运中途不经过换船而直接驶往目的港卸货所签发的提单。转船提单是从装运港装货的轮船不直接驶往目的港，而需在中途港换装另外船舶所签发的底单。提单需注明"transshipped at…"。联运提单是指经过海运和其他运输方式联合运输时由第一程承运人（船公司或其代理人）所签发的包含全程运输的提单。联运提单虽然包括全程运输，但签发提单的各程承运人只对自己运输的一段航程中所发生的货损负责，在后续运程中，提单签发人只是其他实际承运人的代理，这种提单与转船提单性质相同。多式联运提单指货物由海上、内河、铁路、公路、航空等两种或多种运输方式进行联合运输而签的适用于全程运输的提单。这种提单的签发人是多式联运经营人，负责每一段运输。

（5）根据船舶运营方式不同，可分为班轮提单和租船提单。前者是由班轮公司签发给托运人的提单；后者是承运人根据租船合同而签发的提单，由于这种提单受到租船合同的约束，银行接受提单时往往要求提供租船合同。

（6）按提单的签发人不同可分为船东提单和货代提单。前者是船东签发给托运人的提单，收货人持有船东提单就可以直接向承运人提货；后者是指货运代理人签发给托运人的提单，这时货运代理人需要把某一托运人的货物和其他托运人的货物拼成整箱。

（7）其他类型提单。过期提单是指卖方超过提单签发日期后21天或超过了信用证的有效期才提交给银行议付的提单。按照惯例，如果信用证无特别规定，银行将会拒绝接受这种过期提单。倒签提单是指承运人因受托运人请求或其他原因签发提单的日期早于实际转船日期的提单。而预借提单是指因信用证规定的日期和议付日期已到，而货物却未能及时装船，但已经被承运人接管，托运人出具保函让承运人签发的已装船提单。无论是倒签提单还是预借提单，都是不按实际日期签发的提单，因而都是不合法的。

二、其他运输单据

（一）航空货运单

航空货运单是指承运货物的航空承运人（航空公司）或其代理人，在收到承运货物并接受托运人空运要求后，签发给托运人的货物收据，也是承运人与托运人或其代理人的运输合同。航空货运单只是运输合同的证明，不能代表货物的所有权，不能用于提取货物，具有不可转让性。

需要指出的是航空货物运单无论是国内货运还是国际货运都是统一格式。一般而言航空货运单正本三份至少有六份副本。第一份正本由托运人签字承运人存根；第二份由托运人和承运人签字连同货物一起发送给收货人；第三份由承运人签字托运人存根。

（二）海运单

海运单是承运人或代理向托运人签发的表明他已经收到托运人的货物并拟将该货物运往指定目的港、直接交给指定收货人的凭证。海运单也只是运输合同的证明，不能代表货物的所有权，不能用于提取货物，具有不可转让性。

与海运单的性质和作用相类似，国际铁路联运单、邮政收据和特快专递收据等运输单据都只是运输合同的证明，不能代表货物的所有权，不能用于提取货物，具有不可转让性。

第四节　合同中的装运条款

装运条款主要指对装运的细节进行说明，主要讨论装运时间、装运港口、目的港口、发货通知、是否可分批装运和转运等问题。

一、装运时间

装运时间是指发货人必须在一段时间之内或某一截止日之前把合同规定的货物装运发给买方。讨论装运时间时应该考虑以下几个问题：

（1）要考虑货源情况，如需要的数量，现有的库存和运输情况，如舱位、航线、航班和雨季或冰封期。

（2）避免使用如 " immediate/prompt shipment" 等模糊字句，规定应该清晰明了。

（3）装运时间应该具有灵活性，是合情合理的一段时间。比如雨季的装载时间应该长些，工作日因为天气原因会减少。

（4）考虑货物本身的性质。比如烟草的装载就选择天气晴朗的日子。

关于装运时间的规定，通常有以下两种方法：第一种是规定最迟期限装运。例如，"shipment before June 30th" 或 " shipment during March" 或 " shipment during July/August in two equal lots"；第二种是规定是把装船日与截止日联系起来如 " shipment with 30 days after receipt of L/C"。这种方法也需要规定买方的相应义务，如 "L/C must reach the seller not later than…"。第一种方法直接简单；第二种方法则把货款与发货结合起来，所以更加安全灵活。一般而言，信用证应该在装运日之前至少 15 天到达卖方，使其有足够的时间核对信用证，之后才发货，必要时信用证还得修改。信用证的有效期应该是发货日之后的 7 ~ 15 天，给卖方足够的时间到银行提交材料结账。

二、装运港和卸货港

在规定装运港或卸货港，要注意以下几点：

（1）规定装运港或卸货港要清楚明了。有些港口在不同国家却有相同的名称，这时应注明国别或地区，以免引起误解。

（2）选择几个备用港口，增加灵活性。有时候，销售合同订立时还不能确定准确的交货地点。通常的做法是赋予买方之后指定某一具体港口的权利，并且如果他未能做到，则要承担相应的风险和额外的费用。例如，在信用证开出之前，买方从备选港口中选择一港口，由买方从备选港口中选择港口，买方必须在船只预期到达第一备用卸货港口的两天前通知承运人，并承担带来的额外费用。

（3）考虑政治因素。比如某国正经受到国际制裁，或是某些国家的敌意国，那么卖方应该避免经过该国家的港口。

（4）考虑港口的法规、设施和收费种类。避免选择不熟悉的，设施落后和高收费的国家。比如有些国家禁止某种危险商品进入，不允许在其所在地上货卸货；有些国家允许危险商品进入，但必须取得许可证或其他证件；有些国家允许危险商品进入，但商品必须有特殊的包装；而有些国家仅仅允许危险货物在卸货之后收货人立即用装载工具运走。如果违反相关规定，就会受到罚款、起诉甚至进监狱。

三、分批装运和转运

分批装运是指一个合同项的货物分成若干批次装运交付。分批装运和转运涉及买卖双方的权益，在合同中必须有明确的规定，例如：

Partial shipments（not /to be）allowed.（不/允许分批装运。）

Ship 200 M/T during September and 100 M/T during Oct.（9 月份装 200 吨，10 月份装 100 吨。）

Shipment during November and December in two equal lots.（分两批于 11 月及 12 月平均装运。）

转运是指货物从装运港到目的港的运输过程中，从一种运输工具转至另外一种运输工具，或由一种运输方式转至另一种运输方式。合同中也必须有明确的规定，例如：

Transshipment（not/to be）prohibited（不/允许装运）

to be transshipped at…（于……转运）

如果转运是为了货物抵达指定的目的地，那么卖方需要支付转运费用。但如果是因为承运人为了避免可能的障碍，比如冰霜、拥挤、劳务纠纷等，那么装运产生的额外费用应该由买方支付。

《跟单信用证统一惯例》600 号出版物对分批装运和转运规定，可以归纳为以下几条原则：

（1）如果信用证没有禁止分批、禁止转运的规定，应视为可以分批装运和转运。

（2）表明使用同一运输工具并经由同次航程运输的数套运输单据在同一次提交时，只要显示相同目的地，将不视为部分发运，即使运输单据上标明的发运日期不同或装卸港、接管地或发送地点不同。

（3）假如交单由数套运输单据构成，其中最晚的一个发运日将被视为发运日。含有一套或数套运输单据的交单，假如表明在同一种运输方式下经由数件运输工具运输，即使运输工具在同一天出发运往同一目的地，仍将被视为分批装运。

（4）含有一份以上快递收据、邮政收据或投邮证实的交单，假如单据看似由同一块地区或邮政机构在同一地点和日期加盖印戳区签字并且表明同一目的地，将不视为分批装运。

（5）如信用证规定在指定的时间段内分期付款或分期发运，任何一期未按信用证规定期限支取或发运时，信用证对该期及以后各期均告失效。

四、运载工具及运输路线

因为运输路线关系到运费的高低及运输时间。有时候，在合同中要规定运载工具，而指定运载工具往往由买方提出。买方指定运载工具，有的是因为其与船公司有协议，而要求将货物交由该船公司承载；也有的是因某公司的船舶设备优良、船速迅捷，买方要求交其承运可提早收到货物；也有的是因为进口国政府有政策导向，要求用本国船公司的船舶进行运输，以扶持本国的航运事业，因此，买方在合同中就会对运载工具做出规定。例如：Shipment to be effected per American President Line steamer. （货物交给美国总统轮船公司船运。）又如，Shipment should be made per RIL vessel. （货物由 RIL 船公司承运。）

在 CFR 及 CIF 等贸易术语下，由卖方租船订舱，买方若急需货物或货物运输时间不宜太长，就会指定卖方选择航程较短的航线船只装运。例如：Shipment via Suez canal. （经由苏伊士运河运输。）Shipment during October via Panama Canal. （十月份装运，途经巴拿马运河运输。）

五、装运通知

装运通知起到协调进出口双方责任的作用。在卖方承担保险和运输的情况下，对卖方发装运通知的要求不是很严格。但是，如果是在有买方承担保险和运输的情况下，则对卖方发装运通知的要求就显得尤为重要。在 CFR 贸易术语下，卖方负责安排运输，并在装运港将货物装上船，而买方则要自己办理保险。由于买方承担货物在装运港装上船之后的一切风险，因此，对货物装船后可能遇到的风险必须及时办理保险。若有延迟，一旦货物装船后遇险发生损失，就得不到保险公司的赔偿。但是，买方办理保险必须等卖方发出装运通知。按照国际贸易的习惯做法，在装运货物后，发货人应立即（一般在装船后 3 天内）发送装运通知给买方或其指定的人，从而方便买方办理保险和安排接货等事宜。《国际贸易术语解释通则 2010》明确规定，卖方必须给予买方关于货物装船的充分通知。这里的"充分"是指装船通知在时间上是毫不迟疑的，在内容上是详尽的，以满足买方办理保险的需要。

本章小结

国际货物运输的方式有很多种，海洋运输方式是应用最广的国际货物运输方式。依据船公司对船舶经营方式的不同，海洋运输可分为班轮运输和租船运输两大类。

班轮运输的主要特点可概括为"四固定"：航线固定、沿线停靠的港口固定、航行的时间表比较固定和运费率相对固定。班轮运费包括基本运费和附加运费两部分。

合同中的装运条款主要有装运期、装运港和目的港以及运输方式和运输线路，能否分批装运和转船、转运等方面的内容。在业务实践中，分批和转运是最容易出现差错的地方，要予以特别关注。

国际货物运输保险及合同中的保险条款

★学习目标

1. 了解国际货物运输保险的基本原则
2. 理解海上货物运输保险的承保范围
3. 掌握中国货运保险条款和协会货物保险条款
4. 掌握国际贸易合同中保险条款的常见规定方法和主要内容

在国际贸易中，国际货物运输往往需要经历一个较为漫长的过程。在这一过程中，货物有可能因为遭遇自然灾害或意外风险而受到损失。因此，投保国际货物运输保险将是以事先的有限费用规避未来不确定风险的明智选择。

国际货物运输保险的基本类别主要包括海上货物运输保险、陆上货物运输保险、航空货物运输保险和邮包货物运输保险。其中，海上货物运输保险起源最早，可追溯到大约中世纪。海上货物运输保险也是使用频率最高的一种国际货物运输保险，其他货物运输保险都是在海上货物运输保险的基础上发展起来的。因此，本章将着重介绍海上货物运输保险。

第一节　货物保险的基本原则

所谓保险，就是指这样一种合同，当事人一方以收取的保费为代价，承担赔偿另一方当事人的保险标的物可能遭受某些危险或风险导致的损失。这里有两个基本的当事人：投保人和保险人。交付的一定金额的费用就是保费。获得保障的人是投保人，也被称为被保险人，指从保险公司或保险人那里购买保险的那一方。在货运保险中，投保人是卖方还是买方，取决于双方在合同中采用的术语。比如，一个采用 CIF 术语的合同，就是由卖方负责买保险，即使他并不承担运输过程中的主要风险。相反，负责向被保险人赔偿损失的人就是保险人，通常都是保险公司，又叫保险承销商。当承保范围内的损失发生，遭受损失的一方可以向保险人提出索赔。提出索赔的这一方就叫索赔方。在货运保险中，被保险人和索赔人并不一定就是同一方。这大多取决于谁对受损货物拥有保险利益。在保险中涉及的关系人可能还有保险代理人、保险经纪人、保险公证人等。

在法律上，货运保险的投保人和保险人在签订和履行合同时必须遵循以下几个基本原则：

一、保险利益原则

任何事物都可以投保吗？只有可以用货币形式衡量的才能投保。世界上存在着大量类似的风险。被保险人必须与保险标的有保险利益的关系。保险利益的存在是保险法律关系的基本要素。保险利益是指投保人或被保险人对保险标的的所具有的合法的利害关系。保险利益可以表现为某种对保险标的的物的所有权，权利、利益或生存价值，它们必须是基于保险标的的物产生的。投保人或被保险人因保险事故的发生致使保险标的不安全而受损，或因保险事故的不发生而受益，这种利害关系就是保险利益。投保人或被保险人对保险标的的所具有的利害关系必须能被法律认可。我国保险法第 12 条规定，投保人对保险标的的应当具有保险利益，投保人对保险标的的不具有保险利益的，保险合同无效。此原则可以使被保险人无法通过不具有保险利益的保险合同获得额外利益，以避免将保险合同变为赌博合同。

二、最大诚实信用原则

最大诚实信用原则是指保险合同的当事人应在订立合同时完整和准确地告知对方有关保险标的的重要事实，不管对方有没有询问。诚实信用原则对于保险合同尤其重要，因为保险属于无形产品，海上保险承保人在承保时，往往远离船货所在地，对保险标的的难以实地查勘，仅凭投保人的陈述。只有投保人对保险标的的有关情况是最清楚的。当作为保险合同一方的当事人（即投保人）想把风险转嫁出去，他对保险标的的有关情况是最清楚的，投保人的陈述是否完整和准确，对保险人承担的义务关系极大，因此，为了维护保险人的利益，必须要求被保险人坚守诚信。当然，诚信原则是相互的，这一原则也适用于保险人。各国法律规定，违反诚信原则表现为虚假陈述或未披露，对于虚假陈述，不管是有意还是无意为之，由此而签订的保险合同，受到侵害的另一方可解除合同。未披露也区分为无意还是有意欺诈，有意欺诈的构成隐瞒，实质上虚假陈述及未披露重大事实，可能导致保单无效。

三、近因原则

近因是指对保险财产造成损失的最直接原因，这原因不是以时间最接近的标准来衡量的，而应是影响上的最重要的原因。有时候发生的事故是一连串事故，有两种或很多的事故原因，其中有在保险责任中的，也有除外的，那么就要看究竟哪一个是最有影响的原因（最近原因），当发生这种情况时往往会使保险人与被保险人之间引起争议。比如灭火时用水造成财产的损失，再比如一间房子，火灾是由于敌方飞机的空中轰炸引起的，发生损失是因为着火，但是近因是战争（本案例中战争是除外责任），这样就没有任何赔偿。虽然我国保险法及海商法均没有对近因原则进行明文规定，但在国际货物运输保险实践中，近因原则是常用的确定保险人对保险标的的损失是否负有保险责任以及负何种保险责任的一条重要原则。

四、损失赔偿原则

损失赔偿原则是指在保险事故发生而使被保险人遭受损失时，保险人必须在责任范围内对被保险人所受的实际损失进行补偿的一种机制，它使被保险人大致恢复到与损失发生之前相同的财务状况。保险人可以采取金钱支付、修复或替代方式进行赔偿。损失赔偿原则只适用于财产、责任和其他非人身保险。国际货物运输保险合同属于补偿性的财产保险合同，因此，在发生超额保险和重复保险的情况下，保险人只赔偿实际损失，因为保险的目的是补偿，而不能通过保险得利。

五、代位求偿原则

保险代位求偿原则是从补偿原则中派生出来的，只适用于财产保险，是损失赔偿原则的应用。在财产保险中，保险事故的发生是由第三者造成并负有赔偿责任，则被保险人既可以根据法律的有关规定向第三者要求赔偿损失，也可以根据保险合同要求保险人支付赔款。如果被保险人首先要求保险人给予赔偿，则保险人在支付赔款以后，保险人有权在保险赔偿的范围内向第三者追偿，而被保险人应把向第三者要求赔偿的权利转让给保险人，并协助向第三者要求赔偿。反之，如果被保险人首先向第三者请求赔偿并获得损失赔偿，被保险人就不能再向保险人索赔。例如，房东把他的房子向火险保险人投保了金额为 50 万卢比的保险，由于承包商的接线错误导致房屋全部毁损，保险公司须赔偿给房主 50 万卢比。在保险公司全部赔偿之后，就可以取得对承包商的代位求偿权。对代位求偿的几点说明：①无论是全部损失还是部分损失，保险人只有赔偿了被保险人的损失后，才可以取得代位求偿的权利；②定值保险单下，保险人向第三方追偿回来的款项如果少于赔款，则全部归于保险人；如果多于赔偿则余额部分归于被保险人；③被保险人可以取得保险人赔付之前的损失利息，保险人可以取得保险人赔付之后的利息，但也可以根据保险合同约定的方法进行处理。

六、分摊原则

分摊原则也是从补偿原则中派生出来的，适用于两个或以上的保险人对同一风险承保的情形，即重复保险情形。根据保险补偿原则，在发生重复保险赔付责任时，将保险标的的损失赔偿责任在各保险人之间进行分摊，保险人通常言明他们只愿承担的赔偿比例，以避免被保险人获得超过实际损失的赔偿的法律原则。

第二节　海洋货物运输保险的风险与损失

货物在海上运输过程中可能遭遇各种各样的风险，而一旦遭遇风险就可能造成货物的损失。我们可以通过不同的保险条款来回避风险，也就是通过交付保费来购买不同的险别。所以我们需要在投保前对风险及损失等知识认真理解。保险业将这些风险划分为海上风险和外来风险两大类，由风险所带来的损失则分别称之为海损及外来风险损失。

一、海上风险

海上风险是指来自海上的自然灾害和意外事故带来的风险及外来原因带来的风险。

（一）意外风险

在海上保险业务中所指的自然灾害，一般仅指恶劣气候（Heavy Weather）、洪水（Flood）、地震（Earthquake）、海啸（Tsunami）、雷电（Lightning）、流冰（Iceberg）、火山爆发（Volcanic Eruption）等人力不可抗拒的灾害，但又并非包括所有由于自然力所造成的灾害。

在海上保险业务中所指的意外事故，并非囊括所有的海上意外事故，而一般仅指船舶搁浅（Grounding）、触礁（Stranding）、沉没（Sunk）、失踪（Missing）、失火（Fire）、爆炸（Explosion）与流冰或其他物体碰撞（Collision）等引起的有明显海洋特征的重大意外事故。

（二）外来风险

在海上保险业务中所指的外来风险是指除海上风险以外的其他原因所导致的风险，又可划分为一般外来风险和特殊外来风险两大类。

一般外来风险是指在海上运输过程中所发生的偷窃（Theft）、破碎（Breakage）、沾污（Contamination）、渗漏（Leakage）、生锈（Rusting）、串味（Taint of Odour）、钩损（Hook Damage）、短量（Shortage in Weight）、碰损（Clashing）以及受热受潮（Sweating and /or Heating）、淡水雨淋（Fresh and /or Rain Water Damage）等。

特殊外来风险主要是指由于政治、军事或国家法令政策等方面的变化所带来的风险。如由于战争的爆发导致运输途中的货物被扣留而无法交付或由于国家政策的变化导致货物因无法申领进口许可证获许进口。

由一般外来风险和特殊外来风险所造成的损失称为外来风险的损失。在我国，对于外来风险的损失赔付，一般按中国保险条款的附加险条款执行。

二、海上损失

海上损失简称海损，是指货物在海上运输过程中由于遭遇海上风险所造成的损失。根据保险业的惯例，海损还包括与海运相连的陆运及内河运输过程中所遭遇的货物损失。

根据货物损失的程度，海损分为全部损失（Total Loss）和部分损失（Partial Loss）。

（一）全部损失

全部损失简称全损，是指被保险货物全部遭受损失。根据实际情况的不同，全损又可分为实际全损和推定全损两类。

实际全损是指被保险货物在遭遇海上风险后已经完全灭失或变质，不能再被恢复或修理后使用，或者被保险人无可避免地失去了它。比如，某个合同下的一批茶叶被海水浸泡，即使最后到达了目的地，也不可能再有使用价值。当实际全损发生的时候，被保险人在保险人全额赔付后不需要给付保险人委付通知（即放弃保险货物的所有权及利益）。

推定全损是指尽管货物在遭遇海上风险后尚未达到实际全损的状况，但造成实际全损已不可避免，或为了避免实际全损的发生所需支出的费用加上将货物继续运抵目的地所需费用的总和已经超过了货物的保险价值，这样的情形已与实际全损无异，因此称为推定全损。在推定全损的情形下，被保险人可将有关保险标的物的所有权利通过委付通知转给保险人。

（二）部分损失

部分损失是指货物遭遇海上风险后虽发生了损失但又尚未达到全部损失的情形。

根据货物损失的性质而言，部分损失又可分为共同海损（General Average）和单独海损（Particular Average）。

1. 共同海损

共同海损是指当载货的船舶在海上航行过程中，因遭遇海上风险而使船货面临共同的安全威胁时，为了解除这种共同的安全威胁，船方有意识地采取了某些合理的施救措施，并因此造成了某些货物的特殊损失或支出特殊的费用，这样的货物损失和费用，称为共同海损。共同海损将由船方及其他收益方（如货方、运费收入方）根据获救价值按比例分摊，然后再分别向各自的保险机构索赔。

共同海损须具备以下几个条件：一是发生了威胁大家安全且船长无法控制的事件；二是自动和有意地采取的牺牲；三是采取的措施是有效的，必须有船货被救。自愿牺牲的可能是抛弃某些货物，人为使用拖船，主动搁浅，明知使引擎工作会导致损害船舶的情况下也采取措施。

发生共同海损以后，凡属于共同海损范围内的牺牲和费用，均需通过共同海损理算，由有关获救收益方（船、货、运费、燃油）根据获救价值按比例分摊。这种分摊就叫共同海损分摊。共同海损的理算在国际上一般按照1974年或1990年修订的《约克安特规则卫普》办理。在我国，则一般规定按照《中国国际贸易促进委员会共同海损理算暂行规则》（简称"北京理算规则"）办理。

2. 单独海损

单独海损是指仅涉及船或货的一部分的意外损失。既没有达到全部损失的程度，又不具备共同海损的性质的，称为单独海损。这样的损失将由受损方单独承担。如果所遭受的损失属于受损方所投保的险别之列的，可备齐相关的单证向保险公司索赔。

单独海损和共同海损最大的区别在于：

（1）损失的原因不同。单独海损是由海上风险所带来的船舶或货物的直接损失，而共同海损则是船方为了解除船、货遭遇的共同危险而采取合理措施时，给船舶或货物所带来的人为的间接损失。如，当船舶在航行过程中遭遇暴风雪而导致船体发生严重倾斜，面临沉没的危险时，船方为了保持船体的平衡而将部分货物抛入海中。这样的货物损失是为了解除船、货的共同危险所做出的，因此，就属于共同海损。

（2）损失的构成不同。单独海损仅包括由海上风险所导致的货物本身的损失，而共同海损既包括货物损失，又包括为解除船货共同危险所支付的费用。

（3）损失的承担方式不同。单独海损由受损方单独承担；而共同海损则由各受益方按受益比例分摊。

三、海上费用

当货物遭遇海上风险时，不仅可能遭受货物本身的直接损失，而且还可能发生费用方面的支出。这些费用支出只要不超过保险金额，保险公司也会给予赔偿。

海上费用主要分为施救费用和救助费用两种。施救费用是指当被保险货物遭受承保范围内的风险时，被保险人或其代理人或保单受让人为了避免或减少损失的发生而采取相应的施救措施时所发生的费用支出。

"救助"一词指的是向遇险船舶提供援助的做法。救助费用则是指被保险人、其代理人、任何他们雇佣的人以外的第三者对遭受海上风险的货物采取救助措施时，由被救助方向施救方支付的酬谢费。这种费用只要是合适的，救助又有效，就可以根据具体发生的情形作为单独海损或者共同海损进行补偿。救助费用的赔偿常常采取"无效果，无报酬"的原则。

第三节　中国海洋货物运输保险条款

中国人民保险公司的《海洋货物运输保险条款》在 1972 年制定，后经过几次修改，最新的版本是 2009 年版。它是我国进出口公司投保海洋货物运输保险的主要依据，主要规定了保险人的责任范围、除外责任、责任起讫以及被保险人的义务和索赔期限等内容。

一、海洋货物运输保险条款

根据 2009 年版的中国海洋货物运输保险条款的规定，其保险险别主要分为基本险和附加险两大类。基本险可以单独投保，但附加险不能单独投保，投保人只有在投保了任何一种基本险之后，才能加保任何一种或数种附加险。

（一）基本险

依据中国人民保险公司的《海洋货物运输保险条款》的规定，基本险又分为平安险、水渍险和一切险三种基本险别。

1. 平安险

平安险（Free from Particular Average），简称"FPA"，其基本内涵是只负责赔偿因海上风险所造成的全部损失和共同海损，对于单独海损则不予赔偿。依据我国海洋运输条款的规定，平安险的主要承保范围包括：

（1）在运输过程中，由于遭遇恶劣气候、雷电、海啸、地震、洪水等自然灾害所造成的被保险货物的实际全损或推定全损。当被保险人要求赔偿推定全损时，须将受损货物及其权利委付给保险公司。

（2）由于运输工具遭受搁浅、触礁、沉没、互撞、与流冰或其他物体碰撞以及失火、爆炸等意外事故所造成的被保险货物的全部损失或部分损失。

（3）运输工具发生了搁浅、触礁、沉没、焚毁等意外事故，并且在此前后又在海上遭遇了恶劣气候、雷电、海啸等自然灾害所造成的被保险货物的部分损失。

（4）在装卸或转运过程中由于一件、数件或整件货物落海所造成的被保险货物的全部或部分损失。

（5）当被保险货物遭遇承保范围内的危险时，被保险人为防止或减少货损而采取抢救措施时所支付的合理费用，但以不超过该批被救货物的保险金额为限。

（6）由于运输工具遭遇自然灾害或意外事故而需要在中途港口或避难港停靠，由此引起的被保险货物的损失以及在中途港、避难港由于卸货、存仓以及运送货物所产生的特别费用。

（7）由于发生共同海损所引起的牺牲、分摊和救助费用。

（8）依据运输契约中所制定的"船舶互撞责任条款"的规定应由货方偿还船方的损失。

平安险是我国海洋货物运输条款中保险责任最小的一种险别，费率也最低。

2. 水渍险

水渍险（With Particular Average），简称"W. A."，其基本内涵是除了赔偿因海上风险所造成的全部损失和共同海损外，对于单独海损也予以赔偿。因此，保险人在水渍险下的保险责任主要体现在两个方面：

（1）承担平安险项下的全部海损和共同海损的赔偿。

（2）承担被保险货物由于遭遇恶劣气候、雷电、海啸、地震、洪水等自然灾害而造成的部分损失。

3. 一切险

一切险（All Risk），简称"A. R."，是我国海洋货物运输保险条款中承保范围最大的一种基本险别。它的名字只是保险公司的一种承保险别的称谓，并不意味着承保所有风险。比如由于运输延迟、货物内在缺陷或者战争、罢工等原因引起的损失都不在一切险的承保范围。保险人在一切险项下的保险责任主要体现如下：

（1）承担水渍险项下的全部责任。

（2）承担被保险货物在运输途中由于外来原因所遭受的全部或部分损失。

（二）附加险

我国海洋货物运输保险条款中还有附加险条款，可分为一般附加险和特殊附加险两大类。

1. 一般附加险

一般附加险所包括的主要险种有：

（1）偷窃、提货不着险（Theft, Pilferage and Non – delivery, T. P. N. D. ）。是指主要承保在保险有效期内，保险人对保险货物由于被偷窃，整件提货不着所造成的损失负赔偿责任。

（2）淡水雨淋险（Fresh Water Rain Damage, F. W. R. D）。主要承保货物在运输中，由于淡水、雨水以至雪溶所造成的损失，保险公司都应负责赔偿。

（3）短量险（Risk of Shortage）。主要负责赔偿承保的货物因外包装破裂或散装货物发生数量损失和实际重量短缺的损失，但不包括正常运输途中的自然损耗。

（4）混杂、沾污险（Risk of Intermixture & Contamination）。指保险人负责赔偿承保的货物在运输过程中因混进杂质或被玷污，影响货物质量所造成的损失。

（5）渗漏险（Risk of Leakage）。是指保险人负责赔偿承保的流质、半流质、油类货物在运输途中因容器损坏而引起的渗漏损失，或用液体储藏的货物因液体渗漏而引起的腐烂变质造成的损失。

（6）碰损、破碎险（Risk of Clash and Breakage）。主要负责承保金属、木质等货物因震动、颠簸、碰撞、挤压而造成货物本身的损失，或易碎性货物在运输途中由于装卸野蛮、粗鲁、运输工具的颠震所造成货物本身的破裂、断碎的损失。

（7）串味险（Risk of odor）。主要负责赔偿承保的食用物品（如食品、粮食、茶叶、中药材、香料）、化妆品原料等因受其他物品的影响而引起的串味损失。

（8）受潮受热险（Sweating and Heating Risk）。本保险对被保险货物在运输过程中因气温突然变化或由于船上通风设备失灵致使船舱内水汽凝结发潮或发热所造成的损失，负责赔偿。

（9）钩损险（Hook Damage）。保险人负责赔偿承保的货物（一般是袋装、箱装或捆装货物）在运输过程中使用手钩、吊钩装卸，致使包装破裂或直接钩破货物所造成的损失及其对包装进行修理或调换所支出的费用。

（10）锈损险（Risk of Rust）。主要负责赔偿承保的货物在运输过程中由于生锈而造成的损失。但生锈必须是在保险期内发生的，如原装船时就已生锈，保险公司不负责。

（11）包装破裂险（Loss of Damage Caused by Breakage of Packing）。主要负责赔偿承保的货物在运输过程中因搬运或装卸不慎造成包装破裂所引起的损失，以及因保险货物续运安全需要而产生的修补或调换包装所支出的费用。

上述的一般附加险不能单独投保，只能在投保了平安险或水渍险的基础上加保。当已经投保了一切险时，由于上述险别已经包含在一切险的承保范围内，因此，就无须再加保一般附加险了。

2. 特殊附加险

特殊附加险主要承保由于政治原因、军事形势变化等原因引起的损失，以及由于有些国家的行政管理、政策措施等因素引起损失的险别。主要包括以下险种：

（1）战争险（War Risk）是指承保战争或类似战争行为等引起保险货物的直接损失或使用常规武器带来的损失。但是使用原子武器或核武器带来损失和费用并不赔偿。

（2）罢工险（Strikes Risk）是指承保因罢工者，被迫停工工人，参加工潮、暴动和民众斗争的人员采取行动造成保险货物的损失，以及任何人的恶意行为造成的损失。但其负责的损失仅仅是直接损失，对于间接损失是不负责的。

（3）交货不到险（Failure to Delivery risks）是指承保运输货物从装上船舶开始起算，满6个月仍未运抵原定目的地交货，不论什么原因，保险公司均按全损赔付。

（4）进口关税险（Import Duty Risk）是指承保运输货物到达目的港后，因遭受保险责任范围内的损失，而仍需按完好货物缴纳进口关税所造成的损失。

（5）舱面险（On Deck Risk）是指承保有些体积大、有毒性、有污染性、易燃易爆而习惯于必须装在舱面上的货物。

（6）拒收险（Rejection Risk）是指承保运输货物由于在进口港被进口国的政府或有关当局拒绝进口或没收，保险人按该货物的价值予以赔偿。

（7）黄曲霉素险（Aflatoxin Risk）是指承保运输货物在进口港经当地卫生当局检验证明，因含有黄曲霉素，并且超过了进口国对该霉素的限制标准，必须拒绝进口、没收或强制改变用途时造成的损失。

（8）出口货物到香港（包括九龙在内）或澳门存仓火险责任扩展条款，是指为了保障过户银行的利益，出口到港澳的货物，如直接卸到保险单载明的银行所指定的仓库时，加贴这一条款，则延长存仓期间的火险责任。保险期间从货物进入过户银行指定的仓库时开始，直到过户银行解除货物权益或者运输责任终止时计算满30天为止。

特殊附加险也不能单独投保，只能在投保了平安险、水渍险或一切险的基础上加保。此外，根据国际保险业的习惯做法，在已经投保战争险后另加罢工险，无须另行收费，因此，一般可同时投保战争险和罢工险。

二、除外责任

根据中国人民保险公司海洋货物运输条款的规定，保险公司对于下列损失不予赔偿：

（1）由于被保险人的故意行为或过失所造成的被保险货物损失。

（2）属于发货人责任所导致的被保险货物的损失。

（3）在保险责任开始前，被保险货物就已经存在品质不良或数量短差。

（4）由于被保险货物的自然损耗、本质缺陷、固有的特性以及市价涨跌、运输延误等所引起的被保险货物的损失或需额外支出的费用。

（5）在战争险和罢工险条款中所规定的责任范围和除外责任。

三、保险责任的起讫时限

根据中国人民保险公司海洋货物运输条款的规定，保险公司对于平安险、水渍险或一切险所承担的保险责任的起讫时限主要有以下几种情况：

（1）根据中国人民保险公司海洋货物运输条款的规定，保险公司对于"平安险""水渍险"和"一切险"的责任起讫时限沿用国际保险业中惯用的"仓至仓"条款（Warehouse to Warehouse，W/W），即保险公司对被保险货物所承担的保险责任从货物运离保险单所载明的起运地或起运港发货人仓库开始，直到货物抵达保险单所载明的目的地或目的港收货人仓库为止。但是自货物从目的港海轮上卸下时算起满60天，则不论货物是否已进入收货人仓库，保险责任也自行终止。

（2）当被保险货物在保险期内需转运到非保险单所载明的目的地时，保险责任从货物开始被转运时即告终止。

（3）当被保险货物在运抵保险单所载明的目的地或目的港仓库以前的某一仓库就发生了被分配或分派的情况，则该仓库就被视为被保险人的最后仓库，保险公司的保险责任在货物运抵该仓库时即告终止。

保险公司对于战争险的责任起讫时限与基本险不同，不采用"仓至仓"条款，保险公司只承担从货物装上海轮至货物运抵目的港卸离海轮为止的水面风险。如果不卸离海轮或驳船，保险责任以海轮到达目的港当天午夜开始起算满15天为止。

四、保险索赔的时限

保险索赔的时限从被保险货物在最后目的地或目的港全部卸离海轮后起算，最多不超过两年。

被保险人向保险人索赔时，必须提供保险单正本、相关的发票、提单、装箱单、磅码单、货损货差证明、检验报告以及索赔清单等。如果涉及第三方责任，则还必须提供有关第三方责任的单证、函电或证明文件。

第四节　英国伦敦保险协会的海运货物保险条款

在国际海运保险中，各国保险组织都有自己的保险条款，其中英国保险协会所制定的各种保险规章制度，包括海运保险单格式和保险条款，对世界各国有着广泛的影响。目前，世界上有很多国家在海上保险业务中直接采用英国伦敦保险协会所制定的协会货物条款，即《英国伦敦协会海运货物保险条款》，一般简称为《协会货物条款》。

中国的企业以 CIF 或 CIP 术语出口时，一般会选择中国海运货物保险条款中的险别，但是如果国外顾客要求采用协会货物条款中的险别，那也是可以接受的。协会货物条款最早制定于 1912 年，为了适应不同时期法律、判例、商业、航运等方面的变化和发展，需要经常进行补充和修订，较近的一次修订完成于 1982 年。1982 年出台的《伦敦协会货物保险条款》给国际海运保险市场带来了革命性的变化，取代了长期被世界各国视为海上保险圣经的 S. G. 保单，改变了其一直作为保险单附贴条款的地位，在保险市场得到广泛应用。然而随着全球经济一体化的深入发展，国际运输方式发展日新月异，迫切需要改变相应的法律条款以适应时代发展，联合保险委员会从 2006 年起开始进行全方位的调查和咨询，搜集各国专家建议，于 2008 年 11 月 24 日公布了更新的协会货物保险条款，并于 2009 年 1 月 1 日起生效。伦敦保险协会的海运货物保险条款主要有五种：

（1）协会货物条款（A）（Institute Cargo Clauses）（A），简称 I. C. C.（A）；

（2）协会货物条款（B）（Institute Cargo Clauses）（B），简称 I. C. C.（B）；

（3）协会货物条款（C）（Institute Cargo Clauses）（C），简称 I. C. C.（C）；

（4）协会战争险条款（Institute War Clauses Cargo）；

（5）协会罢工险条款（货物）（Institute Strikes Clauses Cargo）。

I. C. C.（A）、（B）、（C）款险可以独立承保，协会战争险和罢工险也具有独立的结构，但是需要与保险公司协商之后独立投保。

一、协会货物 A 款险的承保风险与除外责任

（一）I. C. C.（A）险的承保风险

（1）承保"除外责任"各条款规定以外的一切风险所造成的保险标的损失。

（2）承保依据运送契约及（或）有关适用法律与惯例所理算或认定的共同海损与施救费用，而其发生系为了避免或有关避免"除外风险"以外的任何原因所致之损失。

（3）承保对于被保险人在运送契约之任何"双方过失碰撞条款"下所肇之责任，按照保险单所承保的危险予以理赔。倘船舶运送人依据该条款要求赔偿时，被保险人同意通知保险人，保险人要自备费用为被保险人对该赔偿要求提出抗辩。

（二）I. C. C. （A）险的除外责任

1. 一般除外责任

（1）归因于被保险人故意的不法行为造成的损失或费用。

（2）自然损耗、自然渗漏、自然磨损、包装不足或不当所造成的损失或费用。

（3）被保险标的物的不良、不当包装或配置引起的损害或费用，此种包装或配置已由被保险人或其职员于保险开始生效前完成，且堪能承受正常运输过程中之意外事故（本款所谓的包装，包括货柜内货物积载，且员工并不包括独立承揽人）。

（4）被保险标的物之固有瑕疵或本质所引起的损害或费用。

（5）直接由于延迟所引起的损失或费用，即使该延迟系由承保之危险所致者亦同（依其承保的共同海损条款可予赔付之费用则不在此限）。

（6）当被保险标的物装载于船舶上时，或依正常业务程序，被保险人知道或应知道破产或债务积欠将会妨碍正常航行者，由于船舶所有人、经理人、租船人或船舶营运人之破产或积欠所致之损害或费用。本款不适用于该保险契约已经转让给已买入或已同意买入这批被保险标的物善意受让之索赔者。

（7）由于使用任何原子反应装置物或核子分裂及或融合或其他类似反应或放射性之武器等直接或间接所致或引起的损失或费用。

2. 不适航、不适货除外责任

（1）载运船舶或驳船的不适航或载运船舶驳船不适宜安全运载被保险标的，而此种不适航或不适运原因于被保险标的之装载之时为被保险人已知情者。本规定不适用于该保险契约已经转让给已买入或已同意买入这批保险标的物善意受让之索赔者。

（2）货柜或运输工具的不适安全装载被保险标的物，而此装载系发生于保险生效前由被保险人或其职员完成，且于装货时已知不适载。

3. 战争除外责任

（1）因战争、内乱、革命、造反、叛乱或由此引起的内乱或任何交战方之间的敌对行为。

（2）由上述承保风险引起的捕获、拘留、扣留、禁制或扣押，以及这种行动的后果或任何进行这种行为的企图。

（3）遗弃的水雷、鱼雷、炸弹或其他遗弃的战争武器。

4. 罢工除外责任

I. C. C. （A）不承保下列原因所致损害或费用：

（1）因参与罢工、停工、工潮、暴动或民众骚扰等人员所致者。

（2）因罢工、停工、工潮、暴动或民众骚扰结果引起者。

（3）任何代表人或有关组织因采取以武力或暴动方式，借以直接推翻或影响不论其是否合法成立之任何政府组织的任何恐怖行为所致者。

（4）任何人因政治、意识形态或宗教动机行为所致者。

二、协会货物 B 款险的承保风险与除外责任

（一）I. C. C.（B）险的承保风险

（1）保险标的物的灭失或损坏可合理地归因于下列任何之一者，保险人予以赔偿：

①火灾或爆炸；

②船舶或驳船搁浅、触礁、沉没或颠覆；

③陆上运输工具的倾覆或出轨；

④船舶、驳船或运输工具同除水以外的任何外界物体碰撞；

⑤在避难港卸货；

⑥地震、火山爆发、雷电。

（2）下列原因所致保险标的物的灭失或损坏：

①共同海损牺牲；

②抛货或浪击落海；

③海水、湖水或河水进入船舶、驳船、运输工具、集装箱、大型海运箱或贮存住所。

（3）货物在装卸时落海或摔落造成整件的全损。

（二）I. C. C.（B）险的除外责任

（1）I. C. C.（A）险的除外责任。

（2）对被保险人之外的任何个人或数人故意损害和破坏标的物或其他任何部分的损害要负赔偿责任。

（3）对海盗行为不负保险责任。

（4）任何没有在本险种下被列举出来的承保风险之内的风险。

I. C. C.（B）款险的承保范围与中国保险条款下的水渍险大致相同。

三、协会货物 C 款险的承保风险与除外责任

（一）I. C. C.（C）险的承保风险

（1）只承保重大意外事故，而不承保自然灾害及非重大意外事故，其具体承保的风险有：

①火灾、爆炸；

②船舶或驳船触礁、搁浅、沉没或倾覆；

③陆上运输工具倾覆或出轨；

④船舶、驳船或运输工具同除水以外的任何外界物体碰撞；

⑤在避难港卸货。

（2）因下列原因所致的被保险物的损失或损害：

①共同海损牺牲；

②抛货。

（二）I. C. C.（C）险的除外责任

I. C. C.（C）险的除外责任与 I. C. C.（B）险完全相同。

由此可见，I. C. C.（C）险则类似于我国的平安险，但比平安险的责任范围要小。

四、协会货物罢工险

协会货物罢工险承保由于下列原因引起的保险标的物的灭失或损坏：

（1）罢工者，被迫停工工人或参与工潮、暴动或民变的人员所造成的损失。

（2）任何代表人或有关组织因采取以武力或暴动方式，借以直接推翻或影响不论其是否合法成立之任何政府组织的任何恐怖行为所致者。

（3）任何人因政治、意识形态或宗教动机行为所致的损失。

协会货物罢工险的承保范围与中国保险条款下的罢工险大致相同。

五、协会货物战争险

协会货物战争险承保由于下列原因引起的保险标的物的灭失或损坏：

（1）战争、内乱、革命、造反、叛乱或由此引起的内乱或任何交战方之间的敌对行为。

（2）由上述承保风险引起的捕获、拘留、扣留、禁制或扣押，以及这种行动的后果或任何进行这种行为的企图。

（3）遗弃的水雷、鱼雷、炸弹或其他遗弃战争武器。

协会货物战争险的承保范围与中国保险条款下的战争险大致相同。

第五节　合同中的保险条款

一、国际贸易合同中的保险条款

在国际贸易合同中制定保险条款时，不能采用笼统的规定方法，而应该明确规定由谁投保、投保的险别、投保的金额、所依据的保险条款以及保险费用的承担方等。

国际贸易合同中的保险条款示例：

（1）以 EXW 、 FAS 、 FOB 、 FCA 、 CFR 、 CPT 价格术语成交的合同，其保险条款可定为：

Insurance：To be covered by the Buyers.

（2）以 DAP、DAT 、 DDP 价格术语成交的合同，其保险条款可定为：

Insurance：To be covered by the Sellers.

（3）以 CIF 、 CIP 价格术语成交的合同，其保险条款可定为：

Insurance：To be covered by the sellers for 110% of the total invoice value against All Risks and War Risks，as per and subject to the Ocean Marine Cargo Clause 2009 and Ocean Marine Cargo War Risks Clause 2009 of the People's Insurance Company of China.

二、履约过程中的投保手续

在出口业务中，凡以 CIF 和 CIP 术语成交的合同，依照国际贸易惯例，应由卖方负责办理国际货物运输保险。在我国，一般是由进出口商按逐笔投保的方式办理投保手续。在填制保险单时，应严格按照进出口合同或信用证的规定详细列明被保险人的名称、保险标的的名称、数量、包装、保险险别、保险金额、运输工具的名称及运输路线等内容，以确保缴纳保费后所获得的保险单证能顺利地用于出口结汇。

办理投保的日期不应迟于合同或信用证规定的装船日期。在采用信用证结算的情形下，依据国际商会《跟单信用证统一惯例》（600）的规定，投保金额通常应按 CIF 或 CIP 价格加成 10%。

在进口业务中，按 FCA、FOB、CFR、CPT 术语成交的合同，依据国际贸易惯例，应由买方，即我国进口商自行办理国际货物运输保险。为简化投保手续避免漏保或难以及时办理投保手续等情形的发生，一般采取预约保险的做法，即由进出口商与保险公司提前订立长期的预约保险合同。在其后的每一批进口货物启运时无须再另行填写保单，而仅需将国外的装运通知及时转交保险公司即可视为办妥了投保手续，保险公司将自动承担起对该批货物的保险责任。

三、保险金额的确定和保险费的计算

中国人民保险公司对于承保货物保险金额的确定一般以货物的 CIF（CIP）价再加成10%计算，具体公式为：

$$保险金额 = CIF（CIP）价 \times（1 + 投保加成率）$$

$$保险费 = 保险金额 \times 保险费率$$

$$= CIF（CIP）价 \times（1 + 投保加成率）\times 保险费率$$

【例6-1】 某外贸公司出口一批货物，数量为 200 公吨，价格为 USD 1 200 Per Metric Ton CIF Rotterdam，合同规定卖方应按发票金额加成 10% 投保水渍险和短量险，保险费率分别为 0.2% 和 0.3%，试计算该外贸公司应该支付的保险费。

解：保险金额 = CIF（CIP）价 ×（1 + 投保加成率）

= 1 200 × 200（1 + 10%）

= 264 000（美元）

保险费 = 保险金额 × 保险费率

= 264 000 ×（0.2% + 0.3%）

= 1 320（美元）

答：该外贸公司应支付 1 320 美元的保险费。

四、保险单证

在国际贸易业务中，保险单是保险公司与投保人之间所订立的保险合同。它既是保险公司的承保证明，也是投保人凭以办理保险索赔和保险公司凭以办理保险理赔的书面依据。常用的保险单证有保险单和保险凭证两种。

（一）保险单（Insurance Policy）

保险单又称大保单，是保险公司和投保人之间订立保险合同的正式凭证。所载内容除被保险人，保险标的名称、数量、包装、运输工具、运输路线、投保险别、保险期限、保险金额等内容外，还在背面另附有关保险人责任范围以及保险人和被保险人的权利和义务等方面的详细条款。

保险单除了可作为投保人办理保险索赔和保险公司办理保险理赔的书面依据外，还通常被列为出口商向银行办理出口押汇时所需提交的单证之一。在以 CIF 术语成交的合同中，保险单是卖方必须向买方提交的单证。

（二）保险凭证（Insurance Certificate）

保险凭证又称小保单，是一种简化的保险合同，除在背面没有另附有关保险人责任范围以及保险人和被保险人的权利和义务等方面的详细条款外，其余内容与保险单相同，具有与保险单同等的法律效力。

（三）预约保险单

预约保险单又叫"开口保险单"，一般适用于那些长年有大量进出口业务的外贸公司。保险人与被保险人事先约定保险货物的范围、险别、保险费率或每批货物的最高金额，并在预约保单上载明，但不规定保险的总金额。凡属于预约保险范围的货物，一经启运，保险人即自动按保单所列的条件承保。保险单的有效时间可能持续 6 个月到 12 个月不等，有时也会一直延续下去。

（四）联合凭证

由承保人在投保人或被保险人的相关发票上加注保险编号、承保险别、保险金额及保险机构印章等，这种保单不能转让，现在已经较少使用。

本章小结

国际货物运输保险的基本类别主要包括海上货物运输保险、陆上货物运输保险、航空货物运输保险和邮包货物运输保险。货物在海上运输过程中可能遭遇的风险分为海上风险和外来风险两大类。根据遭遇海上风险后货物损失程度的不同，海损分为全部损失和部分损失两大类。根据造成损失的原因的不同，部分损失又可分为共同海损和单独海损。

中国人民保险公司的《海洋货物运输保险条款》中的保险险别分为基本险和附加险两大类。其中，基本险又分为平安险、水渍险和一切险，它与协会货物保险条款中的 C 险、B 险和 A 险有大致对应的关系。

在国际贸易合同中制定保险条款时，不能采用笼统的规定方法，而应该明确规定由谁投保、投保的险别、投保的金额、所依据的保险条款以及保险费用的承担方等。

在我国，出口业务保险一般由出口商按逐笔投保的方式办理投保手续。在进口业务中，为简化投保手续避免漏保或难以及时办理投保手续等情形的发生，一般采取预约保险的做法。

第七章

国际货款结算

1. 熟悉国际货款结算过程中涉及的支付工具、支付方式
2. 熟悉汇票分类，掌握汇票的内容
3. 掌握信用证的性质、作用及主体内容
4. 熟悉外贸合同中支付条款的内容

在国际贸易活动中，进口商有义务按照约定的付款方式，在既定的时间内，向出口商支付既定金额、既定币种的合同货款。出口商最关心的事就是怎样以及什么时候才能得到他销往国外商品的货款。国内贸易的支付比较简单，它可以是先付款后交货，也可以是在交货后的一小段合情合理的时间内付款。然而，国际贸易的支付要比国内贸易的支付复杂得多。在通信、发货、交货的过程中不可避免地要浪费很多时间。谁来承担这个损失？卖方难道要等六个月才能收到货款？或者说买方在见到货物以前几个月就要支付货款？并且，如果买方拒绝付款，卖方就要花钱打官司，甚至可能血本无归。由于这些原因，制定了国际贸易不同的支付方式。一般来说，在每一笔出口合同中，涉及货款支付的条款包括下列四点：支付时间、支付方式、支付地点和支付货币。各种不同的出口资金融通的方法是这四个方面的不同排列和变体。

第一节 国际贸易支付工具

在国际贸易中，最常使用的支付工具包括货币和票据。前者用于计价、结算和支付；后者用于结算和支付。实际上，销售商几乎从不坚持要求用现金支付的权利，而是乐意用一些票据，如汇票、本票和支票来代替现金支付。其中，最常用的是汇票，本票和支票在国际贸易中只是偶尔使用。

一、汇票

（一）汇票的定义

汇票是由一人向另一人签发的无条件的书面命令，要求接受命令的人在见票时或在指定的或可以确定的将来某一日期，支付一定的金额给特定的人或其指定的人或持票人。汇票的使用程序包括：出票、提示、承兑、付款、背书、拒付及追索。

（二）汇票的内容

汇票示例一、汇票示例二能帮助我们了解汇票的内容，分别见表7-1、表7-2。

表7-1　汇票示例一

编号：1602
汇票金额：10 000 美元纽约，2016 年 1 月 8 日
见票时付汤姆·史密斯或持票人壹万美元整。
（签字）大卫·怀特
此致
费利克斯·布赖恩
伦敦

表7-2　汇票示例二

汇票号：677/96
汇票金额：7 500 美元中国，上海，2015 年 8 月 8 日
凭本汇票（副本未付）于见票后 60 天付上海 A&G 进出口公司或其指定人柒仟伍佰美元整（大写）。
此致
ABC 进出口公司
林登大街 56 号上海 A&G 进出口公司
迈阿密，美国经理（签字）

根据汇票的定义，这两张汇票可以分解成以下几个方面：

（1）无条件的书面命令。

（2）由一人/一方（出票人）签发。

在示例一中是：大卫·怀特，纽约

在示例二中是：上海 A&G 进出口公司

（3）向另一人（受票人）。

在示例一中是：费利克斯·布赖恩，伦敦

在示例二中是：ABC 进出口公司，迈阿密

（4）由给出汇票的人/一方（出票人）签发。

（5）要求接受命令的人/一方（受票人或付款人）。

（6）支付。

（7）在见票时或在指定的或可以确定的将来某一日期。

在示例一中是：在见票时

在示例二中是：见票后 60 天（确定的将来某一日期）

（三）汇票中的当事人

一张汇票主要涉及三个当事人：

出票人：签发命令要求另一人支付一定金额的人。在进出口贸易中，他通常是出口商或出口地银行，并且他也经常是受票人的债权人。

受票人（付款人）：接受命令并将付款的人。在进出口贸易中，他通常是进口商或信用证下的指定银行。还有，当受票人承兑一张远期汇票时，他就成为承兑人。出票人和承兑人必须是不同的人。

受款人：接受付款的人（个人、商号、公司或银行）。出票人和受款人通常是同一个人，在这种情况下，汇票上可能有这样的字句"付款给我们……"。在进出口贸易中，受款人经常就是出口商自己或他指定的银行。受款人也可能是持票人。受款人可以是汇票中的原有受款人，也可以是原有受款人所转让汇票的人。如果一张汇票有这样的指示"付××公司或其指定人"，它意味着汇票可以经受款人（现在是背书人）而转让给新的受款人（被背书人），这样使之成为可以转让的票据。一张汇票可以有多个背书人。

一张汇票中的主要当事人之间的关系可以通过下面的三角形表示出来，如图 7-1 所示。

图 7-1　汇票当事人关系图

（四）汇票的种类

根据不同的标准，汇票可以分为以下几种：

1. 商业汇票和银行汇票

汇票按出票人的不同，分为商业汇票和银行汇票。商业汇票是由工商企业开出的汇票，它经常用于对外贸易的资金融通。银行汇票是由银行开出的汇票，它主要用于汇付。

2. 光票和跟单汇票

光票是指在流转中没有伴随货运单据的汇票。做贸易很少用光票，它通常用于收取佣金、利息、样品费和代垫费用。与之相反，附有提单、保险单、发票等货运单据的汇票就成为跟单汇票，它经常用于进出口贸易中的货款结算。

3. 即期汇票和远期汇票

汇票按付款期限不同，可分为即期汇票和远期汇票。汇票上规定见票后立即付款的称为即期汇票。汇票上规定受票人先承兑，然后在指定的或将来一个可确定的日期付款的，换句话说，要求先承兑后付款的称为远期汇票。

在指定的或将来一个可确定的日期是承兑以后的若干天，分为：

（1）付款人见票后若干天付款，如见票后 30 天或 60 天。

（2）出票后若干天付款，如出票后 90 天付款。

（3）将来某一指定日期，如于 2016 年 7 月 18 日付款。

4. 商业承兑汇票和银行承兑汇票

在商业汇票中，由工商企业出票而以另一工商企业为付款人的远期汇票，经付款人承兑后，就称为商业承兑汇票。如果工商企业出票而以银行为付款人的远期汇票，经付款银行承兑后，就称为银行承兑汇票，由银行承担到期付款的责任。

（五）汇票在对外贸易中的使用

汇票是索款的票据，由出口商开出，提示给进口商，一般都通过银行。可以是见票即付（即期汇票），也可以是提示多少天后再付款（远期汇票）。在后一种情况下，受票人在汇票上写上"承兑"并签上自己的名字。这样出口商便可以将汇票贴现，填送押汇质押书后，立即取款。如果远期汇票到期不付款，公证人就要在汇票上附注受票人拒付字样，并出具拒绝证书，然后再次提交给受票人。显然，这种汇票以及相应的"承兑交单"付款条款对出口商或其银行有一定的风险。

即期汇票条款涉及的风险较少，因为只有在付款后才能取得货物的所有权。即使如此，如果汇票被拒付，货物仍有可能留在发货人手上，或在船上。但是在提示信用证项下的汇票时，发货人受到充分保护，承担风险的只是开证行和保兑行。

二、本票

本票是一人向另一人签发的，保证在见票时或在指定的或可以确定的将来某一日期，支付一定的金额给特定的人或其指定的人或持票人的无条件书面承诺。

本票和汇票的最大区别是汇票的当事人有三个：出票人、受票人和受票人。而本票的当事人只有两个：出票人和受款人。本票的付款人即出票人本人。

本票可以分为商业本票和银行本票。由工商企业或个人签发的称为商业本票，由银行签发的称为银行本票。商业本票有即期和远期之分，银行本票则都是即期的。在国际贸易结算中使用的本票，大都是银行本票。

本票样本，如图 7-2 所示。

图 7-2　本票样单

三、支票

支票是银行存款户对银行签发的授权银行对特定的人或其指定人或持票人在见票时无条件支付一定金额的书面命令。支票样单如图 7-3 所示。

在对外贸易中，出口商不能凭以海外银行为付款人的支票立即议付货款。如果出口商的银行愿意议付，那么出口商就可以立即得到货款，但他要支付贴现的费用。如果出口商的银行不愿意议付，那么出口商只有委托其银行收款，这样就既费时又费钱。

图 7-3　支票样单

第二节　支付方式

对外贸易与国内贸易有所不同。出口商和进口商在交易中都要面对一些危险，因为他们将不可避免地会遇到一方不能履行合同的情况。

对出口商来说，他们有买方违约和买方不支付全部货款的危险。这主要是由以下几个原因引起的：进口商破产，战争爆发，进口商所在国家的政府禁止与出口商所在国的贸易往来，或者他们禁止进口某些商品。另外一个可能的原因是进口商很难获取外汇以支付货款。还有可能是进口商不可靠和根本就不想支付他所答应的货款。

对进口商来说，他们要面临延迟交货或付款很长一段时间才收到货的危险。这可能是由港口拥挤或罢工引起的。出口商的推迟发货以及进口国复杂的结关手续都会给生意带来损失。另外，还有发错货的危险。

为预防这些种种可能的危险，就要采取不同的支付方法。总的来说，国际贸易支付方式可以分为三大类，见表 7-3。

表7-3 国际贸易支付方式

国际贸易支付方式	汇付		信汇
			电汇
			票汇
	托收	付款交单	即期付款交单
			远期付款交单
		承兑交单	
	信用证		

汇付和托收属于商业信用，信用证属于银行信用。"信用"在对外贸易中货物的交接和货款的支付上规定由谁承担付款和提供货物所有权单据的责任问题。在汇付和托收项下，买方负责付款，卖方负责提交装船单据。在信用证交易项下，银行代替买卖双方负责付款和提交单据。

一、汇付

汇付是指付款人通过银行或其他途径将款项汇交收款人。在汇付业务中，通常有四个当事人，即汇款人、收款人、汇出行和汇入行。在对外贸易中，当用下列付款条件进行销售时，通常使用汇付方式支付。这些付款条件是预付现金、订货付款、交货付款和记账交易。

（一）汇付的种类

1. 信汇

信汇是最常使用的汇付方式。买方将款项交给进口地银行，该银行开具付款委托书，通过邮寄交给卖方所在地的进口地银行的支行或往来行，委托其向卖方付款。

信汇费用较少，但速度慢。

2. 电汇

电汇的程序与信汇相似，只不过在电汇中进口地银行向他的支行或往来行发出的付款指示是用电报进行的。这就意味着付款更为迅速，卖方也能及早收到货款，但是买方却因此要负担较高的费用。

3. 票汇

买方从进口地银行购买银行汇票寄给卖方，由卖方或其指定的人持此汇票从出口地银行（汇票的受票人）取款。

（二）汇付的利与弊

在国际贸易中，如果使用汇付时，大多数交易是通过信汇和电汇来完成的。电汇对卖方有利，他可以较快地收到货款，加速资金周转，增加利息收入和避免汇率变动的风险，但买方却要多付电报费用和银行费用。在实际业务中，除非明确规定要使用电汇，买方最好通过信汇付款。有时，当款项的金额较大或因货币市场动荡，使用的结算货币有贬值的可能时，通过电汇付款是买方明智的选择。总之，是用电汇还是信汇要根据实际情况在合同中明确规定。就票汇来说，它是可以转让的，这一点与信汇和电汇不同。

二、托收

当出口商不急需资金，汇票的流通引不起银行的足够注意时，出口商就有可能将这些汇票交给银行托收。出口商委托银行安排汇票的接收和海外支付，然后银行再通过国外分行或代理行来办理此项业务。托收流程如图7-4所示。

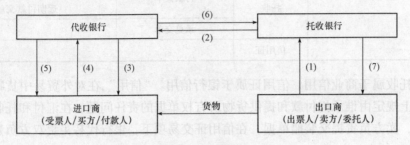

图7-4　托收流程图

说明：

（1）出口商发货，开具汇票连同货运单据交托收行，申请委托代收货款。

（2）托收行将汇票和货运单据寄交进口地代收银行委托代收。

（3）代收行向进口商提示汇票和单据，让其承兑（D/A）或付款（D/P）。

（4）进口商付款（D/P）或背书汇票承兑。

（5）代收行交单。

（6）代收行办理转账并通知托收行款已收妥。

（7）托收行向卖方交款。

托收是指银行为了：第一，取得承兑和/或视情况给予付款；第二，在承兑后和/或视情况在付款后交付商业单据；第三，按其他条件交付单据，而根据所收到的指示来处理有关单据（资金单据和/或商业单据）。托收分为光票托收和跟单托收两种。托收的当事人主要包括委托人、托收银行、代收行和提示行等。光票托收是指资金单据的托收，不附有商业单据（资金单据是指汇票、本票、支票、付款收据或其他用于取得付款的类似凭证，商业单据是指发票、装运单据、所有权单据或其他类似的单据，或一切不属于资金单据的其他单据）。

跟单托收是指：第一，资金单据的托收，附有商业单据；第二，商业单据的托收，不附有资金单据。在国际贸易中，大多采用跟单托收。跟单托收又分为付款交单和承兑交单两种。

当采用托收方式时，进口商/出口商要求通过银行以付款交单（D/P）或承兑交单（D/A）的方式支付货款。付款交单要求转交货运单据即付货款；承兑交单要求进口商承兑出口商开具的汇票才转交货运单据。

（一）付款交单（D/P）

在这种支付方式下，出口商交出单据后指示托收行和代收行在国外的买方付清货款后才交出单据。根据付款时间的不同，付款交单可分为即期付款交单和远期付款交单。

1. 即期付款交单

在这种方式下，卖方开具即期汇票并通过银行向买方提示，买方见票后马上付款，只有付清货款后才能领取单据。这种方式也称为"凭单据现付"。即期付款交单的程序如图7-5所示。

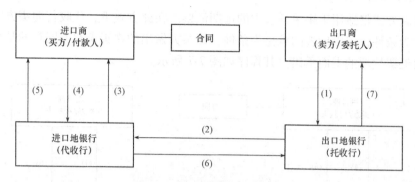

图7-5 即期付款交单程序图

说明：

（1）出口商根据合同规定装货后，开出即期汇票，连同货运单据交托收行，委托代收货款。

（2）托收行将汇票连同货运单据寄交进口地代收银行委托代收。

（3）代收行向买方提示汇票与单据。

（4）进口商付款。

（5）代收行交单给进口商。

（6）代收行办理转账并通知托收行款已收妥。

（7）托收行向出口商交款。

2. 远期付款交单

在这种方式下，卖方开立远期汇票。代收行将此汇票向买方提示汇票和货运单据。买方见票后仅须承兑此票，等汇票到期支付货款。代收行收到货款后，即向他交付单据。远期付款交单的程序如图7-6所示。

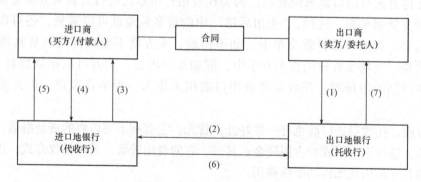

图7-6 远期付款交单程序图

说明：

（1）出口商根据合同规定装货后，开立远期汇票，连同货运单据交托收行，委托代收货款。

（2）托收行将汇票连同货运单据寄交进口地代收银行委托代收。

（3）代收行向进口商提示汇票与单据，让其承兑。进口承兑汇票后，代收行收回汇票和单据。

（4）进口商到期付款。

（5）代收行交单给进口商。

（6）代收行办理转账并通知托收行款已收妥。

（7）托收行向出口商交款。

（二）承兑交单（D/A）

这种付款方式仅适用于跟单托收中的远期汇票。在此方式下，代收行向买方交付单据不以后者付款为条件，仅以后者的承兑为条件，即买方做出的在买卖双方同意的某个将来的日期保证支付汇票款项的书面承诺。其程序如图 7-7 所示。

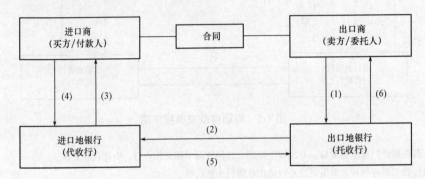

图 7-7　承兑交单程序图

说明：

（1）出口商根据合同规定装货后，开立远期汇票，连同货运单据交托收行，委托代收货款。

（2）托收行将汇票连同货运单据寄交进口地代收银行委托代收。

（3）代收行向进口商提示汇票与单据，进口商在汇票上承兑。代收行收回汇票，同时将货运单据交给进口商。

（4）进口商到期付款。

（5）代收行办理转账并通知托收行款已收到。

（6）托收行向出口商交款。

（三）托收的利与弊

托收支付方式对出口商来说有利。因为托收银行指示代收行直到买方承兑或支付汇票后才向其转交货运单据，这样，"先出后结"中的很多危险就可以避免，还可以防止买方缺乏商业信誉。例如，在付款交单下，如不付款，买方就不能获得代表货物所有权的单据和提取货物。货物所有权仍在卖方手中，假如买方拒付，卖方可以把货物转卖给他人。当面临竞争激烈的市场时，托收常常被出口商用来作为一种争夺客户、扩大销售的竞争手段。

另一方面，托收对进口商也有一些好处：首先，它有利于进口商的资金融通，如能争取到远期付款，还可不占用或少占用资金；其二，它的费用较低，采用托收方式，进口商可以免去像申请开立信用证那样的手续费用。

然而，问题依然存在。在付款交单下，买方或其银行可能拒绝付款赎单，尤其是在行市下跌的时候。在这种情况下，卖方要面临买方不付款或晚付款的危险，虽然他还是货主。在承兑交单下，显著的不利就是买方承兑汇票以后，代表货物所有权的单据就交给了他。如果他在付款前破产或无力支付，卖方就要承担损失。所以，对出口商来说，承兑交单比付款交单要危险。

对外贸易中，只有在进口商的资信良好或以往的生意往来使出口商相信进口商会付款的情况下，才采用托收的支付方式。就卖方的利益来说，即期付款交单比远期付款交单好，而付款交单又比承兑交单好。

为防止托收中的种种弊病，在国际贸易中，人们采用了一种更好的支付方式——信用证。

三、信用证

以上所谈论的种种付款办法，都是在买卖双方彼此有一定信任程度的基础上才采用的。当然，贸易伙伴还有其他的选择。如果进口方的银行对自己的客户进口商的信誉感到满意，它可以开出一张以出口商为受益人的信用证寄给出口方。在信用证上，银行作如下保证：要是出口货物是按信用证条款的规定发运，本行将保证付款不误。信用证条款是以买卖双方签订的销售合约为依据的。银行用开信用证的方式支持了进出口商。现在，信用证已成为国际贸易中获取货款的一种较为安全和迅速的支付方式。

（一）信用证的定义

信用证是为维护买方利益，由银行向卖方开具并签署的一份书面承诺。在信用证中，银行承诺如果卖方确实遵从信用证所提及的各项条件，那么银行就会付款或承兑开给银行的汇票。并且，只有在出具的单证和合同规定相一致时，银行才保证付款给卖方。通过信用证，银行就取代了它的客户（即买方）的付款保证。虽然信用证并不是真正意义上的担保，但其作用却几乎相同。因为只要卖方符合规定的条款，银行就保证向他付款。

（二）信用证的流通

买方和卖方签订合同并同意用信用证支付货款后，买方向银行申请开立信用证。如果银行同意其申请，开证行就开出信用证，然后通知其在国外的分行或代理行，让它转告信用证的受益人（即出口商）。随之，受益人审查信用证的条款。如果信用证条款与销售合同中的条款不符，出口商可要求对之修改。根据 UCP600 规定，信用证一经开出，即为不可撤销信用证，非经有关各方同意，信用证上的条款是不能修改的。

出口商接受信用证后才把出口货物交给承运人，承运人收下货物后出具提单。其他货物单据如商业发票和保险证明等，也由出口商办理。接着出口商开一张以开证行为付款人的汇票，连同信用证和货运单据一起交给本国的银行。通常，出口方银行要审单，如单证相符，银行即凭汇票付款，然后再把信用证和货运单据寄给开证行。银行的责任是对照信用证的条款审单。一旦发现有不符点，则必须改正。改正的办法有：重新开立信用证、制备新单据，或者进行修改。单证不符点有以下情况：信用证逾期；汇票上有错；提单上无"货已装船"的证明；保额不足；发票上的内容与信用证上的内容不符。开证行仔细核对单据无误后，按信用证条款将款项偿还给出口商银行（议付行）。然后，开证行向进口商出示汇票和单据要求其付款或承兑，在付款人付款或承兑后将单据交给进口商，进口商便可凭单据提货。

信用证的流通过程如图 7-8 所示。

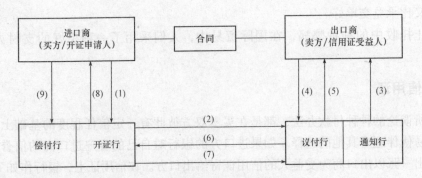

图7-8 信用证使用流程图

说明：

（1）进口商向当地银行（开证行）申请开立信用证，并与开证行签订协议。开证行同意其申请后开出信用证。

（2）开证行将信用证寄给通知行。

（3）通知行将信用证转递受益人（出口商）。

（4）受益人（出口商）经审查信用证认可后，即向买方发货。受益人发货后，备妥信用证规定的单证，送交当地银行议付。

（5）受益人银行（议付行）确认单证后，按照信用证将垫款付给受益人。

（6）议付行将单证寄给开证行索偿。

（7）开证行经审查单据无误后，付款给议付行。

（8）开证行通知进口商付款，赎回单据。

（9）进口商付款并取得货运单据后，即可提货。

（三）信用证相关当事人

（1）开证申请人，即请求银行开立信用证的人，通常为进口商。

（2）开证行，即应进口商申请开立信用证的银行。开证行担负对货物付款的独立责任。

（3）通知行，即受开证行授权将信用证传递到出口商的银行。通知行通常是出口地银行，而且一般是开证行的代理行。它仅对信用证的真实性负责。

（4）受益人，即有权领取信用证下支付货款的人，通常为出口商。

（5）议付行，即买入或贴现收益人按信用证规定提交的汇票和/或单据的银行。它可以在信用证中指定，也可不指定。在汇票的转让过程中，议付行可看作汇票的善意持票人。议付行和通知行通常为一家银行，可在信用证中加以指定。

（6）偿付行，即信用证指定的代开证行对汇票进行付款的银行。大多数情况下，付款行为开证行，也可为其他银行。若信用证结算货币为第三国货币，付款行也可为该国银行。一旦付款行履行付款，款项不可追索。

（7）保兑行，即应开证行请求对信用证加以保兑的银行。信用证一经保兑，保兑行对信用证的议付或付款独立负责。保兑行大多为通知行，亦可为出口商所在国的其他银行。

（四）信用证支付的利弊

使用信用证支付，进出口双方均得到了保障。进口商能获得这样的保证：他所订购的货物一定能符合买卖合同的规定。出口商也得到了保证：出口货款对方一定照付不误。另外，采用信用证方式付款，进口商可以争取到较好的贸易条件，当然要以付出一定数量的银行费用为代价。有了信用证，进口商也可能从银行得到资金融通的便利。当进口商想开立信用证用于支付国外的款项时，他可以向他的银行申请开立任何金额的信用证，他的银行通常不要求他支付该信用证的全部金额，只向他收一笔押金，这样，进口商的资金就不会积压。换句话说，开证行为货款的余额通融资金和提供担保。

就出口商来说，他知道现在这笔出口货款将由进口方银行支付，即便他对外国这家银行的资信不了解也没关系。保兑行对出口商按信用证规定开来的汇票付款后，对出口商没有追索权，即便将来开证行不把钱还给保兑行，后者也不能向出口商追回。由此可见，银行开立的信用证是支付货款的一种最理想的方式。

然而，信用证并不能绝对消除商人遇到的所有危险。由于信用证的开证行是以提交相关单证而不是提交货物为付款条件，因此进口商就有可能成为欺诈行为的受害人。受益人可以凭虚假的、与实际货物不相符的单证或根本就没有货物的单证向银行议付。除了欺诈行为以外，有关人员的错误操作、不恰当的程序、模糊的表述或疏忽也可给有关方带来损失。

（五）信用证范例

信用证的主体内容包含了合同中的主要内容，对单据的要求及银行保证性语句。其内容可进一步分成如下要点：

（1）相关当事人，包括信用证开立申请人、开证行、议付行，付款行等。

（2）关于信用证的说明，如信用证编号、信用证类型、开立日期、信用证金额、信用证到期日及到期地等。

（3）汇票条款，如汇票金额、出票人、受票人、付款日期等。

（4）装运条款，如起运港、目的港、关于分批装运和转运的规定，最迟装运期。

（5）关于货物的说明，包括货物名称、规格、数量、包装、单价、总额等。

（6）单据条款，包括对单据种类的要求、每种单据的份数及每份单据所记载的内容的要求。

（7）议付期限。银行通常不接受迟于运输单据日期 21 天后的交单行为，且交单议付应在信用证有效期内。

（8）开证行保证性条款，保证开证行担负对受益人或汇票持票人的付款责任。

（9）特别条款。

信用证形式不一。尽管国际商会推介了几种信用证标准形式，但仅被开证行采用。大多数银行仅以此为参考，而使用自身的信用证形式。关于信用证使用的法律文本为《跟单信用证统一惯例》。

SWIFT 信用证的式样请参看书后附录。

（六）信用证的种类

在国际贸易中，为满足贸易和支付的不同需求，人们采用不同类型的信用证。根据不同的标准，信用证主要可以划分为以下几种：

1. 不可撤销和可撤销信用证

不可撤销信用证，在信用证有效期内，未经受益人同意，开证行或买方不得对其进行修改、修正或撤销。只要出口商提交的单据与信用证条款一致，开证行一定承担付款责任。所以，在对外贸易中，这种信用证使用得很普遍。

可撤销信用证，如其名称所示，可以在信用证有效期以内的任何时候，不经受益人同意而撤销或修改。在对外贸易中，这种信用证不可能被出口商接受，因为它没有给他提供任何保障。

2. 保兑和不保兑信用证

由另一家银行（通常是通知行）代表开证行保证兑付的信用证叫作保兑信用证。有了保兑信用证，出口商就得到了开证行和保兑行的双重保证付款。保兑行首先对受益人负责。不保兑信用证就是通知行只负责通知受益人，但对单据不愿意承担任何责任的信用证。

3. 可转让信用证

在可转让信用证中，受益人有权指示通知行或议付行，把开具汇票的权利全部转让给另一个人（第二受益人）使用。由第二受益人开具汇票，递交货运单据和领取货款。但第一受益人仍然负有合同内规定的卖方责任。一份可转让的信用证只能转让一次，第二受益人不能再把它转让。

4. 即期和远期信用证

即期信用证与"凭即期汇票付款的信用证"或者"即期汇票信用证"的意思相同。当卖方按照信用证规定把即期汇票连同装船单据向议付行提示时，后者就立即付款。同样，议付行向开证行提交即期汇票和单据时，后者也应立即偿付。当卖方向议付行提示远期汇票和信用证里规定的单据时，后者并不立即付款，而是把汇票和单据转递到开证行，开证行见票时也不立即付款，而仅仅是承兑汇票，然后将之退还给卖方，汇票到期时开证行才付款。这种信用证就称为远期信用证。

5. 循环信用证

如果签订的合同是长期合同，并允许分批交货，或双方想不间断地执行这项合同，买方常常要求开立循环信用证。它通知卖方货已发出，单据已提交，货款已付后，信用证按原来的样子自动开立，下一批货物即可装运，再下一批也这样循环下去。这样，信用证就按同样的条款，在没有必要开立另一信用证的情况下，不断重复使用，直到规定的利用次数或规定的总金额用完为止。循环信用证还可以简化手续和节省费用。

6. 背对背信用证

一份背对背信用证要涉及两张信用证：一张信用证的受益人是出口商，但他并不是真正的供货方；另一张信用证是出口商通过自己的往来银行开给真正的供货方，并以其为受益人。第二张信用证的金额略低，两证的金额差价就是出口商想赚的钱。两张信用证的有效期也相差若干天，目的是出口商有充分时间替换发票。

7. 红条信用证

红条信用证与普通信用证不同的是它含有授权议付行垫付货款给出口商的条款（最初用红字打印）。通过它，受益人可利用进口商的资信情况获得好处。现在，红条信用证主要在进口商在出口商国家有代理的情况下使用。代理人在其中起到购买出口物资的作用。进口商为提供资金往往开具红条信用证。

第三节　其他付款方式

信用证为国际贸易货款的结算提供了一种安全的方式。但是某些情况下，如交易跨越期间较长或业务在非贸易领域操作时，信用证不被采用。此时，保函和备用信用证更为适用。随着国际贸易的发展，一些其他的结算方式也得到发展，其中，保函日渐被更多地采用。

除此之外，各结算方式在多种因素、多种情形下进行了组合运用。其中，有信用证结合托收、信用证结合汇付、跟单托收与预付定金相结合、跟单托收与备用信用证相结合的多种结算方式。

一、保函

保函是银行向第三方开立的一种书面担保凭证，保证申请人在未能按双方协议履行其责任或义务时，由担保银行代其向受益人履行债务清偿责任。保函可应用于多种场合。在此，主要介绍两种形式：投标保函和履约保函。

投标保函是由银行应投标人申请而出具的保证书。银行担保当投标人在投标有效期内撤销投标，或在招标开始前对投标文件进行修改，或中标后不能同业主订立合同或不能支付投标保证金时，担保银行就自己负责付款。

履约保函是指担保银行应委托人的请求，向受益人开立的一种保证文件，保证若委托人不能按时履约，由银行支付给受益人规定款项或采取某些补救措施。在国际贸易实践中，进出口商均可使用履约保函。前者，若进口商不能按时对货物付款，担保银行将对卖方负责；后者，若出口商不能按时交付货物，担保银行将对买方负责赔偿遭受的相关损失。

保函不同于信用证。信用证条件下，开证行对货款的支付负独立责任，但在保函条件下，银行仅在委托人未履行其相关义务时担负责任。信用证条件下，卖方通过议付获取货款，但在保函条件下不可。且信用证条件下，开证行仅处理相关单据，与合同毫不相干；保函条件下，受益人出具委托人未能履约、要求担保银行赔付的书面文件时，银行须查证委托人如何且为何未能履约。这样，银行或许卷入合同纠纷中。诸如美国、日本等一些国家，不允许银行涉及商业纠纷，因此该国家的银行不出具银行保函，而使用备用信用证形式。

二、备用信用证

备用信用证又称为商业文本信用证、担保信用证或履约信用证。它是一种光票信用证。它是指开证行保证在开证申请人未能如期履行其应履行的义务时，由其担负某些责任的凭证。一旦申请人按期履行其相关义务，备用信用证不发生效用。反之，受益人须出具书面文本举证申请人未能如期履约，要求开证行对此加以赔付。备用信用证同银行保函一样，同属银行信用，都向受益人保证支付因开证申请人违约或不付款造成的损失。它们的不同之处在于：保函条件下，若开证申请人未能履行其职责时，开证行自身担负相关损失偿付责任，如此一来，开证行难免牵涉到受益人与开证申请人的纠纷当中；备用信用证条件下，如受益人向开证行提交申请，开证行与申请人和受益人的合同是独立开来的。

在银行保函或备用信用证下，分期付款和延期付款均可使用。分批交货通常使用分期付款。这种情况下，买方应首先支付一定数额的预付款，待分批交付货物后结算余款。延期付款时，买方首先要支付一定数额的定金，余款会在较长一段时间后分期支付。这的确是一种信用交易，通过这种方式，买方能够使用这笔外汇资金做采购。延期付款中涉及利息问题，应在合同中对利息条款加以明确。此外，以上两种方式还可结合使用，即每次分批交货之后，分期付款条件下的分期支付不能马上执行时，可做延期。

三、不同付款方式的结合使用

在国际贸易业务中，对于一笔交易的结算，通常只使用一种结算方式就可以了，但有时也会将两种以上的结算方式结合使用，以收回货款。

常见的不同结算方式的结合有：信用证与汇付结合、信用证与托收结合、跟单托收与预付款相结合、跟单托收与备用信用证相结合等。

（1）信用证与汇付结合。这是指一笔交易的货款，部分用信用证方式支付，余额用汇付方式结算。这种结算方式要求买方按照全部货款的一定百分比开立信用证，其余采取托收形式。使用这种结合形式，合同必须首先订明信用证下使用光票汇票，余款在托收下以跟单汇票结算。装运单据直到付款后才会交给买方。

（2）信用证与托收结合。这是指一笔交易的货款，部分用信用证方式支付，余额用托收方式结算。货物装运前的汇付款被视为预付款。若装运后再行汇付，汇付款常用来结算交货金额变化的余额。

（3）跟单托收与预付款结合。为确保出口商免于承担因托收造成过多的损失，卖方可要求买方支付部分预付款。货物装运后，预付款部分可从待收款上加以扣除。这样一来，若汇票遭到拒付，出口商可将货物运回，相关损失可由预付款加以抵偿。但出口商必须确保货物可从进口国运回。

（4）跟单托收与备用信用证结合。为了避免跟单托收的结汇风险，出口商可要求买方开立备用信用证。当货款被拒付时，卖方可要求开证行开立备用信用证加以偿付。

采用这种结算方式，备用信用证的有效期必须足够长，以便于货款遭到拒付时，买方有足够的时间来要求开证行加以付款。在填写出口托收的申请表时，出口商应让汇出行告知，一旦汇票被拒付，托收行应立即以电传方式告知。

第四节　合同中的支付条款

货款的支付有不同的方式，因此，在合同中必须明确规定使用何种方式及其具体要求。以下是各种支付方式的实例。

一、汇付方式

实例如下：

（1）买方应不迟于 7 月 15 日将 100% 的货款用信汇预付交至卖方。

（2）买方对卖方的支付要以美元电汇至卖方指定的在×国×银行的账号。

二、托收方式

（一）付款交单（即期）条款

（1）买方应凭卖方开具的即期跟单汇票于见票时立即付款，付款后交单。

（2）凭即期汇票和附装货物的提单以净现金支付。

（3）凭即期汇票和所附表明货物发运的提单通过×银行以现金支付。汇票未付清之前，提单不交给买主。

（二）付款交单（远期）条款

（1）买方对卖方开具的见票后××天付款的跟单汇票，于第一次提示时应即予承兑，并应于汇票到期日即予付款，付款后交单。

（2）买方应凭卖方开具的跟单汇票，于提单日后××天付款，付款后交单。

（3）买方应凭卖方开具的跟单汇票，于汇票出票日后××天付款，付款后交单。

（三）承兑交单

买方对卖方开具的见票后××天付款的跟单汇票，于提示时应即承兑，并应于汇票到期日即予付款，承兑后交单。

三、信用证

（一）即期信用证

（1）买方应通过为卖方可接受的银行于装运月份前××天开立并送达卖方不可撤销即期信用证，有效至装运月份后第 15 天在中国议付。

（2）以不可撤销的信用证，凭卖方即期跟单汇票议付，有效期应为装运期后若干天在中国到期。该信用证须于合同规定的装运月份前 30 天到达卖方。

（二）远期信用证

（1）买方应通过为卖方可接受的银行于装运月份前××天开立并送达卖方不可撤销见票后××天付款的信用证，有效至装运月份后 30 天在中国议付。

（2）以不可撤销的信用证，凭卖方开具的见票后××天的跟单汇票议付，有效期限应为装运期后 15 天在中国到期。该信用证须于合同规定的装运月份前 30 天到达卖方。

四、不同方式的结合使用

买方应通过卖方所接受的银行于装运月份前××天开立并送达卖方不可撤销即期信用证，规定 50% 发票金额凭即期光票支付，其余 50% 即期付款交单（或见票后××天付款交单）。100% 发票金额的全套装运单据随附于托收项下，于买方付清发票的全部金额后交单。如买方不付清全部发票金额，则货运单据须由开证银行掌握凭卖方指示处理。

本章小结

　　在国际贸易结算中，大都采用票据作为结算工具。这些结算票据主要包括汇票、本票和支票。其中，最常用的就是汇票。

　　国际贸易中常用的结算方式主要有汇付、托收和信用证三种方式。信用证支付方式在很大程度上解决了进出口双方在付款和交货问题上的矛盾，成为国际贸易中一种主要的结算方式。在业务实践中，各种结算方式应根据实际情况灵活选择。

合同中的一般条款

1. 了解商品检验的时间、地点、检验机构、检验证书、检验的标准和方法
2. 理解争议和索赔的概念、违约的性质和后果
3. 掌握检验检疫、索赔、不可抗力及仲裁条款的订立

在国际贸易实践中，买卖双方要证实交货的商品在品质、数量和包装等方面是否符合合同规定，就要进行商品检验。买卖双方在履行合同时，一方违约给另一方造成损失时，除不可抗力之外，受损方有权向违约方提出赔偿损失的要求。如果买卖双方因各自的权利和义务问题而发生争议时，可采取仲裁的方式来解决。如果 WTO 成员间发生贸易争端，可以通过 WTO 贸易争端解决机制来解决。一般包含检验检疫、索赔、不可抗力及仲裁等内容的条款，通常会涉及订约后可能发生争议的预防和处理，因此交易双方应在合同中订立相关条款。

第一节 商品的检验

一、商品检验的内容和意义

商品检验是指在国际货物买卖过程中，由具有权威性的专门的进出口商品检验机构依据法律法规或合同的规定，对商品的质量、数量、重量和包装等方面的检验和鉴定，同时出具检验证书的活动。商品检验是买卖双方交接货物不可缺少的重要环节。

（一）商品检验的内容

（1）商品品质检验。品质检验亦称质量检验。商品品质检验需要运用各种检验手段，包括感官检验、化学检验、仪器分析、物理测试、微生物学检验等，对进出口商品的品质、

规格、等级等进行检验，以确定其是否符合外贸合同的规定。

品质检验的范围很广，大体上包括外观质量检验与内在质量检验两个方面：外观质量检验主要是对商品的外形、结构、花样、色泽、气味、触感、疵点、表面加工质量、表面缺陷等的检验；内在质量检验一般指有效成分的种类含量、有害物质的限量、商品的化学成分、物理性能、机械性能、工艺质量、使用效果等的检验。同一种商品根据不同的外形、尺寸、大小、造型、式样、定量、密度、包装类型等而有各种不同的规格。

（2）商品数量和重量检验。商品数量和重量检验是按合同规定的计量单位和计量方法对商品的数量和重量进行检验，看其是否符合合同规定。在实务中，商品重量检验允许有一定的合理误差。

（3）商品包装检验。商品包装检验是根据外贸合同、标准和其他有关规定，对进出口商品的外包装和内包装以及包装标志进行检验。包装检验首先核对外包装上的商品包装标志（标记、号码等）是否与进出口贸易合同相符。对进口商品主要检验外包装是否完好无损，包装材料、包装方式和衬垫物等是否符合合同规定要求。对出口商品的包装检验，除包装材料和包装方法必须符合外贸合同、标准规定外，还应检验商品内外包装是否牢固、完整、干燥、清洁，是否适于长途运输和符合保护商品质量、数量的习惯要求。

（4）商品残损检验。商品残损检验主要是对进口受损货物的残损部分予以鉴定，了解致残原因及对商品使用价值的影响，估定残损程度，出具证明，作为向有关各方索赔的依据。商品的残损主要是指商品的残破、短缺、生锈、发霉、虫蛀、油浸、变质等情况。检验的依据包括发票、装箱单、保险单、重量单、提单、商务记录及外轮理货报告等有效单证或资料。

（5）商品卫生检验。商品卫生检验主要是对肉类罐头食品、奶制品、禽蛋及蛋制品、水果等进出口食品检验其是否符合人类食用卫生条件，以保障人民健康和维护国家信誉。

（6）商品的安全性能检验。商品的安全性能检验是根据国家规定和外贸合同、标准以及进口国的法律要求，对进出口商品有关安全性能方面的项目进行的检验，如易燃、易爆、易触电、易受毒害、易受伤害等，以保证安全使用和生命财产的安全。

（二）商品检验的意义

（1）有利于商品出口。从商品出口的角度来看，通过商品检验，卖方能够保证向买方交付合格货物，以此提高自己的信誉。同时商检也为出口合同履行状况起监督和摸底作用。出口商品检验可以为出口商及时发现问题，以便有机会采取补救措施，提高履行合约的质量。因此，出口商应结合自己的企业、商品的总体情况，灵活运用检验方式，使商检成为提高自己合同质量的一个重要环节。

（2）有利于维护进口方的权益。从商品进口的角度来看，进口方可通过行使检验权来保护自己的正当权益，商品检验的结果既为接受合乎质量要求的货物提供了保证，又为可能因货物质量不好而拒收货物或提出索赔要求提供必要的证据，从而可有效防止国际贸易中的欺诈行为，维护进口方的合法权益。

二、商品检验的时间和地点

在国际贸易中，进出口商品检验的时间和地点关系着买卖双方的切身利益。因为它涉及检验权、检验机构以及有关的索赔问题。商品检验的时间和地点的规定，成为合同中商检条款的一个核心问题。主要有以下 5 种做法：

（一）在出口国产地检验

发货前，由卖方检验人员会同买方检验人员对货物进行检验，卖方只对商品离开产地前的品质负责。离产地后运输途中的风险，由买方负责。

（二）在装运港（地）检验

货物在装运前或装运时由双方约定的商检机构检验，并出具检验证明，作为确认交货品质和数量的依据。这种规定，以"离岸品质和离岸数量"为准。

（三）目的港（地）检验

货物在目的港（地）卸货后，由双方约定的商检机构检验，并出具检验证明，作为确认交货品质和数量的依据。这种规定，以"到岸品质和到岸数量"为准。

（四）买方营业处所或用户所在地检验

对于那些密封包装、精密复杂的商品，不宜在使用前拆包检验，或需要安装调试后才能检验的产品，可将检验推迟至用户所在地，由双方认可的检验机构检验并出具证明。

（五）出口国检验，进口国复检

按照这种做法，装运前的检验证书作为卖方收取货款的出口单据之一，但货到目的地后，买方有复验权。如经双方认可的商检机构复验后，发现货物不符合合同规定，且系卖方责任，买方可在规定时间内向卖方提出异议和索赔，直至拒收货物。

上述几种做法，各有特点，应视具体的商品交易性质而定。但对大多数一般商品交易来说，"出口国检验，进口国复验"的做法最为方便而且合理。因为这种做法一方面肯定了卖方的检验证书是有效的交接货物和结算的凭证，同时又确认买方在收到货物后有复验权，这在一定程度上调和了买卖双方在检验问题上的矛盾，符合各国法律和国际公约的规定。我国对外贸易中大多采用这一做法。

三、商品检验机构

（一）国际贸易领域中商品检验机构的种类

（1）国家设立的官方商检机构。官方的检验机构只对特定商品（粮食、药物等）进行检验，如美国食品药物管理局（FDA）。

（2）民间私人或社团经营的非官方机构。国际贸易中的商品检验主要由民间机构承担，民间商检机构具有公证机构的法律地位。比较著名的有：瑞士日内瓦通用鉴定公司（SGS）、日本海外货物检验株式会社（OMIC）、美国保险人实验室（UL）、英国劳合氏公证行（Lloyd's Surveyor）、法国船级社（BV）以及香港天祥公证化验行等。

（3）工厂企业、用货单位设立的化验室、检测室等。

（二）我国商检机构及其职责任务

我国进出口商品检验主要由官方的"中华人民共和国国家出入境质量监督检验检疫局"及其分支机构承担。此外还有各种专门从事动植物、食品、药品、船舶、计量器具等官方检验机构。

1980 年成立的中国进出口商品检验总公司（CCIC）及其分公司根据商检总局的指定，以第三者的身份，从事进出口商品检验和鉴定业务。我国商检机构和一些国外检验机构建立了委托代理关系（如 SGS）或合资检验机构（如 OMIC）。外国检验机构经批准也可在我国设立分支机构，在指定范围内接受进出口商品检验和鉴定业务。

中华人民共和国商品检验及其分支机构，统一按照《商检条例》执行检验任务。主要任务有三条：对重要商品实施法定检验；对所有进出口商品的品质实施监督管理；办理对外贸易公证鉴定业务。

1. 法定检验

进口商品分法定检验商品和非法定检验商品。法定检验是指商检机构依据国家法律法规对重点进出口商品实行的一种强制性检验。法定检验的范围包括：

（1）列入《商检机构实施检验的商品种类表》的进出口商品。

（2）《中华人民共和国食品卫生法》和《进出境动植物检疫法》规定的商品。

（3）对出口危险货物包装容器、危险货物运输设备和工具的安全技术的性能和使用鉴定。

（4）对装运易腐烂变质食品、冷冻品的船只和集装箱等运输工具实施适载检验。

（5）根据国外法规要求强制检验或认证的商品。

（6）对外贸易合同规定由商检局检验出证的进出口商品。

以上范围之外的进口商品为非法定检验商品。

这两类商品在办理报验手续上有所不同。前者到货后，收货人或其代理人必须向口岸或到达站商检机构办理进口商品登记手续，然后按商检机构规定的地点和期限向到货地商检机构办理进口商品报验。

非法定检验进口商品到货后，由收、用货部门直接办理进口通关手续。提货后，可按合同的约定自行检验，若发现问题需凭商检证书索赔的，应向所在地商检机构办理进口商品报验。

2. 监督管理

监督管理是指检验检疫机构通过行政管理手段，对本地区进出口商品的检验检疫工作进行监督管理。其范围包括对一切进出口商品的质量、规格、数量、重量、包装以及生产经营、仓储、运输、安全和卫生要求等进行检验、鉴定。商检机构除依法对规定的进出口商实施检验外，还有权对规定以外的进出口商品进行抽查检验。

3. 鉴定业务

鉴定业务是指商检机构接受对外贸易关系人的申请，或外国检验机构的委托，以公证的态度，对进出口商品进行鉴定，签发鉴定证书，作为申请人办理进出口商品的交接、结算、报关、纳税、计费、理算、索赔、仲裁等的有效证据。鉴定业务与法定检验不同，它不具有强制性。鉴定业务的范围主要包括：进出口商品的质量、数量、重量、包装、海损、商品残损的鉴定；货载衡量、车辆、船舱集装箱等运输工具的清洁、密固和冷藏效能等装运技术的鉴定以及抽取并签发各类样品、签发价值证书等。

四、检验证书及其作用

（一）检验证书

检验证书是商检机构对进出口商品进行检验、鉴定后签发的书面证明文件。商检证书对贸易有关各方履行契约义务、处理索赔争议和仲裁、诉讼举证具有法律效力，也是海关验放、征收关税和优惠减免关税的必要证明。常见的检验证书有：

（1）品质检验证书。品质检验证书是出口商品交货结汇和进口商品结算索赔的有效凭证。法定检验商品的证书，是进出口商品报关、输出输入的合法凭证。商检机构签发的放行单和在报关单上加盖的放行章有与商检证书同等的通关效力，签发的检验情况通知单同为商检证书性质。

（2）重量或数量检验证书。重量或数量检验证书是出口商品交货结汇、签发提单和进口商品结算索赔的有效凭证。出口商品的重量证书也是国外报关征税和计算运费、装卸费用的证件。

（3）兽医检验证书。兽医检验证书是证明出口动物产品或食品经过检疫合格的证件。适用于冻畜肉、冻禽、禽畜罐头、冻兔、皮张、毛类、绒类、猪鬃、肠衣等出口商品。是对外交货、银行结汇和进口国通关输入的重要证件。

（4）卫生健康证书。卫生健康证书是证明可供人类食用的出口动物产品、食品等经过卫生检验或检疫合格的证件。适用于肠衣、罐头、冻鱼、冻虾、食品、蛋品、乳制品、蜂蜜等出口商品。是对外交货、银行结汇和通关验放的有效证件。

（5）消毒检验证书。消毒检验证书是证明出口动物产品经过消毒处理，保证安全卫生的证件。适用于猪鬃、马尾、皮张、山羊毛、羽毛、人发等商品。是对外交货、银行结汇和国外通关验放的有效凭证。

（6）产地证明书。产地证明书是出口商品在进口国通关输入和享受减免关税优惠待遇和证明商品产地的凭证。

（7）残损检验证书。残损检验证书是证明进口商品残损情况的证件。适用于进口商品发生残、短、渍、毁等情况。可作为受货人向发货人或承运人或保险人等有关责任方索赔的有效证件。

（8）价值证明书。价值证明书是进口国管理外汇和征收关税的凭证。在发票上签盖商检机构的价值证明章与价值证明书具有同等效力。

（9）船舱检验证书。船舱检验证书证明承运出口商品的船舱清洁、密固、冷藏效能及其他技术条件是否符合保护承载商品的质量和数量完整与安全的要求。可作货物交接和处理货损事故的依据。

除上述各种检验证书外，还有证明其他检验、鉴定工作的检验证书，如生丝品级及公量检验证书、熏蒸证书、货载衡量检验证书等。

（二）检验证书的作用

（1）作为议付货款的单据之一。

（2）作为证明交货的品质、重量、包装等是否符合规定的依据。

（3）作为对品质、重量、数量、包装等提出异议，拒收、理赔，解决争议的凭证。

（4）作为海关验关放行的凭证。

（5）作为进口国实行关税差别待遇的依据。

五、合同中的商品检验条款

商检条款是国际货物买卖合同中的一项重要内容。其所包含的商检权与当事人的拒收权和索赔权有着直接的联系。当事人依据商检条款，行使相应的商检权。因此应根据平等互利原则与对方协商订立检验条款，从而提高合同的履约率。

（一）出口合同中的商品检验条款

在我国出口合同中，商检条款一般订法为：

双方同意以中国进出口商品检验局所签发的品质/数量检验证书作为信用证项下议付单据的一部分。买主有权对货物进行复检。复检费由买方负担。如发现品质或数量与合同不符，买方有权向卖方索赔，但需提供经卖方同意的公证机构出具的检验报告。索赔期限为货到达目的港××天内。

（二）进口合同中的商品检验条款

在我国进口合同中，商检条款一般订法为：

双方同意以制造厂（或××检验机构）出具的品质及数（重）量检验证明书作为有关信用证项下付款的单据之一。货到目的港经中国进出口商检局复验，如发现品质或数（重）量与本合同规定不符时，除属保险人或承运人责任外，买方凭中国出入境检验检疫机构的检验证书，在索赔有效期内向卖方提出退货或索赔。索赔有效期为××天，自货物卸毕日期起计算。所有退货或索赔引起的一切费用（包括检验费）及损失均由卖方负担。

根据我国《进口商品质量监督管理办法》的规定：对于有些重要的进口货物，可以根据合同的规定，到出口国进行货物装运前的事先检验、监造和监装。最后以货到后的检验为准。

（三）商品检验条款的注意事项

合同中的商检条款主要内容一般包括检验方式、检验内容、检验机构和检验费用等方面的内容。所以，在与外商签订的进出口合同中，需要"科学、明确、具体、合理"地确定这些内容。

（1）确定检验方式。出口检验方式理论上可分为自验、共验、出口商品预先检验、驻厂检验、产地检验、出口商品内地检验与口岸查验、出口商品的重新检验、免验、复验等多种方式。不同的商检机构有不同的要求。所以在与外商签订合同时，事先就要搞清楚，客户所要求的出证机构将会采取哪种方式检验。

（2）确定检验内容。双方商量好每一批货物应检验哪些项目，并将它清楚地写到合同里，这是商检条款的核心内容之一。而且，对该项检验内容进行合同表述的时候要科学、合理和精确。

（3）慎选检验机构。要选择世界公认的、一流的检验机构。这类机构在世界分支机构多、信誉好、技术水平先进、效率高、出证快、权威性强。一般来说比较公正，收费也规范，联系也方便。

（4）明确检验费用由谁承担。出口业务中，商检费用一般由出口商自己承担。但是，当买方提出额外的商检方面的要求时，出口商就得考虑费用该由谁承担的问题了。当然，还要考虑额外的工作所占用的时间和对整个出口流程的影响。

第二节　争议和索赔

一、争议、索赔和理赔的含义

（一）争议

1. 争议的含义

争议（Disputes）是指买卖的一方认为另一方没有履行合同规定的责任与义务所引起的纠纷。

2. 争议引起的原因

（1）卖方违约。如卖方不交货，或未按合同规定的时间、品质、数量、包装条款交货，或单证不符等。

（2）买方违约。如买方不开或缓开信用证，不付款或不按时付款赎单，无理拒收货物，在 FOB 条件下不按时派船接货等。

（3）合同规定不明确。买卖双方国家的法律或对国际贸易惯例的解释不一致，甚至对合同是否成立有不同的看法。

（4）在履行合同过程中遇到了买卖双方不能预见或无法控制的情况，如某种不可抗力，双方有不一致的解释等。

由上述原因引起的争议，概括起来讲就是：是否构成违约，双方对违约的事实有分歧，对违约的责任及其后果的认识不一致。对此，双方应采取适当措施，妥善解决。

（二）索赔与理赔

1. 索赔

索赔（Claim）是指签订合同的一方违反合同的规定，直接或间接地给另一方造成损害，受损方向违约方提出损害赔偿要求。

2. 理赔

理赔（Settlement of Claims）是指违约方受理受损方提出的赔偿要求。可见，索赔和理赔是同一个问题的两个方面。

二、不同法律对违约行为的不同解释

违约（Breach of Contract）是指买卖双方之中任何一方违反合同义务的行为。国际货物买卖合同是对缔约双方具有约束力的法律文件。任何一方违反了合同义务，就应承担违约的法律责任，受损的一方有权提出损害补偿要求。但是，因为不同的法律和文件对于违约方的违约行为及由此产生的法律后果、对该后果的处理有不同的规定和解释，所以我们必须了解和熟悉这方面的知识。

（一）英国的法律规定

英国的《货物买卖法》将违约分为违反要件和违反担保两种。违反要件（Breach of Condition）是指违反合同的主要条款，即违反与商品有关的品质、数量、交货期等要件；在合同的一方当事人违反要件的情况下，另一方当事人，即受损方有权解除合同，并有权提出

损害赔偿。违反担保（Breach of Warranty）是指违反合同的次要条款，受损方只能提出损害赔偿，而不能解除合同。至于在每份具体合同中，哪个属于要件，哪个属于担保，该法并无明确具体的解释，只是根据"合同所作的解释进行判断"，这样，在解释和处理违约案件时，难免带有不确定性和随意性。

（二）《联合国国际货物销售合同公约》规定

《联合国国际货物销售合同公约》则对违约的后果及其严重性进行判断，将违约分为根本性违约和非根本性违约。根本性违约（Fundamental Breach）是指违约方的故意行为造成的违约，如卖方完全不交货，买方无理拒收货物、拒付货款，其结果给受损方造成实质损害。如果一方当事人根本违约，另一方当事人可以宣告合同无效，并可要求损害赔偿。非根本性违约（Non-fundamental Breach）是指违约的状况尚未达到根本违反合同的程度，受损方只能要求损害赔偿，而不能宣告合同无效。

三、进出口合同中的索赔条款

买卖双方可根据交易的需要在合同中订立或不订立索赔条款。订立索赔条款通常有两种方式：

（一）异议和索赔条款

该条款针对卖方交货品质、数量或包装不符合合同规定而订立。主要内容包括索赔依据和索赔期限、赔偿损失的办法和金额。

（1）索赔依据主要是指双方认可的商检机构出具的检验证书。

（2）索赔期限主要是指受损方向违约方提出索赔要求的有效期限。如逾期提出索赔，违约方可不予理赔。至于索赔期限究竟以多长时间为宜和采用何种办法计算索赔期的起讫，则应根据商品的性质、港口条件、检验货物的可能性及所需时间等加以明确。实务中对索赔期限的起算时间通常有下列几种规定方法：

①货物到达目的港后某某天起算。②货物到达目的港卸至码头后某某天起算。③货物到达买方营业处所后某某天起算。④货物到达用户所在地后某某天起算。⑤货物经检验后某某天起算。

（3）索赔的办法和索赔金额。一般对此问题只作笼统规定，主要是由于违约的原因通常较复杂，在订立合同时很难进行预计。

（二）罚金条款

该条款针对当事人不按期履约而订立。如卖方未按期交货或买方未按期派船、开证等。主要内容是在合同中规定：如有一方未履约或未完全履约，应向对方支付一定数量的约定金额，即罚金或违约金以补偿对方的损失。罚金的支付并不意味着解除违约方继续履行合同的义务。因此，违约方除支付罚金外，仍应履行合同义务，如因故不能履约，则另一方在收受罚金之后，仍有权索赔。

违约金的起算日期有两种方法：一种是按合同规定的交货期或开证期终止后立即起算；一种是规定优惠期，指在合同规定的有关期限终止后再宽限一段时间，在优惠期内免于罚款，优惠期届满即开始起算。

各国在法律上对罚金条款的解释和规定存在差异，实务中应引起重视。如英美法系国家的法律，只承认损害赔偿，不承认带有惩罚性的罚金。所以在与这些国家进行贸易时，应注意约定的罚金额的合法性。

罚金条款常用于大宗商品或成套设备的合同中。

第三节　不可抗力

一、不可抗力概述

不可抗力（Force Majeure）是指买卖合同签订后，不是由于当事人一方的过失或故意，发生了当事人在订立合同时不能预见，对其发生和后果不能避免，并且不能克服的事件，以致不能履行合同或不能如期履行合同。遭受不可抗力事件的一方，可以据此免除履行合同的责任或推迟履行合同，对方无权要求赔偿。

可见，不可抗力是一种免责条款，即免除由于不可抗力事件而违约的一方的违约责任，也是一项法律原则。对此，在国际贸易中不同的法律、法规等各有自己的规定（我国的相关规定见表8-1）。1980年《联合国国际货物销售合同公约》在其免责一节中作了如下规定："如果他能证明此种不履行义务，是由于某种非他所能控制的障碍，而且对于这种障碍，没有理由预期他在订立合同时能考虑到或能避免或克服它或它的后果。"该《公约》指明了一方当事人不能履行义务，是由于发生了他不能控制的障碍，而且这种障碍在订约时是无法预见、避免或克服的，可予免责。

表8-1　我国《合同法》对不可抗力的规定

我国《合同法》对不可抗力的规定
不可抗力是指不能预见、不能避免并不能克服的客观情况。所谓"不能预见"是指以一般人的预见能力为标准而不能预见；"不能避免"是指当事人已尽了最大努力仍然不能避免某种事件的发生；"不能克服"是指当事人在事件发生以后，已尽到了最大的努力，仍不能克服事件所造成的后果使合同不能得以履行。 　　因不可抗力不能履行合同的，根据不可抗力的影响，部分或者全部免除责任，但法律另有规定的除外。 　　当事人一方因不可抗力不能履行合同的，应当及时通知对方，以减轻可能给对方造成的损失，并应当在合理期限内提供证明。

又如，英美法称"合同落空"，意思是说合同签订以后，不是由于双方当事人自身的过失，而是由于事后发生了双方当事人意想不到的根本性的不同情况，致使订约的目的受到挫折，因而未能履行合同义务，当事人可以据此免除责任。大陆法称"情势变迁"或"契约失效"，意思是说，不属于当事人的原因，而发生了当事人预想不到的变化，致使合同不可能履行，或对原来的法律效力需做相应的变更。综上所述，虽然国际贸易公约和各国的法律法规对不可抗力的叫法不统一，解释也不一致，但其基本精神则大体相同，主要包括以下几点：

（1）意外事故必须是发生在合同签订以后。

（2）不是由于合同当事人双方自身的过失或疏忽而导致的。

（3）意外事故是当事人双方所不能控制的、无能为力的。

引发不可抗力事故的原因通常包括两种情况：一种是自然原因引起的，如水灾、旱灾、暴风雪、地震等；另一种是社会原因引起的，如战争、罢工、政府禁令等。国际上对自然原因引起的各种灾害在解释上都较一致，但对社会原因引起的意外事故的解释存在较大分歧。因此，哪些意外事故应视作不可抗力，应由买卖双方在合同的不可抗力条款中约定。

二、进出口合同中的不可抗力条款

不可抗力条款的内容，主要包括不可抗力事件的范围、不可抗力事件的处理原则和方法、事件发生后通知对方的期限和通知方式以及出具事件证明的机构等。

（一）不可抗力事件的范围

不可抗力事件的范围较广，哪些意外事故构成不可抗力，哪些不能构成，买卖双方在交易磋商时应达成一致意见。而且对不可抗力条款的表述应明确具体。

关于不可抗力事件的范围，应在买卖合同中订明。通常有下列三种规定办法：

1. 概括规定

即在合同中不具体规定不可抗力事件的范围，只作概括的规定。例如：如果由于不可抗力的原因导致卖方不能履行合同规定的义务时，卖方不负责任，但卖方应立即电报通知买方，并须向买方提交证明发生此类事件的有效证明书。

2. 具体规定

即在合同中明确规定不可抗力事件的范围，凡在合同中没有订明的，均不能作为不可抗力事件加以援引。例如：如果由于战争、洪水、火灾、地震、雪灾、暴风的原因致使卖方不能按时履行义务时，卖方可以推迟这些义务的履行时间，或者撤销部分或全部合同。

3. 综合规定

即采用概括和列举综合并用的方式。在我国进出口合同中，一般都采取这种规定办法。例如：如果是战争或其他人力不可控制的原因，买卖双方不能在规定的时间内履行合同，如此种行为或原因在合同有效期后继续存在三个月，则本合同的未交货部分即视为取消，买卖双方的任何一方不负任何责任。

（二）不可抗力的法律后果

发生不可抗力事件后，应按约定的处理原则和办法及时进行处理。不可抗力引起的法律后果有两种：一种是解除合同；另一种是延期履行合同。至于什么情况下可以解除合同，什么情况下只能延期履行合同，应视事件的原因、性质、规模及其对履行合同所产生的实际影响程度而定。

（三）不可抗力事件的通知期限、方式

不可抗力事件发生后如影响合同履行时，发生事件的一方当事人应按约定的通知期限和通知方式，将不可抗力事件情况如实通知对方，如以电报通知对方，并在15天内以航空信提供事故的详尽情况和影响合同履行的程度的证明文件。对方在接到通知后，应及时答复，如有异议也应及时提出。

（四）不可抗力事件的证明

在国际贸易中，当一方援引不可抗力条款要求免责时，必须向对方提交有关机构出具的证明文件，作为发生不可抗力的证明。在国外，一般由当地的商会或合法的公证机构出具。

在我国，由中国国际贸易促进委员会或其设在口岸的贸促分会出具。

（五）援引不可抗力条款和处理不可抗力事件应注意的事项

当不可抗力事件发生后，合同当事人在援引不可抗力条款和处理不可抗力事件时，应注意如下事项：

（1）发生事故的一方当事人应按约定期限和方式将事件情况通知对方，对方也应及时答复。

（2）双方当事人都要认真分析事件的性质，看其是否属于不可抗力事件的范围。

（3）发生事件的一方当事人应出具有效的证明文件，以作为发生事件的证据。

（4）双方当事人应就不可抗力的后果，按约定的处理原则和办法进行协商处理。处理时，应弄清情况，体现实事求是的精神。

第四节　仲　裁

一、仲裁的含义和特点

（1）仲裁的含义。仲裁（Arbitration）又称公断，是指买卖双方在争议发生之前或发生之后，签订书面协议，自愿将争议提交双方所同意的第三者予以裁决。由于仲裁是依照法律所允许的仲裁程序裁定争端，因而仲裁裁决是最终裁决，具有法律约束力，当事人双方必须遵照执行。

（2）仲裁的特点。国际贸易中，双方在履约过程中有可能发生争议。由于买卖双方之间的关系是一种平等互利的合作关系，所以一旦发生争议，首先应通过友好协商的方式解决，以利于保护商业秘密和企业声誉。如果协商不成，则当事人可按照合同约定或争议的情况采用调解、仲裁或诉讼方式解决争议。

诉讼是一方当事人向法院起诉，控告合同的另一方，一般要求法院判令另一方当事人以赔偿经济损失或支付违约金的方式承担违约责任，也有要求对方实际履行合同义务的。其特点是诉讼是当事人单方面的行为，只要法院受理，另一方就必须应诉。但诉讼方式的缺点在于立案时间长、诉讼费用高，异国法院的判决未必是公正的，各国司法程序不同，当事人在异国诉讼比较复杂。

与诉讼相比，仲裁方式具有解决争议时间短、费用低、能为当事人保密、异国执行方便等优点。且仲裁是终局的，对双方都有约束力。因此，在国际贸易实践中，仲裁是最被广泛采用的一种方式。

二、仲裁协议的形式和作用

（一）仲裁协议的形式

仲裁协议是表明双方当事人愿意将他们的争议提交仲裁机构裁决的一种书面协议。仲裁协议有两种形式：

（1）双方当事人在争议发生之前订立的，表示一旦发生争议应提交仲裁，通常为合同中的一个条款，称为仲裁条款。

（2）双方当事人在争议发生后订立的，表示同意把已经发生的争议提交仲裁的协议，往往通过双方函电往来而订立。

（二）仲裁协议的作用

（1）仲裁协议表明双方当事人愿意将他们的争议提交仲裁机构裁决，任何一方都不得向法院起诉。

（2）仲裁协议也是仲裁机构受理案件的依据，任何仲裁机构都无权受理无书面仲裁协议的案件。

（3）仲裁协议还排除了法院对有关案件的管辖权，各国法律一般都规定法院不受理双方订有仲裁协议的争议案件，包括不受理当事人对仲裁裁决的上诉。

三、仲裁协议的内容

仲裁协议的内容一般应包括仲裁地点、仲裁机构、仲裁程序、仲裁裁决的效力及仲裁费用的负担等。

（一）仲裁地点

仲裁地点通常是指在哪个国家仲裁。因为仲裁地点与仲裁适用的程序和合同争议所适用的实体法密切相关，规定在哪个国家仲裁实际上就意味着适用该国的仲裁法和实体法。所以仲裁地点是协议中最为重要的一个问题。由于当事人对本国的法律和仲裁程序较为了解，一般都希望将仲裁地点定在本国。而且适用不同国家的法律，仲裁结果往往也可能不同。

我国进出口贸易合同中的仲裁地点一般采用下列三种规定方法：

（1）力争规定在我国仲裁。

（2）有时规定在被诉方所在国仲裁。

（3）规定在双方同意的第三国仲裁。

（二）仲裁机构的选择

国际贸易中的仲裁，可由双方当事人在仲裁协议中规定在常设的仲裁机构进行，也可以由当事人双方共同指定仲裁员组成临时仲裁庭进行仲裁。当事人双方选用哪个国家（地区）的仲裁机构审理争议，应在合同中做出具体说明。

（1）常设仲裁机构。世界上许多国家和一些国际组织都设有专门从事国际商事仲裁的常设机构，如国际商会仲裁院、英国伦敦仲裁院、英国仲裁协会、美国仲裁协会、瑞典斯德哥尔摩商会仲裁院、瑞士苏黎世商会仲裁院、日本国际商事仲裁协会以及中国香港国际仲裁中心等。我国的常设仲裁机构为中国国际经济贸易仲裁委员会和海事仲裁委员会。

仲裁机构不是国家的司法部门，而是依据法律成立的民间机构。

（2）临时仲裁庭。临时仲裁庭是专为审理指定的争议案件而由双方当事人指定的仲裁员组织起来的，案件审理完毕后即自动解散。因此，在采取临时仲裁庭解决争议时，双方当事人需要在仲裁条款中就双方指定仲裁员的办法、人数、组成仲裁庭的成员、是否需要首席仲裁员等问题做出明确的规定。

（三）仲裁程序法的适用

在买卖合同的仲裁条款中，应订明用哪个国家（地区）和哪个仲裁机构的仲裁规则进行仲裁。各国仲裁机构的仲裁规则对仲裁程序都有明确规定。按我国仲裁规则规定，基本程序如下：

（1）申请仲裁。申请人应提交仲裁协议和仲裁申请书，并附交有关证明文件和预交仲裁费。仲裁机构立案后，应向被诉人发出仲裁通知和申请书及附件。被诉人可以提交答辩书或反请求书。

（2）成立仲裁庭。当事人双方均可在仲裁机构所提供的仲裁员名册中指定或委托仲裁机构指定一名仲裁员，并由仲裁机构指定第三名仲裁员作为首席仲裁员，共同组成仲裁庭。如果用独任仲裁员方式，可由双方当事人共同指定或委托仲裁机构指定。

（3）仲裁审理。仲裁审理案件有两种形式：一种是书面审理，也称不开庭审理，根据有关书面材料对案件进行审理并做出裁决，海事仲裁常采用书面仲裁形式。另一种是开庭审理，这是普遍采用的一种方式。

（4）做出仲裁裁决。裁决是仲裁程序的最后一个环节。裁决做出后，审理案件的程序即告终结，因而这种裁决被称为最终裁决。根据我国仲裁规则，在仲裁过程中，仲裁庭认为有必要或接受当事人之提议，可就案件的任何问题做出中间裁决或者部分裁决。中间裁决是指对审理清楚的争议所作的暂时性裁决，以利于对案件的进一步审理；部分裁决是指仲裁庭对整个争议中的一些问题已经审理清楚，而先行做出的部分终局性裁决。这种裁决是构成最终裁决的组成部分。

仲裁裁决必须于案件审理终结之日起45天内以书面形式做出仲裁裁决书，除由于调解达成和解而做出的裁决书外，应说明裁决所依据的理由，并写明裁决是终局的和做出裁决书的日期地点，以及仲裁员的署名等。

（四）仲裁裁决的效力

仲裁裁决的效力主要是指由仲裁庭做出的裁决，对双方当事人是否具有约束力，是否为终局性的，能否向法院起诉要求变更裁决。进出口中的仲裁条款一般都规定仲裁裁决是终局的，对争议双方都有约束力，任何一方都不得向法院提出诉讼。但是有些国家则规定允许向上一级仲裁庭或法院上诉。即使向法院提起诉讼，法院也只是审查程序，而不审查裁决本身是否正确。即便如此，双方当事人在签订仲裁条款时仍应规定：仲裁裁决是终局的，对双方都有约束力。

（五）仲裁费用的负担

通常在仲裁条款中明确规定出仲裁费用由谁负担。一般规定由败诉方承担，也有的规定由仲裁庭酌情决定。

本章小结

在国际贸易中，买卖双方常常会因各种原因而发生争议，有的还可能导致索赔、仲裁和诉讼等情况的出现。为了在合同履行过程中尽量减少争议或在争议发生时能妥善解决，交易双方通常都要在合同中订立一些预防争议及发生争议时如何处理的条款，如检验检疫、索赔、不可抗力及仲裁。我们应重视此类条款的签订。

国际贸易合同的履行

★学习目标

★学习目标

1. 熟悉进出口业务流程
2. 掌握信用证审核方法、改证事项
3. 熟悉主要结汇单据

对大多数国家来说，进出口贸易是最重要的国际经济活动。每个国家都必须进口本国所不生产的货物和商品，还得创收外汇来支付货款，还要出口本国的制成品和富余的原料。因此，进出口贸易是同一件事物的两个方面，两者对国内市场都能产生有利影响。进口货物使国内产品有了竞争，而出口则为厂商的产品提供了更广阔的市场，有助于降低单位成本。无论是进口还是出口，其作用都是控制国内市场的价格。

但是，可能出于某些因素，政府不得不对对外贸易加以限制。为了保护国内某一产业，或者由于需要外汇用于购买更为重要的物资，政府可能要控制进口或以关税来制约进口。同样，为了保留发展中的国内产业所需要的某一种特殊的原料，出口也会受到限制。

这些因素意味着进出口贸易受许多手续的限制，诸如报关和外汇审批，而国内的零售及批发业务则不受此限制。这说明对外贸易的程序比国内贸易的程序要复杂得多，后者需要专门的知识以及受过良好训练的人才。

通常，在出口贸易和进口贸易中，要涉及下列所有或大部分机构：①出口公司；②装运港或机场的运输经纪人；③出口商所在国家的铁路部门（在有些情况下）；④出口商所在国家的陆路承运人（在有些情况下）；⑤港口当局；⑥船务公司（对海洋运输）；⑦航空公司（对空运）；⑧保险公司或经纪人；⑨出口商银行；⑩进口商银行；⑪进口商所在国家的铁路部门（在有些情况下）；⑫进口商所在国家的陆路承运人（在有些情况下）；⑬装卸港或机场的运输经纪人；⑭进口公司。

出口和进口贸易中要涉及很多事项，这包括：

（1）运输经纪人或货物转运商负责制单并安排货物的空运、海运、铁路运输或公路运输。如果有专门人才的话，这些服务将由供货商自己的出口部门实施。

（2）航空公司、船运公司、铁路公司或货物承运人实际运输货物。

（3）如果进出口贸易是用信用证或汇票来支付，那就要涉及进口商银行和出口商银行。

（4）海关和征税官员要检查货物和核查进出口许可证，征收关税和/或增值税。

（5）如果进口国要求，商会将颁发原产地说明书。

（6）保险公司为运输中的货物保险。

（7）如果要起草特殊合同的话，还需要律师。

很多进口或出口交易是通过出口商在国外的代理商或经销商来安排的。在这种情况下，进口商从本国的公司那里购买货物，而那家公司自己也进口货物。交易还可以通过进口商的购买代理商或代表进口商的购买商行来进行，或通过出口商国家的出口行来进行。在这些情况下，出口商把货物直接销售给他本国的公司，该公司再出口货物。

进出口贸易的过程非常复杂，往往要花很长一段时间才能完成一笔交易。在一宗出口或进口贸易中，要经历各种各样复杂的程序。从开始到结束，一笔交易一般要经历四个阶段：进口或出口的准备—商务谈判—执行合同—解决纠纷（如果有的话）。每个阶段包含一些具体的步骤。既然进口贸易和出口贸易是同一件事物的两个方面，一个国家的进口就是另外一个国家的出口，那么我们结合进出口贸易步骤的流程图来说明进出口贸易的一般程序。

第一节　出口合同的履行

不同的国家有不同的经济政策或不同的经济体系。因此，任何想做外贸出口的人都必须事先了解外贸出口流程。按 CIF 贸易术语成交，以信用证结算的出口合同履行流程如图 9-1 所示。

一、出口准备阶段

对出口贸易来说，最困难的是迈出第一步。任何一个想把货物销售到国外的出口商首先都必须做很多的市场调研。市场调研是指对某一产品在某一特定市场条件下进行调研。具体到出口贸易的市场调研，指的是对国外某一特定市场的调查研究以确定其市场需求及供货的方式。出口商必须了解有哪些国外公司可能使用其产品，有哪些公司对在其所在国销售和分销其产品感兴趣。其必须考虑是否有赚取利润的可能，还必须调查那些国家的市场结构和总的经济形势。如果经济不景气，对产品的需求量通常会下降。那么，此时出口商的产品价格就会受到影响。市场调查主要包括：

（1）对国家或地区的调查研究。拥有不同政治和经济制度的国家或地区对待对外贸易的态度会大不相同。出口商应调查它们的政治、金融和经济状况，以及关于外贸、外汇管制、海关关税和商业惯例的政策、法律和制度，还有它们在对外贸易（如进出口商品的结构、数量、金额、贸易对象及贸易管制等方面）的情况。这些因素对选择正确的销售市场有很大影响。

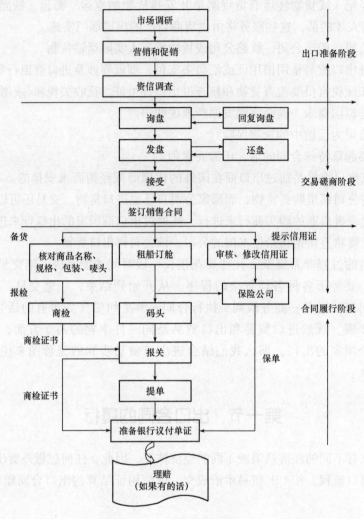

图 9-1　出口交易流程图

（2）对商品市场调查的研究。调查有关某一特定的商品在国外市场的生产、消费、价格及其变化趋势以及主要进出口国等情况，以便正确制定出口商品的价格和恰当地处理其他交易条件。

（3）对客户的调查研究。在国际贸易中，资信情况比在国内贸易中更为重要。出口商应了解买方或进口商的声誉，他们的经营规模，付账方式以及贸易活动方面的信息。显然，拥有良好信誉和资信状况的客户能促进出口交易。出口商可以从各种途径获取这些信息，如买方提供的证明人、参与银行、各种行业公会和咨询机构。通过对客户的调查研究，出口商可以更好地确定潜在的顾客。

除了通过外部渠道进行市场调研来收集信息或数据以外，出口商也可以主动到国外市场去推销产品，常用的方法有促销的印刷品、售货点广告、包装、赞助、展列室、商品交易会和展览会、公共宣传、公共关系等。

二、交易磋商阶段

如果外国公司有兴趣购买出口商的产品，就应安排磋商。交易的磋商在销售合同的签订及以后的履行中起着重要的作用，它与交易双方的经济利益密切相关。

不管磋商用什么方法开始，它通常包括下列四个环节：询盘、发盘、还盘、接受和签订销售合同。其中，发盘和接受是达成协议和签订合同不可缺少的环节。

（一）询盘

询盘是要求提供有关交易的信息，如商品的价格单、目录、样品以及贸易条件。询盘可由进口商或出口商发出。收到询盘以后，作为惯例卖方应立即回复以便开始交易磋商。

（二）发盘和还盘

发盘是卖方向买方提出的签订合同的建议。换句话说，它指的是发盘人向受盘人提出的交易条件。按此条件，发盘人愿意同受盘人进行交易。发盘有两种：一种是实盘，另一种是虚盘。对发盘表示接受但载有添加限制或其他更改的答复，即为拒绝该项发盘并构成还盘。

（三）接受

接受指的是受盘人声明或做出其他行为表示无条件同意一项发盘。发盘接受以后，合同随即达成。

（四）签订销售合同

发盘接受以后，买卖双方通常要求签订书面的销售合同或销售确认书以规定各自的权利和义务。销售合同或销售确认书包含一些一般的条款，还包含一些随商品而变化的特定的条款。但是以下这些条款是不可缺少的：买卖双方的姓名，商品的品名、品质和规格、数量、包装、单价、总价，支付、发货日期、装运、保险、商检、索赔和仲裁。销售合同或销售确认书一般一式两份，买卖双方各执一份。

三、合同的履行阶段

当出口合同为 CIF 合同，采用信用证为付款方式时，出口合同的履行一般包括备货，报检，催证、审证和改证，租船和订舱，装运、报关手续、保险、制单结汇和理赔等。

（一）备货

合同签订以后，出口商的主要任务就是为装船而准备货物，并按合同条款对货物进行核查。货物的品质、规格、数量、唛头和包装必须与合同或信用证一致。备货时间应结合船期安排。

（二）报检

如有国家或者合同的规定要求，出口商的货物应经有关的商检部门检验并获取商品检验证。通常，只有在取得商检部门发给的合格检验证书后，货物才能被海关放行。

（三）催证、审证和改证

在国际贸易中，银行信用证被广泛用来支付货款。在执行合同时，其中一个重要步骤就是卖方应催促买方按时开立信用证。按照合同规定，买方应该及时开立信用证，但有时由于各种原因而迟开。为了安全收汇，卖方应催促买方开证。当卖方收到信用证后，必须对照合同逐条进行严格审核。审核时发现的任何不符点，应该立即联系买方尽快修改，以保证合同的顺利进行。

1. 催开信用证

若以信用证付款，出口商要等到信用证开立后才发货。有时有必要提示买方开立信用证，买方亦应该在规定时限前开立。因财务困难或其他原因，买方有可能不能及时开证。卖方有必要事先定明信用证开立时限，同时可保留因延迟开立的索赔权。

2. 审核信用证

收到信用证后，受益人应仔细审核信用证是否与合同及《跟单信用证统一惯例》相符。若存有不符点，卖方应让进口商要求开证行加以改正。出口商确认信用证的有效完备并接受信用证后，才会发货。审核信用证一般分为大体审核和对具体条款的审核。

（1）总体审核。信用证应由与我国有经济往来关系的国家或地区的银行开立，该国家或地区的分行开立信用证依旧适用此原则。信用证必须与两贸易国缔结的商务协议相一致，且信用证不可附有不赞同或政治倾向性条款。

通知行应审核信用证印章或检验信用证以辨真伪，若信用证直接递交卖方，卖方应呈交通知行做此项审核。若通知行担心开证行的信用状况，可据此采取一些措施，诸如要求开证申请人对此信用证由一家声誉良好的银行对付款行加以保兑，在信用证中加入保证付款的条款；若成交货物量较大，可规定分批装运、分批付款，且在信用证中加入电报求偿条款。

受益人也应注意审核信用证是否存在矛盾性条款，如在不允许转运却要求以联运提单议付；以 CIF 价格成交，却要求提交保险单议付等情形。

最后，受益人应审核信用证当中是否存在拼写错误，其是否可履行全部要求（如某些情况下信用证含有指定船只装运的要求），或者其是否可在规定时间内备齐全部单据议付等。

（2）审核单个条款。这意味着我们应逐条审核信用证条款。信用证受益人应审核信用证中关于货物名称、规格、数量、包装、单价、装运时间、装运港和卸货港等列在合同中的相关条款，特别应注意以下内容：

第一，信用证有效期。例如：有效期截止到 2016 年 6 月 15 日，在中国议付有效。这一规定给出了有效期及议付在出口商所在国相关行即可。除非有更加稳妥的方式来避免相关风险，否则受益人不接受其他条款是十分明智的。

第二，信用证金额。信用证金额及结算货币应与合同一致。若合同中货物数量有增减性条款，信用证也应据此对数量条款和结算金额灵活调整，汇票及发票金额不应超过合同金额。

第三，装运时间、信用证有效期、装运单据的交单期限。装运时间应与合同规定一致。若信用证中规定装运时限长于合同，是可接受的。如因信用证延迟开立、备货延误等种种原因不能如期装运，卖方可要求将期限进行延展。

信用证应有有效期的规定。通常为装运日之后的 10～15 天，以便卖方有足够的时间备齐单据，交单议付。

有些时候信用证会规定运输单据的交单时限。当信用证有效期较长，买方想及时获取装运单据提取货物时，需加以规定交单期限。常见规定方法为：海运提单日期后若干天交付相关装运单据。若无此项规定，应在装运后 21 天内提交单据。交单期限订立太过短暂，如 2 天，也是不可接受的。

第四，转运和分批装运。如销售合同规定允许转运和分批装运而信用证无相关描述，则可视为允许。

第五，开证申请人和受益人。出口商应确保开证申请人和受益人在信用证中规定及拼写正确无误。发货人某些时候可能是分支代理机构或其他制造商，而非出口商本身。此时可使用可转让信用证，免去了受益人和发货人不同的麻烦。

第六，审核各条款。出口商还应逐一审核信用证中其他条款，确信其是否与合同一致，是否成为交易履行不必要的障碍，或与其他条款相矛盾。

第七，付款时间。付款时间应与合同规定一致。若为远期信用证，销售合同规定买方担负利息损失，信用证也应有相同规定。

3. 修改信用证

一旦出口商发现信用证存在某些瑕疵，他应考虑是否有必要加以修改。若此瑕疵是结算货款的障碍或妨碍合同履行，出口商应及时要求进口商通知开证行加以修改。如此瑕疵不太重要，合同依旧可以顺利履行，则不必要加以修改以便节约时间、节省费用。

信用证如需多处修改，出口商应一次性提出。进口商可接受或拒绝修改。一旦做出修改，修改内容将成为信用证的一部分并应按此执行。

（四）租船和订舱

收到有关的信用证后，出口商必须马上与轮船代理人或船运公司联系租船和订舱，并按照进口商的装船要求准备装运。如果货物数量较大，需要整船载运的要办理租船手续；如果数量不大，办理订舱手续即可。

（五）报关手续

在货物装运以前，还应完成一些报关手续。按照规定，必须把填写有出口货物细节的报关单以及合同副本、发票、装箱单、重量单、商检证书和其他相关文件提交给海关办理报关。货物一装上船，船运公司或轮船代理人将开出提单，它是证明货物已装上船的收据。

（六）保险

出口贸易要面临很多危险，例如，船舶可能沉没、货物可能在运输途中受损、外汇兑换率可能有变动、买主可能违约或者政府部门突然宣布禁运等。为避免旅途中的风险，通常要为出口货物购买保险。所购买的险种应根据货物的种类和环境来选择。如果出口商购买了保险，他就会得到相应险种下损失的赔偿。

（七）制单结汇

出口货物在装运后，进出口商应立即按照信用证的规定，编制各种单据，并在信用证规定的交单有效期内，交银行办理议付结汇手续。其要求的单据包括商业发票、提单、保险单、装箱单、重量单、商检证书。在有些情况下，还包括领事发票、原产地证书等。编制的单据必须正确、完整、简明和整洁。开证行只有在审核单证与信用证完全相符后才付款。单证中有任何不符点都可能遭到银行的拒付。

四、解决争议

有时尽管进出口双方小心谨慎地履行合同，但抱怨和索赔仍不可避免。这些抱怨和索赔可能由各种各样的原因引起，例如交货数量或多或少、发错货、粗劣的包装、质量低劣、样品和实际到货之间存有差异、发运的迟误等。根据具体情况，索赔可向出口商、进口商、保险公司或运输公司提出。一旦发生争议时，明智的选择是仲裁比诉讼好，而调解又比仲裁好。

第二节　进口合同的履行

上一节我们已经学习了出口贸易的一般程序，并从出口商的角度简要地了解了其各个阶段和步骤。因为已经熟悉了出口贸易的流程，我们就可以很容易理解进口商是如何进行进口贸易的。毕竟，出口贸易和进口贸易是同一件事物的两个方面。在进口贸易中，你竭力争取的交易条件很可能与你在出口贸易中争取的恰恰相反。不管你是作为出口商还是进口商，交货术语的意思是一样的。不管是谁出于什么样的目的，提单还是提单。我们前面获得的关于出口贸易的知识同样适用于进口贸易。具备了出口程序的基本知识后，我们可以很容易地掌握进口程序的要点，从而能顺利地进行进口贸易。

按 FOB 成交，以信用证来结算，进口交易流程如图 9-2 所示。

进口贸易的一般程序如下：

（1）进行市场调查。

（2）制订某一商品的进口计划。

（3）向海外可能的卖方发出询盘。

（4）比较、分析所收到的报盘或时价。

（5）还盘并决定最有利的报盘。

（6）签订购买合同。

（7）向银行申请开立信用证。

（8）如果是 FOB 合同的话，要租船订舱接运货物。

（9）收到装运通知后，让保险公司为货物投保。

（10）如有必要要申请商检。

（11）申请办理海关手续，让货物清关。

（12）委托承运人将货物从港口运至最终用户的仓库。

（13）解决争议（如果有的话）。

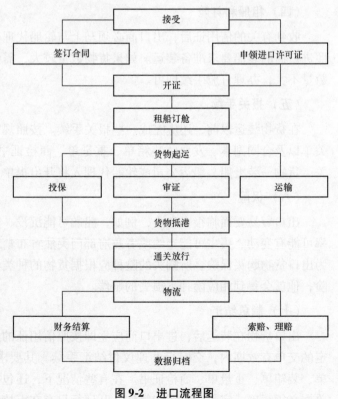

图9-2　进口流程图

第三节 主要进出口单证

国际贸易对单证的要求非常高，从某种程度上讲国际贸易是一种单证的交易，或"象征性交易"，因为货运单证代表着对货物的所有权。例如，在信用证下，买方只有在获取货运单证后才能提货，而卖方也只有在交出货运单证后才能得到货款。要使用什么样的单证以及如何仔细、准确地完成这些单证非常值得我们注意。一般来说，每一笔销售合同都要规定各种需要的单证。忽略这些单证将会产生严重的后果，这在国际贸易中并不罕见。所以，对进出口双方来说，遵守这些规定是很有必要的。商业发票、提单、保险单、装箱单、重量单等通常称为货运单证。此外，货运单证还包括买方所要求的以及与对进口货物征收关税有关的单证，如形式发票、领事发票、产地证书、价值证书、检验证书等。在国际贸易中，商业发票、提单和保险单是最主要的单证。在每一宗进出口货物中，它们都是必不可少的。本小节将主要涉及商业发票、形式发票、提单、装箱单、重量单、检验证书和保险单。

一、商业发票

（一）商业发票的定义

发票是卖方对其所出售的货物开出的包括各项细节的清单，并作为向买方收取货款的凭证。买卖双方出于不同的目的需要这些清单。发票的形式多种多样，有商业发票、银行发票、领事发票、海关发票和形式发票等。在这些发票中，商业发票使用得最为广泛，几乎每一宗货物都要求提供商业发票，作为证明货物已发送的单证之一。当我们提到为发送货物开具的发票时，一般指的就是商业发票。商业发票是一种载有买方必须付款的所售货物的识别情况的单证。所有商业发票应写明债务人的名称和地址、支付条款、商品名称、价格。此外，该发票还应写明运输方式。

（二）商业发票的内容

商业发票的形式各不相同，但不管采取什么样的形式，其内容必须与合同完全相符。通常商业发票概括了合同条款，并表明货物根据合同条款已经装运。商业发票首先包括卖方和买方的姓名和地址；其次，包括运出货物的细目：所有包件的重量、数量和唛头；再次，包括货物的单价和总价。发票也要说明装运的港口和日期以及价格条款，如到岸价和付款方式，如即期汇票或信用证项下的即期汇票。最后，发票必须由卖方授权的雇员签字，甚至还可能要写上进出口许可证的号码。

有时发票上的价格可以分解为诸如材料费、加工制造费和包装运输费。对发票细节的要求取决于进口国家的规章制度。有些国家要求发票提供更为详细的价格分解。

有些国家政府对商业发票有特殊的规定，比如要求商业发票翻译成当地使用的语言或者在重量上使用国际公制还有其他的一些计量方法。有的海关当局和一些管制机构还强调不同文件的一致性。因此，商业发票、领事发票、保险证明书和提单上的号码和标号必须绝对一致。

商业发票的原件必须有卖方的签名，通常就是发货人。缩写"E. & O. E."代表着"错误和遗漏例外"，通常印在发票的页脚，意思是如果有错误和遗漏，发货人将随时准备修改。

（三）商业发票的作用

商业发票主要是为买方、卖方和海关当局提供一种出口交易的记录。出口商、出口商银行、付款行、卸货港的接收代理、出口国和进口国的海关都要使用商业发票。进口商用它来检查所运送给他的货物是否与相应的合同中的条款一致；银行需要它连同提单和保险单来进行议付；海关需要它来计算关税（如有的话）；出口商和进口商需要它来记账。在没有汇票的情况下，商业发票还可以取代汇票来收取钱款。

（四）商业发票范例

为了理解和正确填写商业发票，一般要包含以下几点：

①顾客的姓名；②办公地址、发货地址；③发票号码（便于记录）；④订单号码；⑤参照号码和/或各项商品的货名；⑥数量；⑦各项商品的价格；⑧各项商品的总价和所有商品的总价；⑨所同意的折扣数量及条件；⑩支付运费、保险费和成本的方法；⑪发货地址；⑫包、件或箱的数量；⑬包、件或箱上的标志；⑭其他事项。

商业发票样本请参见本书附录。

二、形式发票

（一）形式发票的定义

Proforma 在拉丁文里是"为了形式"的意思。形式发票是指一种凭证只作为临时单证使用，数天后由开具的最终发票所代替。从外观上看，除注明"形式"字样外，它和一般的商业发票一样，包含一些常规项目，如唛头、货物的件数、货名、数量、质量、价格等。但是，从本质上讲，形式发票是一种不同形式的发票，因为它涉及的是一种"假设"的销售，就像是签订了合同，实际发生了一样。形式发票不是正式文件，而是一种对买卖双方都无约束力的文件。

（二）形式发票的作用

需要使用形式发票的原因有多种。其中最重要的是进口商需要形式发票，以便适应其本国现行的规章制度。进口商还可事先要求得到形式发票来了解进口货物的有关情况，申请进口许可证或信用证。

在许多国家，特别是第三世界的发展中国家，对外贸易由国家严格管制。这些国家的政府通常都实行进口许可证制度或进口配额制。进口商必须申请必要的进口许可证或外汇，未经批准进口许可证或未经配给外汇，进口商就不得进口任何货物。他们的申请往往须提供外国出口商签发的非正式发票——形式发票为依据，该发票须列明商品名称、规格、单价等。

进口商索取形式发票事实上是在询盘，而出口商寄送形式发票实际上是在报盘。假如出口商要报实盘，他必须在信内指明其有效期内寄送形式发票。

三、提单

（一）提单的定义

最重要的货运单证就是提单。它是轮船公司签发的单证，既代表承运货物的收据，又代表承运人和托运人之间的运输合同。它也是代表货物所有权的证件，它给予持有人或受让人提货的权利。

（二）提单的内容

提单首先是发货人与轮船公司之间的一种合同；其次，是收到货物的收据；再次，是所有权的一种单证。提单不止包含货物的详细说明——数量、重量和箱包的唛头——它还包含许多其他的情况：发货人的姓名和运载船舶的船名、装运港及目的港、运费率、收货人姓名（除非是指示提单，同支票相像）以及装运日期，从合同的观点来看，装运日期极为重要。

提单也可能包含许多其他条款。在发货人按 CIF 或 CFR 条款出售货物时，提单上注明"运费已付"。有些提单允许转船，转船的意思是货物可在一些中间港从一艘船转到另一艘船。对发货人来说，有一点也很重要，那就是提单应该是"清洁"的，而不是"不清洁"的；也就是说轮船公司对实际装运的货物，对其数量及情况，不应注有任何条件。这是因为发货人的信用证可能一定要求清洁提单，就像要求"已装船提单"而不同意"备运提单"一样。有时在签发提单以前，先给发货人一张大副收据，因为签发提单得花一些时间。

（三）提单的作用

提单有三个重要的作用。首先，它是轮船公司签发的并给发货人的收据；同时，它也是轮船公司和托运人之间的运输合同的证明；另外，提单还是证明货物所有权的单证，因为提单的法定拥有者即是货物的拥有人。

出于这个原因，提单可以用来把货物从一个拥有者手中移交到另一人手中。当出口商填单时，他们在提单上的"收货人"一栏空白处写上买方的姓名。这就意味着收货人如提单上所示是货物的法定拥有者。另外，出口人可在"收货人"一栏中填上"凭指定"。在"凭指定"下面填上代理人的姓名和地址。然后，进口国的代理人可以背书提单把它交给买方。通过这种方法，进口商就可以把货物转交给顾客。这就意味着不同的收货人必须有分别开立的提单，并且几件货物不能同时开在一张提单上。

（四）提单的使用

提单是海运出口交易的核心文件。发货人一旦获取关于货物的所有细节，他就开始填写由轮船公司提供的表格。然后，将它送到船上，在那里轮船公司的官员核查货物是否"情况良好"，并在装运货物越过船舷时在提单上签字。在货物准备装运之前，提单必须在轮船公司或他们的代理手中。

当货物装上船后，轮船公司的官员或代理人在提单上签上"货物已收到，表面情况良好"的字样。换句话说就是，货物必须与提单上所记载的严格一致，没有差异。包装箱必须没有损伤，如有包装袋的话，不能有破损和污点。盛装液体的桶不能有凹痕和渗漏。货物的件数和种类必须与提单上一致。

如果提单上所记载的与货物的实际情况有所不同，轮船就要在提单上批注损伤或损毁。这样，提单就不再是清洁提单了，进口商银行就不会接受这样的提单，出口商银行也不能获取货款。所以，应极力避免不清洁提单，出口商必须保证他们的货物到达码头时"情况良好"。

有时，货物的某些缺陷是难免的。比如，木材的末端开裂、化学品引起包装的掉色。在这些情况下，出口商必须同进口商在提单的某些条款上达成一致。这些条款必须在出口合同签订以前就达成共识，出口商还应该把这些达成共识的条款通知给相关银行。

提单通常制成一套三至四份原件。发货人可以要求多签发几份以便存档。船方保留一份提单。其他的几份送交给出口商或直接给相关银行。用这些可转让的提单来获取货款。它们被送到进口国的买方和代理商手中。

接下来当轮船抵达时，提单和其他货运单据就送交到轮船公司。然后轮船公司就将可转让的提单和船上的复本提单作比较。这样，进口商就可以表明其对货物的所有权并从船上提取货物。

近几年来，在单据的简化方面取得了很大的进展。提单经常被非海运的其他运输方式所使用的类似的、不可转让的运输单据所替代。这些单据被称为海运单、班轮运单、货运收据等。

（五）提单范例

为能理解和制定提单，单据中一般应包含以下几点：

①货物发送方式；②运费支付者；③发货人姓名；④收货人姓名；⑤船只名称及装运港；⑥卸货港；⑦包装的类型、数量和唛头、毛重和计量；⑧其他方面。

如果一份提单有以上内容，叫作"略式提单"。然而，如果一份提单的背面还有有关轮船公司和发货人权利和义务的条款，称为"全式提单"。在国际贸易中，全式提单比略式提单运用得更为广泛。

提单样本请参见本书附录。

四、装箱单

装箱单是国际贸易中卖方售货时出具的单据，是说明所发运货物的数量、种类、毛重，每件货物的法定净重以及标志和号数的单证。它用于补充发票的不足之处，以便收货人在货物抵达目的港后区分和核对货物并向海关申报货物。这样可以加快货物的清关过程。并且，在货物失落或受到损坏时，可凭装箱单向保险公司索赔。

装箱单样本请参见本书附录。

五、重量单

重量单是对外贸易中卖方售货时出具的单据，是说明每件货物的毛重和净重的证件。它用于补充发票的不足之处，以便收货人在货物到达目的港后核对货物和加快办理海关手续。装箱单和重量单通常合二为一，做成一个单证。

六、检验单

检验单或检查报告是表明货物的数量、质量或其他因素的单证。它由制造商、商会、检查人员或政府机构签发。它主要有两方面的作用：①作为质量和数量的单证，它可确定所运货物是否与合同中规定的相一致，是拒绝付款、索赔或理赔时的一个重要证明；②它是议付货款的货运单证之一。

按对外贸易合同规定，进口商品在抵达后必须接受检验部门的检验，收货人、用户或货运代理商应在抵达港/站及适当时间内向商检部门申请商检。商检后，将发给申请人一份检查报告或检验单。在出示附在海关申报单上盖有商检部门印章后，海关将检查和放行这些商品。

根据合同或法律规定，出口商品必须接受检查，制造商、供应商在货物装运前应申请商检。经检查，如果货物符合标准，商检部门将为他们签发检验单为货物结关。

如果用船只和集装箱来运输供出口的易腐烂的谷类食物、油、粮食和冷冻产品时，运输工具和集装箱的填装部门应向所在港的商检机构申请检验他们的货舱、箱和集装箱。经检验后，如果他们符合运输的技术条件，商检机构将签发检验单允许他们运输货物。

检验单样本请参见本书附录。

七、保险单和保险凭证

货物投保后，将由保险公司签发保险单或保险凭证。保险单和保险凭证是主要的装运单证。在信用证下，保险单还可作为抵押担保从银行获得垫付货款。

保险单是保险人签发的一种具有法律效力的单证，它严格规定了一笔保险业务的条款和条件——被保险人姓名、保险货物名称、保险金额、载货船只名称、承保险别、保险期限和可能发生的免责事项。它也是保险人和被保险人之间订立的书面契约。

保险凭证是一种保险证明，实际上是简化的保险单。它包含保险单上的必要项目，但它并不列出保险人和被保险人的权利和义务，它们应以正式的保险单详细的保险条款为准。保险凭证与保险单具有同样的效力。

保险单样本请参见本书附录。

本章小结

本章按照进出口合同签订后的履约顺序，对进出口合同履行的各个环节分别进行了介绍。同时介绍了主要进出口贸易单证。

跨境电子商务实务

1. 了解跨境电子商务的发展历史
2. 熟悉并掌握跨境电子商务的概念
3. 了解跨境电子商务的特点

第一节　跨境电子商务概述

一、跨境电子商务的概念

跨境电子商务（Cross-border Electronic Commerce），简称跨境电商，是指分属不同关境的交易主体，通过电子商务平台达成交易、进行支付结算，并通过跨境物流送达商品、完成交易的一种国际商业活动。

跨境电子商务可以分为广义的跨境电子商务和狭义的跨境电子商务。

从狭义上看，跨境电商实际上基本等同于跨境零售。跨境零售指的是分属于不同关境的交易主体，借助计算机网络达成交易，进行支付结算，并采用快件、小包等行邮的方式通过跨境物流将商品送达消费者手中的交易过程。跨境电商在国际上流行的名称叫 Cross – border 和 E – commerce，其实指的都是跨境零售，从海关来说跨境电商通常等同于在网上进行小包的买卖，基本上是针对消费者。从严格意义上说，随着跨境电商的发展，跨境零售消费者中也会含有一部分碎片化小额买卖的 B 类商家用户，但现实中这类小 B 商家和 C 类个人消费者很难区分，也很难界定小 B 商家和 C 类个人消费者之间的严格界限。所以，从总体来讲，这部分针对小 B 的销售也归属于跨境零售部分。

从广义上看，跨境电商基本等同于外贸电商，是指分属不同关境的交易主体，通过电子商务的手段将传统进出口贸易中的展示、洽谈和成交环节电子化，并通过跨境物流送达商品，完成交易的一种国际商业活动。从更广意义上看，跨境电商指电子商务在进出口贸易中的应用，是传统国际贸易商务流程的电子化、数字化和网络化。它涉及许多方面的活动，包括货物的电子贸易、在线数据传递、电子资金划拨、电子货运单证等内容。从这个意义上看，在国际贸易环节中只要涉及电子商务应用都可以纳入这个统计范畴内。

二、中国跨境电子商务的发展

1999 年阿里巴巴实现用互联网连接中国供应商与海外买家后，中国对外出口贸易就实现了互联网化。在此之后，共经历了三个阶段，实现从信息服务到在线交易、全产业链服务的跨境电商产业转型。

（1）跨境电商 1.0 阶段（1999—2003）。跨境电商 1.0 时代的主要商业模式是网上展示、线下交易的外贸信息服务模式。跨境电商 1.0 阶段第三方平台主要的功能是为企业信息以及产品提供网络展示平台，并不在网络上涉及任何交易环节。此时的盈利模式主要是通过向进行信息展示的企业收取会员费（如年服务费）。在跨境电商 1.0 阶段的发展过程中，也逐渐衍生出竞价推广、咨询服务等为供应商提供一条龙的信息流增值服务。

在跨境电商 1.0 阶段中，阿里巴巴国际站平台以及环球资源网为典型代表平台。其中，阿里巴巴成立于 1999 年，最初，阿里巴巴中国供应商只是互联网上的黄页，将中国企业的产品信息向全球客户展示，定位于 B2B 大宗贸易。买方通过阿里巴巴平台了解到卖方的产品信息，然后双方通过线下洽谈成交，所以当时的大部分交易是在线下完成的。环球资源网 1971 年成立，前身为 Asian Source，是亚洲较早的贸易市场资讯提供者，并于 2000 年 4 月 28 日在纳斯达克证券交易所上市，股权代码 GSOL。

在此期间还出现了中国制造网、韩国 EC21 网、Kellysearch 等大量以供需信息交易为主的跨境电商平台。跨境电商 1.0 阶段虽然通过互联网解决了中国贸易信息面向世界买家时遇到的难题，但是依然无法完成在线交易，对于外贸电商产业链的整合仅完成信息流整合环节。

（2）跨境电商 2.0 阶段（2004—2012）。2004 年，随着敦煌网的上线，跨境电商 2.0 阶段来临。这个阶段，跨境电商平台开始摆脱纯信息黄页的展示行为，将线下交易、支付、物流等流程实现电子化，逐步实现在线交易平台。

相比较第一阶段，跨境电商 2.0 更能体现电子商务的本质，借助于电子商务平台，通过服务、资源整合有效打通上下游供应链，包括 B2B（平台对企业小额交易）平台模式，以及 B2C（平台对用户）平台模式两种模式。跨境电商 2.0 阶段，B2B 平台模式为跨境电商主流模式，通过直接对接中小企业商户实现产业链的进一步缩短，提升商品销售利润空间。2011 年敦煌网宣布实现赢利的目标，2012 年持续赢利。

在跨境电商 2.0 阶段，第三方平台实现了营收的多元化，同时实现后向收费模式，将"会员收费"改为以收取交易佣金为主，即按成交效果来收取百分点佣金。同时还通过平台上营销推广、支付服务、物流服务等获得增值收益。

（3）跨境电商 3.0 阶段（2013 至今）。2013 年成为跨境电商重要转型年，跨境电商全产业链都出现了商业模式的变化。随着跨境电商的转型，跨境电商 3.0"大时代"随之到来。

跨境电商 3.0 具有大型工厂上线、B 类买家成规模、中大额订单比例提升、大型服务商加入和移动用户量爆增五方面的特征。与此同时，跨境电商 3.0 服务全面升级，平台承载能力更强，全产业链服务在线化也是 3.0 时代的重要特征。

在跨境电商 3.0 阶段，用户群体由草根创业向工厂、外贸公司转变，且具有极强的生产设计管理能力。平台销售产品由网商、二手货源向一手货源好产品转变。

3.0 阶段的主要卖家群体正处于从传统外贸业务向跨境电商业务的艰难转型期，生产模式由制造工厂向智造之都转变，对代运营和产业链配套服务需求较高。另一方面，3.0 阶段的主要平台模式也由 C2C、B2C 向 B2B、M2B 模式转变，批发商买家的中大额交易成为平台主要订单。

三、跨境电子商务的特点

与传统国际贸易相比，跨境电子商务呈现五大新特征：全球性、无形性、匿名性、即时性和无纸化。

（1）全球性。跨境电子商务以电子商务平台为依托进行国际贸易，决定了其全球化的特征。与传统贸易相比，跨境电子商务突破了地域限制和时间限制：一国卖方可以通过互联网发布产品和服务的相关信息，与另一国买方进行交流、磋商，进而达成交易；一国买方也可以通过互联网寻找卖家，进行询价议价、支付结算，最终买到物美价廉的产品或服务。跨境电子商务的全球性特征能最大限度地给全球卖家和买家带来信息资源的共享，但同时也存在一定的支付风险和结算风险。任何人只要掌握了一定的网络基础知识，何时何地都可以将信息输入网络，进行网上交易。例如，一家较小的中国外贸公司，通过电子商务平台就可以向任何国家的贸易或消费者提供产品和服务，只要他们进入了互联网并有相应的需求。这在很大程度上为国际贸易提供了便利，但同时给国家的税收带来了一定的麻烦。

（2）无形性。传统的对外贸易主要是进行实物交易，而随着网络的发展，一些数字化产品和服务（如电子图书、电影、版权等）交易越来越多，数字化传输是在全球化网络大环境下进行的，具有无形性。而跨境电子商务是基于网络发展起来的，必然会具有网络的无形性特征。以书籍交易为例，在传统对外贸易中，是以一本书（即交易实物）为标的物进行买卖，而在跨境电子商务交易中，一国买家只需购买该书网上的数据权即可获取相应的信息，方便快捷。电子商务的无形性特征给一国的税务机关和法律部门带来了新的考验，其交易记录体现为数据代码形式，使得相关部门很难界定该项交易活动，也就无法进行有效监督和征收税款。

（3）匿名性。由于跨境电子商务的全球化和无形性，交易双方主体可以随时随地地利用网络进行交易，而且利用电子商务平台进行交易的消费者出于规避交易风险的目的，通常不暴露自己的真实信息，如真实姓名和确切的地理位置等。但这却丝毫不影响他们顺利地进行交易，网络的匿名性给消费者提供这样的便利条件，允许他们这样做。跨境电子商务的匿名性造成消费者权利与义务的极其不对称，消费者可以在虚拟的网络环境中享受最大的权利和利益，却承担最小的责任与义务，有的甚至想方设法逃避责任，使得他们无法通过查找交易人获知真实的所得利益和交易情况，也就无法计算应缴税款，对纳税人进行合法征税。

（4）即时性。传统对外贸易中，交易双方交流多数是通过邮件、传真等方式，信息的发送与接收存在着不同程度的时间差，而且传输过程中还可能遇到一定的障碍，使得信息无法流畅即时地进行传递，这在一定程度上会影响国际贸易的进行。不同于传统对外贸易模式，跨境电子商务对于信息的传输是即时的，也就是说无论实际地理位置相距多远，卖家发送信息与买家接收信息几乎是同时进行的，不存在时间差，这在一定程度上等同于传统对外贸易的面对面交流磋商。对于一些数字化商品（如软件、电影等）的交易，下单、付款、交货、结算都可以通过网络瞬间完成，给交易双方带来了极大的便利。

跨境电子商务的即时性特征减少了传统对外贸易中的中间商环节，使出口商直接面对最终消费者，提高了贸易的效率，但也隐藏了法律危机。在税收领域表现为：由于电子商务活动的即时性，买卖双方可以随时开始、变动和终止交易活动，增加了贸易的随意性、降低了贸易的有效性。这使得税务机关无法查证双方交易的真实情况，监督无效，给税务机关征收税款带来了一定困难。

（5）无纸化。在传统对外贸易中，从询价议价、磋商、订立合同到货款结算都需要一系列的书面文件，并以之为交易的依据。而在电子商务中，交易的主体主要使用无纸化的操作形式，这是跨境电子商务不同于传统贸易的典型特征。卖方通过网络发送信息，买方通过网络接收信息，整个电子信息的传输过程实现了无纸化。无纸化的交易方式一方面使信息传递摆脱书面文件的限制，更加有效率；另一方面也造成了法律制度的混乱。因为现行的法律法规多数是以"有纸交易"为出发点，并不适应跨境电子商务的"无纸化"交易。

跨境电子商务以"无纸化"交易方式代替了传统对外贸易中的书面文件（如书面合同、结算单据等）进行贸易往来，在这种无据可查的情况下，税务机关无法获知纳税人交易的真实情况从而增加了税务当局获取纳税人经营状况的难度，使得其中很大一部分税收流失，不利于国际的税收政策。例如，印花税作为各国普遍征收的传统税种之一，必须以交易双方提供的书面合同为依据进行征税。但是在"无纸化"的电子商务环境下，没有物质形态的法律合同和书面凭证，因而国家对印花税的征收便无章可循，无法可依。

四、跨境电子商务所面临的挑战

目前我国跨境电子商务的发展主要面临以下四个主要挑战。

（1）个人隐私和跨境数据传输方面的障碍。电子支付是国际电子商务进行的必要条件，也是交易得以实现的重要环节。如果没有第三方支付，就没有电子商务产业的今天。跨境电子商务也离不开第三方支付。因此，第三方支付的国际化已然成为占领未来消费者市场的重要条件。

目前，海外买家欺诈是中国中小外贸商户的心头大患。对于交易安全问题，调查显示，一半以上的受访商户表示担心与海外客户交易时遭遇欺诈，27%的商户担忧他们现在使用的支付系统在进行跨境交易时不够安全。此外，25%的商户认为未来三个月内海外买家拒绝支付的风险将增加。

对于电子商务应用在国际贸易中的安全性问题，应该紧随电子商务技术的进步，继续完善法律及各种规范性的措施，而各行业及相关机构应通过制定各种行业规范和完善认证体系来促进电子商务在国际贸易中安全性问题的解决。

（2）信用评价和标识亟待统一。电子商务具有虚拟的特点，它不仅具有传统商务活动的风险，而且还具有自己独特的开放性、全球性、低成本、高效率特点。交易双方的行为、市场中介的行为等都具有极大的不确定性，不守信用的行为在电子商务领域中更加突出。

而建立电子商务信用保障体系还存在着许多制约因素，如缺乏信用意识和信用道德规范、企业内部电子商务信用管理制度不健全、信用中介服务落后、缺乏有效的法律保障和奖惩机制等。跨境交易更加要求完善的、跨区域、跨文化、跨体制的信用体系来应对更加复杂的交易环境，对当事人的商业信用提出了更高的要求。

（3）国家还没有制定有关电子商务和跨境电子商务的法规。电子商务对国际贸易法律方面的冲击，主要是因为现今应用于国际贸易的法律不健全而造成的。我国相关法律制度的制定远远滞后于信息产业的发展。对于跨境电子商务服务业，目前我国只有《互联网信息服务管理办法》《电子签名法》等几部相关法律法规，对于跨境电子商务涉及的交易、税收以及消费者权益保障等方面都没有专门的规范和标准。所以，目前一个迫切需要解决的问题是制定一些相应的电子商务法律，以解决电子商务中发生的各种纠纷。

鉴于电子商务发展受到技术、信用水平的缺陷和法律滞后的阻碍，电子商务立法既要以确定性的安排弥补技术和信用的不足，又要给其发展创造宽松自由的环境。构建这样的法律体系需要制定新的法律，也要合理解释原有法律和创造有利于电子商务发展的配套法律规范。

2014年7月23日，海关总署公布了第56号公告《关于跨境贸易电子商务进出境货物、物品有关监管事宜的公告》，该文件是对前段时间电子商务跨境试点进行的总结梳理，文件明确了跨境电子商务的合法地位，强调后续口岸海关对电商需要进行系统监管的思路，这意味着跨境电子商务将从试点走向推广。

（4）跨境电子商务人才缺乏。与在国内利用电商相比，跨境电商支付、物流等要复杂得多，中小外贸企业在发展电商时也面临诸多风险。我国中小外贸企业由于规模小、实力不强、发展空间小，难以吸引相对紧缺的技术高、能力强的高级电子商务人才。而电子商务人才短缺，严重阻碍我国中小外贸电商的发展。培养人才这方面亟须进一步加强。

五、跨境电子商务的人才需求

与传统的国际贸易、外语外贸、电子商务、国际商务等专业的人才相比较，跨境电商人才不仅要掌握电子商务的专业知识，还需要有较高的外语水平，能及时熟练地与境外客户进行在线交流与谈判、能管理好多版本语言网店的相关知识；要掌握一定的国际贸易知识，能使用各种国际结算与国际物流方式；还要能对许多国家地区的文化、习俗、法律法规等有所了解，能利用最新的国际网络营销工具，基于社交媒体等与客户沟通交流，并进行境外客户的需求分析等。

从对相关企业的调研分析看，企业对跨境电商人才的职业能力要求主要包括：国际贸易技能、电子商务技能、国际物流操作技能、跨境网络营销技能、外语沟通表达技能以及较高的职业素养等。

（1）国际贸易技能。跨境电商企业的业务范围是全球化的，从业务性质上说属于国际贸易的范畴，这势必要求跨境电商的从业人员具有相当的外贸业务能力，即能熟知进出口与外贸业务流程及相关法律法规，熟练填制各种外贸单证，及时回应并处理境外客户订单与贸易纠纷；能积极开展网上交易、利用 EDI 通关、报检、退税，能跟进国际物流、国际保险、国际结算的相关业务处理。

（2）电子商务技能。跨境电商是利用互联网搭建的电子商务平台开展业务活动，从业务媒介上来看属于电子商务的范畴，这又要求从业人员具有一定的电子商务能力。要具有较强的电子商务信息检索、搜集、制作与发布的能力；较强的产品与服务的市场营销、网络营销能力；较强的小型电子商务系统的运行与维护的能力；初步的电子商务项目计划与管理能力；初步依据组织现状，制订跨境电子商务实施计划的能力；了解跨境电子商务的发展现状与发展趋势的能力。

（3）国际物流操作技能。跨境电商要求从业人员能够掌握跨境采购管理、供应链管理等知识，能熟练地运用现代物流技术从事国际物流业务的管理工作，了解各国的海关通关规则，为进出口货物选择合理运输方式，能跟踪签订运输合同或代理货物运输合同，并为跨境电商采购或销售的货物办理进出口货物的合理投保。

（4）跨境网络营销技能。跨境电商的人才培养要满足全球多层次的跨境网络营销技能需求。所以要求从业人员具有国际市场调研、预测的能力，能精准地进行跨境的网络策划、信息采集和大数据分析，最终在网络上开拓国际市场。

（5）外语沟通表达技能。跨境电商平台上的工作人员在操作过程中通常会涉及大量的外文资料，这就需要具备较高的外语读写能力，并能根据各国的商务礼仪处理往来函电；能在网络平台上用英语准确地描述产品细节，用外语流利地进行客户服务。同时，还要大量地收集了解境外各国的市场、文化、经济等现实发展环境的信息，能更地道地表达与沟通。

（6）职业素养。跨境电商业务面临的是来自全球各地的客户，是基于互联网的全球性开放贸易。跨境电商人员要面临多变复杂的国际商务环境，这势必要求从业人员具有灵活应变的处事谈判能力、良好的沟通应变能力、踏实认真的工作态度与积极合作的团队意识，能了解国际商务法律法规，妥善处理跨境业务纠纷等。

第二节　全球主要的跨境电商平台

目前，全球主要的跨境电商平台有全球速卖通（AliExpress）、Wish、敦煌网（DHgate）、eBay 和亚马逊（Amazon）等。

一、全球速卖通（**AliExpress**）

全球速卖通（AliExpress）于 2010 年 4 月正式上线，是阿里巴巴为帮助中小企业接触海外终端，拓展利润空间而全力打造的融订单、支付、物流于一体的外贸在线交易平台，被广大卖家称为"国际版淘宝"。目前全球速卖通是中国最大的 B2C 交易平台，覆盖 220 多个国家和地区。每天海外买家的流量超过 5 000 万。

从 2010 年上线截止到 2014 年，AliExpress 每年成交额保持 300% 到 500% 增长，在线商品数量已达到亿级，订单成功覆盖全球 220 多个国家和地区，平台卖家 20 多万个，注册的速卖通账号包含未开店的已接近 200 万个。其中 2014 年"双十一"，第一次参加全球化

"双十一"，24小时创下684万笔交易订单，当天有效订单覆盖211个国家和地区。而流量排名方面，购物类网站排名：巴西、俄罗斯、土耳其第二，美国第五，西班牙第一，印尼第六。访问流量每月大概在600M，占流量总和的50%左右。

全球速卖通首页如图10-1所示。

图10-1　全球速卖通首页

二、Wish

Wish于2011年12月1日成立于美国旧金山硅谷，是一款基于移动端App的商业平台。刚开始Wish平台主要是用来交流和分享，并不涉及商品交易。2013年3月，Wish在线交易平台正式上线，移动APP于同年6月推出，当年年经营收益即超过1亿美元，成为跨境电商移动端平台的黑马。

Wish目前有1亿的注册用户，日活跃用户为120万，为家庭年收入在6.5万美元的人群。Wish的主力消费群年龄是15～28岁，男女占比3∶7。Wish商户的比例是亚洲对欧洲81∶19的比例，也就是说Wish是中国卖家的天下。Wish客户在美国的下载量排名第一。而其创始人被评为最佳创始人之一，APP被评为Android最好用的APP。

在Wish平台上对比其他电商平台如亚马逊、eBay、全球速卖通等，Wish不会通过关键词额外收取费用来向用户推荐商品。Wish的系统是通过买家行为等数据的计算，判断买家的爱好、感兴趣的产品信息，并且选择相应的产品推送给买家。

Wish手机应用页面如图10-2所示。

图10-2　Wish首页

三、敦煌网（DHgate）

敦煌网（www.dhgate.com）成立于 2004 年，是
中国第一个 B2B 跨境电子商务平台，致力于帮助中国中小企业通过电子商务平台走向全球
市场。目前拥有 120 多万家国内供应商，3 000 多万种商品，遍布全球 224 个国家和地区，
达到 1 000 万买家在线购买的规模。每小时 10 万买家实时在线采购，每 3 秒产生一张订单。
敦煌网主要采取佣金制、增值服务、广告等赢利模式。敦煌网的优势项目为手机和电子
产品。

敦煌网秉承"促进全球通商、成就创业梦想"的使命，帮助中国中小企业大规模
地应用电子商务平台，为中国外贸寻求新的发展空间，开拓外贸交易的新局面。敦煌
网为中小企业客户带来的价值体现在两个方面：一是"只为成功付费"，使中小企业只
为成功的交易支付小额佣金，不需投入高额年费；二是敦煌网为企业客户寻找海外买
家，为客户解决海外营销的难题。同时，敦煌网拥有专业的外籍员工营销团队在 Face-
book、YouTube、Google 等做推广，利用最新的热门营销手段——社区化营销为中小企
业客户找到海外买家。

敦煌网首页如图 10-3 所示。

图 10-3　敦煌网首页

四、eBay

eBay 是一个可让全球民众上网买卖物品的线上拍卖及购物网站。于 1995 年 9 月 4 日由
Pierre Omidyar 以 Auctionweb 的名义创立于加利福尼亚州圣荷西。eBay 的最初目的是帮助创
始人的女友在全美寻找 Pez 糖果盒爱好者进行交流，没想到 eBay 很快就受到很多人欢迎。
成立之初，eBay 就定位于全球网民买卖物品的线上拍卖及购物网站。

在 eBay 平台上卖家发布的产品主要有两种销售方式："拍卖"和"一口价"。不同的销
售方式 eBay 向卖家收取的费用不同，通常情况是按照"刊登费"加上"成交费"计算，
即：产品发布费用和成交佣金。

eBay 首页如图 10-4 所示。

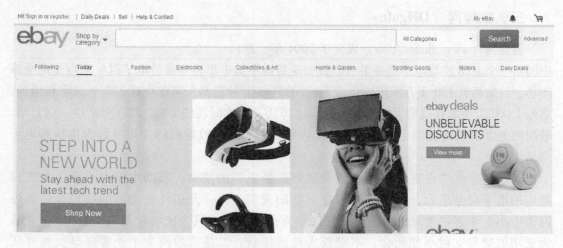

图 10-4　eBay 首页

五、亚马逊（Amazon）

亚马逊（Amazon）是美国最大的电子商务公司，成立于 1995 年，初期定位是一个销售书籍和音像制品的网络书店，1997 年转变成为最大的综合网络零售商。2015 年，在全球市值最高的 20 家互联网公司排名中位列第四位。

Amazon 目前主要经营全新、翻新以及二手的商品，包括图书、数码、家居、厨具、家用电器、美妆、食品、酒水、母婴、玩具、服装、汽车用品等。

亚马逊首页如图 10-5 所示。

图 10-5　亚马逊首页

本章小结

开展国际贸易现在还可以通过互联网和广泛地利用各种 B2B、B2C 的平台。利用跨境电子商务平台会有很多好处，我们应该熟悉国际上主要的跨境电商平台，并初步学会应用这些平台来推广国际贸易商品。

附　录

序号	单据名称	英文名称	单据类别	页码
1	信用证	Letter of Credit，L/C	信用证单证	
2	商业发票	Commercial Invoice		
3	装箱单	Packing List		
4	提单	Bill of Lading，B/L		
5	提单附带声明和证明	Declaration and Certificate Appended to the B/L		
6	保险单	Insurance Policy		
7	保险公司声明	Underwriter's Declaration		
8	汇票	Bill of Exchange		
9	受益人证明书	Beneficiary's Certificate		
10	一般原产地证书	Certificate of Origin，C. O.		
11	销售确认书	Sales Confirmation	报检、托运、报关单据	
12	销售合同	Sales Contract		
13	出境货物报检单	Application of Inspection		
14	货物出运委托书	Transport Entrustment		
15	进出口货物报关单	Import and Export Declaration		
16	投保单	Application for Export/Import Insurance		
17	普惠制原产地证明书表格 A	Generalized System of Preferences Certificate of Origin Form A	其他单据	
18	出口许可证	Export License		
19	订单	Purchase Order		
20	海关发票	Customs Invoice		

1. 信用证

MT 700		ISSUE OF A DOCUMENTARY CREDIT
SENDER		HSBC BANK PLC, DUBAI, U. A. E.
RECEIVER		HANGZHOU CITY COMMERCIAL BANK, HANGZHOU, CHINA
SEQUENCE OF TOTAL	27:	1 / 1
FORM OF DOC. CREDIT	40A:	IRREVOCABLE
DOC. CREDIT NUMBER	20:	FFF07699
DATE OF ISSUE	31C:	150225
APPLICABLE RULES	40E:	UCP LATEST VERSION
DATE AND PLACE OF EXPIRY	31D:	DATE 080330PLACE IN U. A. E.
APPLICANT	50:	DIM TRADING CO. , LTD. 16 TOM STREET, DUBAI, U. A. E.
BENEFICIARY	59:	ZHEJIANG JINYUN IMPORT & EXPORT CO. , LTD. 118 XUEYUAN STREET, HANGZHOU, P. R. CHINA
AMOUNT	32B:	CURRENCY USD AMOUNT 54450. 00
AVAILABLE WITH/BY	41D:	ANY BANK IN CHINA, BY NEGOTIATION
DRAFTS AT…	42C:	60 DAYS AFTER SIGHT
DRAWEE	42A:	HSBC BANK PLC, NEW YORK
PARTIAL SHIPMENT	43P:	PROHIBITED
TRANSSHIPMENT	43T:	ALLOWED
PORT OF LOADING/ AIRPORT OF DEPARTURE	44E:	CHINESE MAIN PORT
PORT OF DISCHARGE	44F:	DUBAI, U. A. E.
LATEST DATE OF SHIPMENT	44C:	150325
DESCRIPTION OF GOODS AND/OR SERVICES.	45A:	4500 PIECES OF LADIESJACKET, SHELL: WOVEN TWILL 100% COTTON, LINING: WOVEN 100% POLYESTER, ORDER NO. DIM768, AS PER S/C NO. ZJJY0739 STYLE NO.　QUANTITY　UNIT PRICE　　AMOUNT L357　　　　2250PCS　　USD12. 10/PC　　USD27225. 00 L358　　　　2250PCS　　USD12. 10/PC　　USD27225. 00 AT CIF DUBAI, U. A. E.
DOCUMENTS REQUIRED	46A:	
		+ COMMERCIAL INVOICE SIGNED IN TRIPLICATE.
		+ PACKING LIST IN TRIPLICATE.
		+ CERTIFICATE OF CHINESE ORIGIN CERTIFIED BY CHAMBER OF COMMERCE OR CCPIT.

		+ INSURANCE POLICY/CERTIFICATE IN DUPLICATE ENDORSED IN BLANK FOR 110% INVOICE VALUE, COVERING ALL RISKS OF CIC OF PICC (1/1/1981) INCL. WAREHOUSE TO WAREHOUSE AND I. O. P AND SHOWING THE CLAIMING CURRENCY IS THE SAME AS THE CURRENCY OF CREDIT. + FULL SET (3/3) OF CLEAN "ON BOARD" OCEAN BILLS OF LADING MADE OUT TOORDER MARKED FREIGHT PREPAID AND NOTIFY APPLICANT. + SHIPPING ADVICE SHOWING THE NAME OF THE CARRYING VESSEL, DATE OF SHIPMENT, MARKS, QUANTITY, NET WEIGHT AND GROSS WEIGHT OF THE SHIPMENT TO APPLICANT WITHIN 3 DAYS AFTER THE DATE OF BILL OF LADING.
ADDITIONAL CONDITION	47A:	+ DOCUMENTS DATED PRIOR TO THE DATE OF THIS CREDIT ARE NOT ACCEPTABLE. + THE NUMBER AND THE DATE OF THIS CREDIT AND THE NAME OF ISSUING BANK MUST BE QUOTED ON ALL DOCUMENTS. + TRANSSHIPMENT ALLOWED. + SHORT FORM/CHARTER PARTY/THIRD PARTY BILL OF LADING ARE NOT ACCEPTABLE. + SHIPMENT MUST BE EFFECTED BY 1 ×40' FULL CONTAINER LOAD. + THE GOODS SHIPPED ARE NEITHER ISRAELI ORIGIN NOR DO THEY CONTAIN ISRAELI MATERIALS NOR ARE THEY EXPORTED FROM ISRAEL, BENEFICIARY'S CERTIFICATE TO THIS EFFECT IS REQUIRED. + ALL PRESENTATIONS CONTAINING DISCREPANCIES WILL ATTRACT A DISCREPANCY FEE OF USD60. 00 PLUS TELEX COSTS OR OTHER CURRENCY EQUIVALENT. THIS CHARGE WILL BE DEDUCTED FROM THE BILL AMOUNT WHETHER OR NOT WE ELECT TO CONSULT THE APPLICANT FOR A WAIVER
CHARGES	71B:	ALL CHARGESOUTSIDE DUBAI ARE FOR ACCOUNT OF BENEFICIARY INCCOUNT.
PERIOD FOR PRESENTATION	48:	WITHIN 5 DAYS AFTER THE DATE OF SHIPMENT, BUT WITHIN THE VALIDITY OF THIS CREDIT.
CONFIRMATION INSTRUCTION	49:	WITHOUT
REIMBURSING BANK	53A:	HSBC BANK PLC, NEW YORK
INFORMATION TO PRESENTING BANK	78:	ALL DOCUMENTS ARE TO BE REMITTED IN ONE LOT BY COURIER TO HSBC BANK PLC, TRADE SERVICES, DUBAI BRANCH, P O BOX 66, HSBC BANK BUILDING 312/45 Al SUQARE ROAD, DUBAI, UAE.

2. 商业发票

上海进出口贸易公司

SHANGHAI IMPORT & EXPORT TRADE CORPORATION.

1321ZHONGSHAN ROAD SHANGHAI, CHINA

COMMERCIAL   INVOICE

TEL : 021-65788877

FAX : 021-65788876

INV NO : TX0522

DATE: JUN. 01. 2006

S/C NO : TXT264

L/C NO: XT173

TO:

TKAMLA CORPORATION

6-7,KAWARA MACH OSAKA

JAPAN

FROM ____SHANGHAI PORT_____ TO _____OSAKA PORT_____

MARKS & NO	DESCRIPTIONS OF GOODS	QUANTITY	U/ PRICE	AMOUNT
T. C TXT264 OSAKA C/NO. 1-66	CHINESE GREEN TEA ART NO.555 ART NO.666 ART NO.777 Packed in 66 cartons	100 KGS 110 KGS 120 KGS	CIF OSAKA USD 110.00 USD 100.00 USD 90.00	USD 1100.00 USD 1100.00 USD 10800.00 USD 32800.00

TOTAL AMOUNT: SAY US DOLLARS THIRTY TWO THOUSAND EIGHT HUNDRED
ONLY.

WE HEREBY CERTIFY THAT THE CONTENTS OF INVOICE HEREIN ARE TRUE
AND CORRECT.

3. 装箱单

GUANGDONG HUANYA IMPORT AND EXPORT CO. , LTD.
118 XUEYUAN STREET, GUANGZHOU, P. R. CHINA
TEL：0086 – 020 – 86739178　　　FAX：0086 – 020 – 86739178

PACKING LIST		

To：	DIM TRADING CO. , LTD. 16 TOM STREET, DUBAI, U. A. E.	Invoice No. :	GH08018
		Invoice Date：	APR. 10, 2008
		S/C No. :	GDHY0 739
		S/C Date：	FEB. 15, 2 015

From：	GUANGZHOU, CHINA	To：	DUBAI, U. A. E.
Letter of Credit No. :	FFF07 699	Issued By：	HSBC BANK PLC, DUBAI, U. A. E.
Date of Issue：	FEB. 25, 2015		

Marks and Numbers	Number and kind of package Description of goods	Quantity	Package	G. W	N. W	Meas.
DIM S/C No. : ZJJY0 739 Style No. : L357/L358 Port of destination： DUBAI, U. A. E. Carton No. : 1 – 502	LADIESJACKET Style No. L357 Style No. L358 PACKED IN9 PCS/CTN, SHIPPED IN40' FCL.	2 250 PCS 2 268 PCS	250 CTNS 252 CTNS	2 500 KGS 2 520 KGS	2 250 KGS 2 268 KGS	29. 363 M^3 29. 597 M^3
TOTAL：		4 518 PCS	502 CTNS	5 020 KGS	4 518 KGS	58. 96 M^3
SAY TOTAL：	FIVE HUNDRED AND TWO CARTONS ONLY.					

4. 提单

BILL OF LADING

1)SHIPPER Shanghai Hongya Imp. & Exp. Corp. Room 705E, 668, Beijing (E) Road, Shanghai, China		10)B/L NO. DDL 478388	
		CARRIER	
2)CONSIGNEE To the Order of Shippers			
3)NOTIFY PARTY Kamlar Trading Stores, P.O.Box 108, Dubai		中国远洋运输（集团）总公司	
4)PLACE OF RECEIPT	5)OCEAN VESSEL Dingyuan	CHINA OCEAN SHIPPING (GROUP) CO.	
6)VOYAGE NO. V.380	7)PORT OF LOADING Shanghai, China	ORIGINAL	
8)PORT OF DISCHARGE Dubai	9)PLACE OF DELIVERY	Combined Transport BILL OF LADING	

11)MARKS	12) NOS. & KINDS OF PKGS.	13)DESCRIPTION OF GOODS	14) G.W.(kg)	15) MEAS(m³)
ARYA NTS0505 DUBAI NO. 1-660	660 cartons	"ARYA" BRAND VACUUM FLASKS FREIGHT PREPAID ON BOARD We hereby certify that the shipment is effected in 40' H.Q. containers.	15120 kg	84.780 m³

16)TOTAL NUMBER OF CONTAINERS OR PACKAGES(IN WORDS)	Say Six Hundred and Sixty Cartons Only				
FREIGHT & CHARGES	REVENUE TONS	RATE	PER	PREPAID	COLLECT
PREPAID AT	PAYABLE AT		17)PLACE AND DATE OF ISSUE SHANGHAI 17-Jul-05		
TOTAL PREPAID	18)NUMBER OF ORIGINAL B(S)/L 3 (THREE)		(21) 上海中远货运服务代理有限公司 COSCO SHANGHAI CONTAINER SHIPPING AGENCY CO., LTD		
LOADING ON BOARD THE VESSEL					
19)DATE 17-Jul-05	20)BY 上海中远货运服务代理有限公司 COSCO SHANGHAI CONTAINER SHIPPING AGENCY CO., LTD AS AGENT for to Carrier, China Ocean Shipping (Group) Co.		AS AGENT for to Carrier, China Ocean Shipping (Group) Co.		

提单背面:
Delivered to the National Bank of Dubai PJSC., Dubai
Shanghai Hongya Imp. & Exp. Corp.

奇少贵

提交份数: 3份

5. 提单附带声明和证明

DECLARATION TO BILL OF LADING

TO WHOM IT MAY CONCERN:

RE: 384 CTNS OF MEN'S VELVET SHOES UNDER B/L NO. KGDA—6 009 T CY – CY L/C NO. AN – 29 854

(1) NAME OF VESSEL: SUI SUN; PREVIOUS NAME: SUI SUN.

(2) NATIONALITY OF VESSEL: CHINA.

(3) OWNER OF VESSEL: CHINA OCEAN SHIPPING (GROUP) COMPANY

(4) THE VESSEL WILL CALL OR PASS THROUGH THE FOLLOWING PORTS ENROUTE TO SAUDI ARABIA:

1. HONG KONG;　　2. SINAGPORE;　　3. BOMBAY;　　4. DUBAI.

WE DECLARE THAT THE INFORMATION PROVIDED IN RESOIBSES TO (1) TO (4) ABOVE IS CORRECT AND COMPLETE AND THAT THE VESSEL SHALL NOT CALL AT OR ANCHOR ON ANY OTHER PORTS THAN THE A. M. ENROUTE TO SAUDI ARABIA.

WRITTEN ON THE 16th DAY OF DEC. , 20 XX.

SWORN TO BEFORE ME ON THE 16 DAY OF DEC. , 20 XX.

PACIFIC INTERNATIONAL LINES (PTE) LTD

× × ×

AS CARRIER

CERTIFICATE

TO WHOM IT MAY CONCERN:

RE: 384 CTNS OF MEN' S VELVET SHOES UNDER B/L NO. KGDA—6 009 T CY – CY L/C NO. AN – 29 854

THIS IS TO CERTIFY THAT THE SHIPMENT HAS BEEN EFFECTED BY REGULAR LINE VESSEL.

WRITTEN ON THE 16th DAY OF DEC. , 20 XX.

PACIFIC INTERNATIONAL LINES (PTE) LTD

× × ×

AS CARRIER

6. 保险单

中国平安保险股份有限公司
PING AN INSURANCE COMPANY OF CHINA，LTD.

NO. 1000005959   货物运输保险单
CARGO TRANSPORTATION INSURANCE POLICY

NO. S OF ORIGIAL：TWO（2）

被保险人：CHINA NATIONAL LIGHT IND. PRODUCTS I/E CORP.，NINGBO BRANCH

    中国平安保险股份有限公司根据被保险人的要求及其所交付约定的保险费，按照本保险单背面所载条款与下列条款，承保下述货物运输保险，特立本保险单。

This policy of Insurance witnesses that PING AN INSURANCE COMPANY OF CHINA，LTD.，at the request of the insured and in consideration of the agreed premium paid by the Insured，undertakes to insure the undermentioned goods in transportation subject to the conditions of policy as per the clauses printed overleaf and other special clauses attached hereon.

保单号赔款偿付地点

Policy No. _____   Claim Payable at NEW YORK IN USD

发票或提单号

Invoice No. or B/L No.   INVOICE NO. GMS – 025

运输工具查勘代理人

per conveyance S. S. MAYER V. 225 Survey By：PICC NEW YORK BRANCH，

                                 666 CARNEY STREET, NEW YORK, U. S. A.

                                 001 – 65432143

起运日期

Slg. onor abt.   AS PER B/L

自经至

From NINGBO _____   VIA To   NEW YORK _____

保险金额

Amount Insured

USD 55 000. 00 （SAY US DOLLARS FIFTY – FIVE THOUSAND ONLY.

保险货物项目、标记、数量及包装：承保条件

Description，Marks，Quantity &Packing of Goods：   Conditions：

PEN, AS PER INV. NO. GMS – 025, 110 CTNSCOVERING MARINE INSTITUTE CARGO CLAUSES（ALL RISKS）AND WAR CLAUSES AND SRCC FOR 110% INVOICE VALUE.

签单日期

Date：OCT. 15th, 2015 _____   For and on behalf of

PING AN INSURANCE COMPANY OF CHINA，LTD.   authorized signature 签名

7. 保险公司声明

DECLATATION

POLICY NO.　HK11/P×× 200224 WUHAN, CHINA DEC. 12, 20× ×

DOC. CREDIT NUMBER：× ×IM27 110 TOC

(1) NAME OF ITS INSURANCE COMPANY：THE PEOPLE'S INSURANCE COMPANY OF CHINA

(2) ADDRESS OF ITS PRINCIPAL OFFICE：410 FU CHENG MEN NEI DA JIE, BEIJING, CHINA

(3) COUNTRY OF ITS INCORPORATION：THE PEOPLE'S REPUBLIC OF CHINA

WE CERTIFY THAT THE SAID COMPANY HAS A DULY QUALIFIED AND APPOINTED AGENT IN SAUDI ARABIA WHOSE FULL NAME AND ADDRESS APPEARS BELOW：

CARGO SURVEYING AGENT：

ARABIAN INSPECTION & SURVEY CO., LTD.

8th FLOOR, NEW BAKHASHAB BUILDING BAB MECCA,

P. O. BOX：832, SAUDI ARABIA

WRITTEN ON THE 12th DAY OF DEC., 20× ×

SWORN TO BEFORE ME ON THE 12th DAY OF DEC., 20× ×

THE PEOPLE'S INSURANCE COMPANY OF CHINA

×××

（GENERAL MANAGER）

8. 汇票

号码	汇票金额	中国上海	年 月 日
No.　4728294	Exchange for　US$35,000.00	Shanghai, China,	SEP. 08　19 93

见票

At　45 DAYS AFTER　日后（本汇票之副本未付）付交　sight of this FIRST of Exchange (Second of Exchange, being unpaid) Pay to the order of　BANK OF CHINA SHANGHAI BRANCH

金　额
the sum of　US DOLLARS THIRTY FIVE THOUSAND ONLY.

against shipment of :- SPORTING GOODS UNDER INVOICE NO. 4728294

此致
To　CHENG HO INTERNATIONAL CO., LTD.

313 MOUNT STREET DAYTON, OHIO

上海市文教体育用品进出口公司
SHANGHAI STATIONERY & SPORTING GOODS
IMPORT & EXPORT CORP.

俞培林

9. 受益人证明书

BENEFICIARY'S　CERTIFICATE

DATE：NOV. 22, 2015

TO：GLOBAL RESOURCE CO. LTD.　（0081 − 2 − 7871082）

ATTN：MR. KAZI ROLL,

RE：100 CASES（100 M/T）ANTIMONY INGOT UNDER S/C NO.　008 AND L/C NO.　YJQ00 819 AND B/L NO.　JX00 918 AND INVOICE NO. M198 AND INVOICE VALUE USD 400, 000.

WE HEREBY CERTIFY THAT ONE SET OF NON − NEGOTIABLE SHIPPING DOCUMENTS PLUS ONE ORIGINAL BILL OF LADING HAVE BEEN SENT BY EMS DIRECTLY TO APPLICANT WITHIN 7 DAYS AFTER SHIPMENT.

WUZHOU METALS & MINERALS IMPORT & EXPORT CORPORATION

XIAOMO

10. 一般原产地证明书

1. Exporter GUANGZHOU KNITWEAR AND MANUFACTURE GOODS IMPORT AND EXPORT TRADE CORPORA-TION. 321, BEIJING ROAD, GUANGZHOU, CHINA	Certificate No. 201 586 **CERTIFICATE OF ORIGIN** **OF** **THE PEOPLE'S REPUBLIC OF CHINA**
2. Consignee YOUNG TRADING CORPORTION. 88 MARSHALL AVE DONCASTER VIC 3 180, CANADA	
3. Means of transport and route FROM GUANGZHOU TO MONTREAL PORT BY SEA	5. For certifying authority use only
4. Country/region of destination CANADA	

6. Marks and numbers	7. Number and kind of packages; description of goods	8. H. S code	9. Quantity	10. Number and date of invoices
YOUNG MONTREAL C/NO. 1 – 360	THREE HUNDRED AND SIXTY (360) CARTONS COTTON DISH TOWELS ******************	8 204. 666 1	G. W. 19. 911 KGS 33 350 PAIRS	NO. MN8 866 SEP 28th, 2015

11. Declaration by the exporter The undersigned hereby declares that the above details and statements are correct; that all the goods were produced in china and that they comply with the rules of origin of the People's Republic of China. GUANGZHOU KNITWEAR AND MANUFACTURE GOODS IMPORT AND EXPORT TRADE CORPORATION. (STAMP) OCT 15th, 2015, GUANGZHOU 张三 --------------------------- Place and date, signature and stamp of certifying authority	12. Certification It is here by certified that the declaration by the exporter is correct. 中国国际贸易促进委员会 单据证明专用章 (粤) CHINA COUNCIL FOR THE PROMOTION OF INTERNATIONAL TRADE. (STAMP) OCT 16th, 2015, GUANGZHOU 李四 --------------------------- Place and date, signature and stamp of certifying authority

11. 销售确认书

中国国际纺织品进出口公司江苏分公司
CHINA INTERNATIONAL TEXTILES I/E CORP.　JIANGSU BRANCH
20 RANJIANG ROAD，NANJING，JIANGSU，CHINA

销售确认书编号 NO.：CNT0 219

SALES CONFIRMATION 日期 DATE：MAY 10th，2015

OUR REFERENCE：IT123 JS

买方

BUYERS：TAI HING LOONG SDN，BHD，KUALA LUMPUR.

地址

ADDRESS：7/F，SAILING BUILDING，NO.50 AIDY STREET，KUALA LUMPUR，MALAYSIA

电话传真

TEL：060 – 3 – 74 236 211　　　　　　FAX：060 – 3 – 74 236 212

兹经买卖双方同意成交下列商品，订立条款如下：

　　THE UNDERSIGNED SELLERS AND BUYERS HAVE AGREED TO CLOSE THE FOLLOWING TRANSACTION ACCORDING TO THE TERMS AND CONDITIONS STIPULATED BELOW：

DESCRIPTION OF GOODS	QUANTITY	UNIT PRICE	AMOUNT
		CIF SINGAPORE	
100% COTON GREY LAWN	300 000 YARDS	@ HKD 3.00 PER YARD	HKD 900 000.00

（船运）

装运 SHIPMENT：　　　DURING JUNE/JULY，2015 IN TRANSIT TO MALAYSIA

付款条件 PAYMENT：　IRREVOCABLE SIGHT L/C

保险 INSURANCE：　　TO BE EFFECTED BY SELLERS COVERING WPA AND WAR RISKS FOR 10% OVER THE INVOICE
　　　　　　　　　　VALUE

买方（签章）THE BUYER　　　　　　　　　卖方（签章）THE SELLER

TAI HING LOONG SDN，BHD，KUALA LUMPUR.　　中国国际纺织品进出口公司江苏分公司

　　　　　　　　　　　　　　　　　　　　　CHINA INTERNATIONAL TEXTILES I/E CORP.　JIANGSU BRANCH

12. 销售合同

SALES CONTRACT

NO.: GDHY0739   DATE: FEB. 15, 2015

THE SELLER: GUANGDONG HUANYA IMPORT & EXPORT CO., LTD.

                118 XUEYUAN STREET, GUANGZHOU,

                P. R. CHINA

THE BUYER: DIM TRADING CO., LTD.

                16 TOM STREET, DUBAI,

                U. A. E.

This Contract is made by and between the Buyer and Seller, whereby the Buyer agree to buy and the Seller agree to sell the undermentioned commodity according to the terms and conditions stipulated below:

Commodity & specification	Quantity	Unit price	Amount
Ladies Jacket (6204320090)		CIF Dubai, U. A. E.	
Style No. L357	2 250pcs	USD 12. 00/pc	USD 27 000. 00
Style No. L358	2 250pcs	USD 12. 00/pc	USD 27 000. 00
Shell: Woven twill 100% cotton			
Lining: Woven 100% polyester			
As per the confirmed sample of Jan. 30, 2008 and Order No. DIM768			
TOTAL	4 500pcs		USD 54 000. 00
TOTAL CONTRACT VALUE: SAY U. S. DOLLARS FIFTY-FOUR THOUSAND ONLY.			

Size/color assortment for Style No. L357:   Unit: piece

Size	S	M	L	XL	Total
White	180	360	450	180	1 170
Red	180	360	360	180	1 080
Total	360	720	810	360	2 250

Size/color assortment for Style No. L358:   Unit: piece

Size	S	M	L	XL	Total
White	180	360	450	180	1 170
Blue	180	360	360	180	1 080
Total	360	720	810	360	2 250

More or less 5% of the quantity and the amount are allowed.

    PACKING: 9 pieces of ladies jackets are packed in one export standard carton, solid color and solid size in the same carton.

    MARKS:

Shipping mark includes DIM, S/C no., style No., port of destination and carton No.

Side mark must show the color, the size of carton and pieces per carton.

TIME OF SHIPMENT:

Within 60 days upon receipt of the L/C which accord with relevant clauses of this Contract.

PORT OF LOADING AND DESTINATION:

From Guangzhou, China to Dubai, U. A. E.

Transshipment is allowed and partial shipment is prohibited.

INSURANCE: To be effected by the seller for 110% of invoice value covering All Risks as per CIC of PICC dated 01/01/1981.

TERMS OF PAYMENT: By irrevocable Letter of Credit at 30 days after sight, reaching the seller not later than Mar. 5th, 2015 and remaining valid for negotiation in China for further 15 days after the effected shipment. In case of late arrival of the L/C, the seller shall not be liable for any delay in shipment and shall have the right to rescind the contract and /or claim for damages.

DOCUMENTS:

+ Signed Commercial Invoice in triplicate.

+ Full set of clean on board ocean Bill of Lading marked "freight prepaid" made out to order of shipper blank endorsed notifying the applicant.

+ Insurance Policy in duplicate endorsed in blank.

+ Packing List in triplicate.

+ Certificate of Origin certified by Chamber of Commerce or CCPIT.

INSPECTION:

The certificate of Quality issued by the China Entry – Exit Inspection and Quarantine Bureau shall be taken as the basis of delivery.

CLAIMS:

In case discrepancy on the quality or quantity (weight) of the goods is found by the buyer, after arrival of the goods at the port of destination, the buyer may, within 30 days and 15 days respectively after arrival of the goods at the port of destination, lodge with the seller a claim which should be supported by an Inspection Certificate issued by a public surveyor approved by the seller. The seller shall, on the merits of the claim, either make good the loss sustained by the buyer or reject their claim, it being agreed that the seller shall not be held responsible for any loss or losses due to natural cause failing within the responsibility of Ship – owners of the Underwriters. The seller shall reply to the buyer within 30 days after receipt of the claim.

LATE DELIVERY AND PENALTY:

In case of late delivery, the Buyer shall have the right to cancel this contract, reject the goods and lodge a claim against the Seller. Except for Force Majeure, if late delivery occurs, the Seller must pay a penalty, and the Buyer shall have the right to lodge a claim against the Seller. The rate of penalty is charged at 0. 5% for every 7 days, odd days less than 7 days should be counted as 7 days. The total penalty amount will not exceed 5% of the shipment value. The penalty shall be deducted by the paying bank or the Buyer from the payment.

FORCE MAJEURE：

The seller shall no thold responsible if they, owing to Force Majeure cause or causes, fail to make delivery within the time stipulated in the Contract or cannot deliver the goods. However, in such a case, the seller shall inform the buyer immediately by cable and if it is requested by the buyer, the seller shall also deliver to buyer by registered letter, a certificate attesting the existence of such a cause or causes.

ARBITRATION：

All disputes in connection with this contract or the execution there of shall be settled amicably by negotiation. In case no settlement can be reached, the case shall then be submitted to the China International Economic Trade Arbitration Commission for settlement by arbitration in accordance with the Commission's arbitration rules. The award rendered by the commission shall be final and binding on both parties. The fees for arbitration shall be borne by the losing party unless otherwise awarded.

This contract is made in two original copies and becomes valid after signature, one copy to be held by each party.

Signed by：

THE SELLER：

GUANGDONG HUANYA IMPORT & EXPORT CO. , LTD.

张立 *Jack chang*

THE BUYER：

DIM TRADING CO. , LTD

13. 出境货物报检单

中华人民共和国出入境检验检疫
出境货物报检单

报检单位（加盖公章）：中韩合资大连海天服装有限公司　　　　　　　　　　　　　* 编号_____

报检单位登记号：联系人：电话：报检日期：2015 年 3 月 16 日前

发货人	（中文）中韩合资大连海天服装有限公司					
	（外文）DALIAN HAITIAN GARMENT CO. LTD					
收货人	（中文）/					
	（外文）WAN DO APPAREL CO. LTD					

货物名称（中/外文）	H. S. 编码	产地	数/重量	货物总值	包装种类及件数
男女羊套衫毛		CHINA	1 300 PCS	USD 14 300.00	260 CTNS
LADY'S JUMPER	6 201. 931 0		1 300 PCS	USD 14 300.00	
MAN'S JUMPER	6 202. 131 0				

运输工具名称号码		DAIN/431 E	贸易方式	加工贸易	货物存放地点	大连
合同号		9911113	信用证号	/	用途	
发货日期	2015. 03. 26	输往国家（地区）	韩国	许可证/审批号		/
启运地	DALIAN，CHINA	到达口岸	仁川	生产单位注册号		/
集装箱规格、数量及号码			1X40'：EASU9608490			

合同、信用证订立的检验检疫条款或特殊要求	标记及号码	随附单据（画"√"或补填）	
QUALITY INSPECTION CERTIFICATES ISSUED BY CCIC	与发票同	□合同√ □信用证 □发票√ □换证凭单 □装箱单	□厂检单 □包装性能结果单 □许可/审批文件 □ □

需要证单名称（划"√"或补填）		* 检验检疫费
□品质证书√ □重量证书 □数量证书 □兽医卫生证书 □健康证书 □卫生证书	□动物卫生证书 □植物检疫证书 □熏蒸/消毒证书 □出境货物换证凭单 □通关单	总金额 （人民币元） 计费人 收费人

报检人郑重声明：	领取证单
1. 本人被授权报验。	日期
2. 上列填写内容正确属实，货物无伪造或冒用他人的厂名、标志、认证标志，并承担货物质量责任。 签名：×××	签名

14. 货物出运委托书

Shipper（发货人）ZHENGCHANG TRADE CO. LTD NO. 168 XUSHI ROAD HUZHOU ZHEJIANG				D/R NO.（编号）	
Consignee（受货人）TO ORDER OF THE BANK OF GOOD COLOMBO				集装箱货物托运单	
Notify Party（通知人）ELECTRADE THE FIRST STREET COLOMBO – 11 SRI LANKA					
Pre – Carriage By（前程运输）　　place of Receipt（收货地点）					
Ocean Vessel（船名）　Voy No.（航次）　Port of Loading （装货港）SHANGHAI，CHINA					
Port of Discharge（卸货港）COLOMBO, SRI LANKA		Place Of Delivery（交货地点）		Final Destination（目的港）COLOMBO，SRI LANKA	
Container No.（集装箱号）	Seal No. （封志号）Marks &Nos. （标志与号码）E. L. E.HZ0 114Colombo C/No. 1 – 280	No. of Containers or Pkgs（箱数或件数）280 CTNS	Kind of packages；description of goods（包装种类与货名）CANDLE LAMPSFREIGHT PREPAID	Gross Weight（毛重/千克）3 500 KGS	Measurement（尺码/立方米）21. 00 M³
TOTAL NUMBER OF CONTAINERS OF PACKAGES（IN WORDS）集装箱数或件数合计（大写）			SAY TWO HUNDRED AND EIGHTY CARTONS ONLY		
Freight & Charges（运费与附加费）	Revenue Tons（运费吨）	Rate（运费率）	Per（每）	Prepaid（运费预付）	Collect（到付）
Ex Rate（兑换率）	Prepaid at预付地点 SHANGHAI	Payable at（到付地点）		Place of Issue（签发地点）SHANGHAI，　CHINA	
	Total Prepaid（预付总额）	No. of Original B（S）L（正本提单份数）THREE			
Service Type on Receiving CY CFS DOOR	Service Type on Delivery CY CFS DOOR		Reefer – Temperature Required（冷藏温度）	F	C
Type Of Goods（种类）	Ordinary，　Reefer，　Dangerous，　Auto（普通）　（冷藏）　（危险品）（裸装车辆）Liquid ，　　Live animal，　Bulk（液体）　　（活动物）　（散货）			危险品	ClassProperty
可否转船 NO		可否分批 NO			
装期 FEB. 28th, 2015		有效期 MARCH 20th , 2015			
金额 USD 24 820. 00					
制单日期 JAN – 16th, 2015					

15. 进出口货物报关单

数据中心统一编号：000000000868859201

中华人民共和国海关出口货物报关单

预录入编号：53042013004511 申报现场：蛇口海关（5304）　海关编号：53042013C

出口口岸（5304）蛇口海关		备案号		出口日期	申报日期 20130206
经营单位		运输方式 水路运输	运输工具名称 9220304/1303N		提运单号 A330206903
发货单位		贸易方式 一般贸易 （0110）	征免性质（101）一般征税		结汇方式 电汇
许可证号	运抵国（地区）（133）韩国		指运港 （133）韩国		境内货源地（44199）东莞
批准文号	成交方式 FOB	运费		保费	杂费
合同协议号 XFM20130204-01	件数 500	包装种类 纸箱		毛重（千克）3459	净重（千克）2870
集装箱号 XINU1441125 * 1(1)	随附单证 出境货物通关单				生产厂家
标记唛码及备注 备注： 随附单证号：441930213016957000					

项号	商品编号	商品名称、规格型号	数量及单位	最终目的国（地区）	单价	总价	币制	征免
1	84145990.50	散热风扇 散热用 螺丝固定 3.12瓦 无 牌 DF0802024SEM2M型	50000台 50000台	韩国 （133）	0.5850	29250.00	（502）美元	照章征税

税费征收情况

录入员　录入单位 8800000283977	兹申明以上申报无讹并承担法律责任	海关审单批注及放行日期（签章）	
报关员		审单	审价
单位地址	申报单位签章	征税	统计
邮编　　电话	填制日期 2013.02.06	查验	放行

1/1

```
★0000000000764190162★
```

数据中心统一编号: 000000000764190162

中华人民共和国海关进口货物报关单

```
★5216201111161010324★
```

预录入编号: 161010324　　申报现场: 埔沙田办(5216)　　海关编号: 5216201111161010324

进口口岸 (5216) 埔沙田办	备案号		进口日期 20120512	申报日期 20120518
经营单位 (4419962538) 东莞市巨昇进出口有限公司	运输方式 水路运输	运输工具名称 正华118/521601105110	提运单号 ZA884-06	
收货单位 广州市　　　有限公司	贸易方式 一般贸易 (0110)	征免性质 (101) 一般征税	征税比例	
许可证号	启运国(地区) 瑞士 (14)	装货港 巴塞尔BASEL (CHBSL)	境内目的地 (44199) 东莞	
批准文号	成交方式 CIF	运费	保费	杂费
合同协议号 SPGS0410	件数 1	包装种类 其它	毛重(千克) 10880	净重(千克) 10640
集装箱号 GLDU2004449＊1(1)	随附单证 入境货物通关单, 进口许可证		用途 进口原料深加工	
标记唛码及备注				

备注: T/T[装卸口岸]海腾码头
随附单证号: 441900110007748000, 1144002133

项号	商品编号	商品名称, 规格型号	数量及单位	原产国(地区)	单价	总价	币制	征免
1	84621010.00	三胜肽 (SYN®-AKE)原液5000千克/5000kg 号:PEGA-304040/		瑞士 (14)	30000.0000	30000.00	(502) 美元	照章征税

税费征收情况

录入员	录入单位	兹声明以上申报　　　　负责		海关审单批注及放行日期(签章)	
8600000232599				审单	审价
报关员		报关专用章		征税	统计
单位地址		东莞市　　　　有限公司		查验	放行
邮编	电话	填制日期 2011.05.13			
```

1/1 (埔)

```
★1076079 70★
```
```

16. 投保单

出口货物运输保险投保单				
发票号码			投保条款和险别	
被保险人	客户抬头		（　　　）	PICC CLAUSE
			（　　　）	ICC CLAUSE
			（　　　）	ALL RISKS
			（　　　）	W. P. A. /W. A.
			（　　　）	F. P. A.
			（　　　）	WAR RISKS
	过户		（　　　）	S. R. C. C.
			（　　　）	STRIKE
			（　　　）	ICC CLAUSE A
			（　　　）	ICC CLAUSE B
			（　　　）	ICC CLAUSE C
保险金额	USD （　　　　　　）		（　　　）	AIR TPT ALL RISKS
	HKD （　　　　　　）		（　　　）	AIR TPT RISKS
	（　　　） （　　　　　）		（　　　）	O/L TPT ALL RISKS
启运港			（　　　）	O/L TPT RISKS
目的港			（　　　）	TRANSHIPMENT RISKS
转内陆			（　　　）	W TO W
开航日期			（　　　）	T. P. N. D.
船名航次			（　　　）	F. R. E. C.
赔款地点			（　　　）	R. F. W. D.
赔付币别			（　　　）	RISKS OF BREAKAGE
正本份数			（　　　）	I. O. P.
其他特别条款				
以下由保险公司填写				
保单号码		费率		
签单日期		保费		

投保日期：　　　　　　　　　　　　　　　　　　　　　　投保人签章：

17. 普惠制原产地证明书表格 A

1. Goods consigned from (Exporter's business name, address, country) CHINA NATIONAL METALS AND MINERALS EXP & IMP CORP. JIANGSU BRANCH, 201 ZHUJIANG ROAD, NANJING, JIANGSU, CHINA	Reference No. 4 501 290/WJG09/099 GENERALIZED SYSTEM OF PREFERENCES CERTIFICATE OF ORIGIN (Combined declaration and certificate) FORM A Issued in THE PEOPLE'S REPUBLIC OF CHINA
2. Goods consigned to (Consignee's name, address, country) ALEXANDER FRASER AND SON LTD. FRANKLAND MOORE HOUSE 185/187 HIGH ROAD, CHADWELL HEATH, ROMFORD, ESSES. RM62 NR.	(country) See Notes. overleaf
3. Means of transport and route (as for as known) BY VESSEL FROM SHANGHAI TO TORONTO CANADA VIA HONG KONG	4. For official use

5. Item number	6. Marks and numbers of packages	7. Number and kind of packages; description of goods	8. Origin criterion (see Notes overleaf)	9. Gross weight or other quantity	10. Number and date of invoices
1	N/M	16 (SIXTEEN) CRATES POLISHED MARBLE TILES ***********************************	"P"	N. W. : 30 MT	M049418 APRIL 16th, 2015

11. Certification: It is hereby certified, on the basis of control carried out, that the declaration by the exporter is correct.	12. Declaration by the exporter The undersigned hereby declares that the above details and statements are correct; that all the goods were Produced in　　　　CHINA --- (country) And that they comply with the origin requirements specified for those goods in the Generalized System of preferences for goods exported to CANADA --- (importing country)
NANJING：APRIL 16th, 2015 小莫 --- Place and date, signature and stamp of certifying authority	NANJING：APRIL 17th, 2015 刘芳 --- place and date, signature of authorized signatory

18. 出口许可证

中华人民共和国出口许可证
EXPORT LICENCE OF THE PEOPLE'S REPUBLIC OF CHINA　No. 3302699

1. 出口商： Exporter　4403781380172 深圳市恒皓通贸易有限公司	3. 出口许可证号： Export licence No. 13-AF-101331
2. 发货人： Consigner　4403781380172 深圳市恒皓通贸易有限公司	4. 出口许可证有效截止日期： Export licence expiry date 2013年09月20日
5. 贸易方式： Terms of trade　　一般贸易	8. 进口国（地区）：韩国 Country/Region of purchase
6. 合同号： Contract No.　LYG130319	9. 付款方式： Payment　　汇付
7. 报关口岸： Place of clearance　南京海关	10. 运输方式： Mode of transport　海上运输
11. 商品名称：碳化硅 Description of goods	商品编码： Code of good 2849200000

12. 规格、等级 Specification	13. 单位 Unit	14. 数量 Quantity	15. 单价 USD Unit price	16. 总值 USD Amount	17. 总值折美元 Amount in USD
颗粒状 80%MIN	千克	*40,000.0	*0.6500	*26,000	$26,000
18. 总　　计 Total	千克	*40,000.0		*26,000	$26,000

19. 备　　注 Supplementary details	20. 发证机关签章 Issuing authority's stamp & signature

19. 订单

Eurosa Furniture（Kunshan）Co. Ltd. Purchase order

NO. 4002401

PO#：——————— Date：2015. 5. 16

Supplier：		Changlong Wood Industry		Phone：	0435 – 8752222	
				Fax：	0435 – 8752567	
No#	Item #	Description	Units	Qty		
1	918711	Lola side chair	piece	520		
2	918712	Lola arm chair	piece	260		
3	918780	Lola Bench	piece	180		
4						
5						
Total					960	
Order date		2015. 5. 16		Odd numbers		
Delivery date		2015. 07. 05		Receiver		
Terms		according to invoices		payment date		
Bank Info				Tax#		

Remarks：

1. A grade birch wood, no knots and wormholes, MC 8% ~ 10% ;

2. Chair legs cannot glued;

3. Sanding is good and can be used for finishing directly;

5. legs bottom beveling 2. 5 mm;

6. legs rounded corner R2;

pls acknowledge your receipt and confirm back to us, thanks!

Supplier（signature & stamp） Person in charge： Approval：

20. 海关发票

			Page
◆◆ Canade Customs d Revenue Agency	Angece des douances et du revenu du Canada	**CANADA CUSTOMS INVOICE** **FACTURE DES DOUANES CANDIENNIES**	of ae

1. Vendor(name and address)-Vendeur(nom at adresse)	2. Date of direct shipment to Canada-Date of expedition direct versle Canda
出口商名称地址 ZHEJIANG 314409 CHINA	SEP.20TH ,2006 3. Otherreferences (include purchaser's order No.) Autres referenes(include le n'de commande de Tacheteur) PURCHASE ORDER NO.:2401

4. Consignee(name and address)-Destinataire(nom et adresse)	5. Purchaser's name and address(if other than consignee0 nom et adresse de Tacheleur(s'il differe du destinataire)
收货人名称地址	SAME AS CONSIGNEE

	6. Country of transshipment-Pays de transbordement N/A
	7. Country of origin of goods　IF SHIPMENT INDLUDES GOODS OFDIFFRENT ORIGINS pays oorigine des marchandises ENTER ORIGINS AGINST ITEMS IN 12 **CHINA**　SIL EXPEDITION COMPRESSED DESMARCHANDISES D'ORIGINS DIFFERENTES,PRECISEZ LEUR ROVENANCEEN12

8. Transportaion:Give mode and place of direct shipment to Canada Transport:Precisez mode et point dexpedition directe vers le Canada	9. Conditions of sale and terms of payment (Le,sale,consignment shipment,leased goods,etc) Conditions de venle el modalites de paiement (p.ex,vente,expedition en consignation,location de marchandises,etc) **FOB SHANGHAI BY L/C AT SIGHT**
FROM SHANGHAI CHINA TO TORONTO CANADA BY SEA	10. Currency of settlement -Devises du paiement USD

11. Number of pakages Nombre de colis	Specification of commodities(kind of pacages,marks and numbers,general description and characteristics.Le,grade.quality) Designation des articles (naure des colis,marques et numeros.description generale et caraclensliques,p.ex,classe,qualite)		13. Quantify (state unit) Quantile (precisez Tunite)	Selling price-Prix de vente		
				14. Unit price Prix unilaire	15.	Total
388ROLLS	PO# ITEM#/NAME, DESIGN#/NAME, COLOUR NAME, CONTENT, WIDTH	**UPHOLSTERY FABRIC** (70 PCT POLYESTER 30 PCT VISCOSE) AS PER PO2401 ASTRA JACQUARDS 150 CM WIDTH 7403.70 METERS AT USD 2.80 PER METER FREE ON BOARD ANY CHINA PORT WE HEREBY CERTIFY THAT THIS SHIPMENT DOES NOT CONTAIN ANY NON MANUFACTUREL WOODEN MATERIAL,DUNNAGE,BRACING MATERIALS,PALLETS,CRATING OR OTHER NON-MANUFACTURED WOODEN PACKING MATERIALS.	7403.70M	$2.80		$20,730.36

18. If any of fields 1 to 17 are included on an attached commercial invoice,check this box Si tout renseignement relativement aux zones 1à 17 figure sur une u des faclures commerciales ci-attachès,cochez cette case Commercial Invoice No./ N" de la faclure commerciale　2401	☑	16. Total weight-Poids total		17. Invoice total Total de la facture
		Net 3331.67KGS	Gross - Brut 3478.17KGS	$20,730.36

19. Exporter's name and address(if other than vendor) Nom et adresse de Texportateur(sil diffère du vendeur) THE SAME AS VENDOR	20. Originator(name and address)- Expédieur dorigine(nom at adresse) 出口商 ZHEJIANG 314409 CHINA Kendy

21. CCRA ruling(if aplicable)-Désicion de rAgence(s'il y a fieu) N/A	22. If fields 23 to 25 are not applicable,check this box Si les zones 23 à 25 sont sans objet,cochez cett case	X

23. If included in field 17 indicate amount: Si compris dans le total à la zone 17. Précisez:	24. If not included in field 17 indicate amount: Si compris dans le total à la zone 17. Précisez:	25. Check(if applicable): Cochez(s il y a lieu):
(i) transpotation charges,expenses and insurance from the place of direct shipment to Canada Les frais de transport, dépenses et assurances à partir du point dexpédition direcle vers le Canada N/A	(i) transpotation charges,expenses and insurance from the place of direct shipment to Canada Les frais de transport, dépenses et assurances à partir du point dexpédition direcle vers le Canada N/A	(i) Royalty payments or subsequent proceeds are paid or payable by the purchaser Des redevances ou roduits ont été ou seront versés par Tacheteur
(ii) Costs for construction,erection and assembly incurred after importation into Canada Les coûtis de construction,dérection et dassemblage après importation au Canada N/A	(ii) Amounts for commissions other than buying commissions Les commissions autres que celles versées pour Tachat N/A	(ii) The purcheser has supplied goods or sevices for use in the production of these goods L'acheteur a fourni des marchandises ou des services pour la production de ces marchandises
(iii) Export packing Le coût de Temballage dexportation N/A	(iii) Export packing Le coût de Temballage dexportation N/A	

Dans ce formulaire.Ioules les expressions dèsignant des personnes visent à la fois les hommes et les femmes.

c11 (00) Printed in Canada - Imprimé au Canada　　　　　　　　　　　　　　　　　A466

参考文献

[1] Thomas E. Johnson & Donna L. Bade. Export Import Procedures and Documentation (Revised and updated fourth edition). American Management Association, New York, 2010.

[2] Kenneth D. Weiss. Building an Import/Export Business. (Fourth Edition). John Wiley & Sons, Inc., Hoboken, New Jersey, 2008.

[3] ICC. International Standard Banking Practice ISBP 745. 2013.

[4] 蔡玉彬, 龙游宇. 国际贸易理论与实务[M]. 3 版. 北京: 高等教育出版社, 2012.

[5] 张靓芝. 国际贸易实务(英文版)[M]. 北京: 对外经济贸易大学出版社, 2013.

[6] 周瑞琪, 王小鸥, 徐月芳. 国际贸易实务(英文版)[M]. 3 版. 北京: 对外经济贸易大学出版社, 2015.

[7] 易露露, 方玲玲, 陈原. 国际贸易实务双语教程[M]. 3 版. 北京: 清华大学出版社, 2011.

[8] 陈宝珠. 国际贸易实务(双语版)[M]. 大连: 大连理工大学出版社, 2008.

[9] 张圣翠. 国际商法[M]. 6 版. 上海: 上海财经大学出版社, 2012.

[10] 田运银. 国际贸易实务精讲[M]. 6 版. 北京: 中国海关出版社, 2014.

[11] 帅建林. 国际贸易实务(英文版)[M]. 3 版. 北京: 对外经济贸易大学出版社, 2015.

[12] 中国国际贸易学会商务专业培训考试办公室. 进出口货物贸易单证实务[M]. 北京: 中国商务出版社, 2012.

[13] 秦定. 国际贸易合同实践教程(双语教材)[M]. 北京: 清华大学出版社, 2006.

[14] 李鹏博. 揭秘跨境电商[M]. 北京: 电子工业出版社, 2015.

[15] 翁晋阳, Mark, 管鹏, 等. 再战跨境电商[M]. 北京: 人民邮电出版社, 2015.

[16] 丁晖. 跨境电商多平台运营[M]. 北京: 电子工业出版社, 2015.

[17] 肖旭. 跨境电商实务[M]. 北京: 中国人民大学出版社, 2015.

[18] 张炳达, 顾涛. 海关报关实务[M]. 3 版. 上海: 上海财经大学出版社, 2015.

[19] 雨果网.《2013—2014 年中国跨境电商产业研究报告》NO. 1: 中国跨境电商发展的三个时代. http://www.cifnews.com/Article/11604, 2015.

[20] 艾瑞咨询. 2014 年中国跨境电商行业研究报告简版. http://www.iresearch.com.cn/report/2294.html, 2014.

[21] 贝多罗. 贝多罗观点: 跨境电商存在的问题与挑战. http://mt.sohu.com/20150906/n420541268.shtml, 2015.

[22] 中华人民共和国海关总署网站 http://www.customs.gov.cn/

[23] 海关信息网 http://www.haiguan.info/

[24] 中国人民财产保险股份有限公司网站 http://www.piccnet.com.cn/

[25] 全球银行间金融电讯协会 http://www.swift.com/

[26] 马士基航运公司网站 htttp://www.maerskline.com/

[27] 阿里巴巴国际站 htttp://www.alibaba.com/

[28] 全球速卖通网站 http://www.aliexpress.com/

[29] 国际商会网站 http://www.iccwbo.org/

[30] 锦程物流网 http://www.jctrans.com/

[31] 中国国际贸易仲裁委员会网站 http://www.cn.cietac.org/

[32] 全关通信息网 http://www.qgtong.com/

[33] 中华人民共和国商务部网站 http://www.mofcom.gov.cn/

[34] 维基百科网站 http://en.wikipedia.org/